T0100621

Modeling and Simulation in Science, Engineering and Technology

Circuit Simulation with
SPICE OPUS

Theory and Practice

Tadej Tuma
Árpád Bűrmen

Birkhäuser
Boston • Basel • Berlin

Tadej Tuma
University of Ljubljana
Faculty of Electrical Engineering
Tržaška cesta 25
SI-1000 Ljubljana, Slovenia
tadej.tuma@fe.uni-lj.si

Árpád Bűrmen
University of Ljubljana
Faculty of Electrical Engineering
Tržaška cesta 25
SI-1000 Ljubljana, Slovenia
arpadb@fides.fe.uni-lj.si

ISBN 978-0-8176-4866-4   e-ISBN 978-0-8176-4867-1
DOI 10.1007/978-0-8176-4867-1

Library of Congress Control Number: 2009926302

Mathematics Subject Classification (2000): Primary 95Cxx; Secondary 94C05, 94C03

Printed on acid-free paper

Birkhäuser Boston is part of Springer Science+Business Media (www.birkhauser.com)

Contents

And all amid them stood the Tree of Life,
High eminent, blooming Ambrosial Fruit
Of vegetable Gold; and next to Life
Our Death the Tree of Knowledge grew fast by,
Knowledge of Good bought dear by knowing ill.

John Milton, Paradise Lost (4,218)

Abbreviations

ASCII	American standard code for information interchange
BDF	Backward differentiation formula
BJT	Bipolar junction transistor
BSIM	Berkeley short-channel IGFET model
CMOS	Complementary metal oxide semiconductor
EKV	Enz–Krummenacher–Vittoz
FET	Field effect transistor
FM	Frequency modulation
IC	Integrated circuit
IGFET	Insulated gate field effect transistor
Inf	Infinity
JFET	Junction field effect transistor
KCL	Kirchhoff current law
KVL	Kirchhoff voltage law
LTE	Local truncation error
LTI	Linear time-invariant
MESFET	Metal semiconductor field effect transistor
MN	Modified nodal
MOSFET	Metal oxide semiconductor field effect transistor
NaN	Not a number
NR	Newton–Raphson
ODE	Ordinary differential equation
PFD	Phase frequency detector
PLL	Phase-locked loop
PSRR	Power supply rejection ratio
RMS	Root mean square
SOI	Silicon on insulator
SPICE	Simulation Program with Integrated Circuit Emphasis
STAGSOI	Southampton Thermal Analogue silicon on insulator
THD	Total harmonic distortion
TM	Typical mean
UFSOI	University of Florida silicon on insulator
VCO	Voltage-controlled oscillator

WO	Worst one
WP	Worst power
WS	Worst speed
WZ	Worst zero

About SPICE OPUS and This Book

To most people who are new to SPICE, it may be interesting to note that the circuit simulation program and the topic of the following 400+ pages was created as a tool for analyzing integrated circuits (ICs). As a matter of fact, in the 1960s and 1970s the IC industry was a rising star on the high-tech horizon. It quickly became obvious that designing ICs by trial and error was just too tedious and expensive. Prediction of circuit behavior using mathematical models of circuit components and digital computers seemed to be a far better approach.

In 1967 William Howard, at the time a recent post-graduate student of Donald Pederson, created a computer program for the analysis of the DC operating point of an integrated circuit and named it BIAS. This research at the University of California, Berkeley continued and the next result was a program named SLIC (Simulator for Linear Integrated Circuits). It was capable of solving the operating point of a circuit and then simulating the corresponding linearized circuit in the frequency domain. CANCER (Computer Analysis of Nonlinear Circuits, Excluding Radiation) was developed in parallel with SLIC and appeared in 1970. It is considered the predecessor of the Simulation Program with Integrated Circuit Emphasis (SPICE).

The proprietary nature of CANCER's code was the reason that triggered the development of SPICE. Lawrence Nagel, a graduate student at UC Berkeley, created SPICE1 as a result of his graduate study. The program was first released in 1971. Dr. Nagel continued his work on SPICE until version SPICE2A. SPICE2 [33] was first released in 1975, and the development eventually led to version 2G6 released in 1983, which is the basis of many commercial simulators. Among other features, SPICE2 introduced dynamic memory management, which makes it possible for the program to be scalable with respect to circuit size.

The clumsy FORTRAN source code of SPICE2 was rewritten in the C programming language, which became the de facto standard for software development. The rewrite was done in the 1980s by Thomas Quarles, and his efforts resulted in SPICE3 [36] (released in 1985). SPICE3 had many improvements in device models and analysis techniques, albeit it also contained many bugs and memory leaks which made it useless when the analysis and the postprocessing commands were run in a loop with several thousand iterations. The last released version of SPICE3 was 3F5. More on the history of SPICE can be found in [34, 35, 39, 50].

The need for a stable and memory leak-free simulator was the trigger for the development of SPICE OPUS [51], which began in the 1990s. These features were a prerequisite for implementing optimization algorithms in the simulator. The name SPICE OPUS is an acronym for SPICE engine for OPtimization UtilitieS. Until 2000 there were several releases that supported only the Windows operating system. In 2000 SPICE OPUS 2.0 was released with a completely redesigned graphical front-end. Version 2.0 was also available for the Linux operating system. SPICE OPUS has over 10,000 users worldwide in the areas of research, education, and industry.

Due to its public domain licensing, SPICE spread like wildfire. The source code of SPICE2 and SPICE3 was used as a basis or model for many commercial simulators like PSPICE [42], ISSPICE [25], HSPICE [22], SPECTRE [49], and many others.

This book is intended for a very wide audience ranging from undergraduate students to IC designers and simulator developers. Not all chapters are intended for all readers. Some readers might find a certain chapter irrelevant to their work, while others may find in the same chapter just what they are looking for.

The first chapter deals with the formulation of the circuit equations and basic mathematical notions required to understand circuit modeling and simulation. This chapter is intended mostly for students and simulator developers.

The second chapter is a kind of tutorial for SPICE OPUS. Everyone who wants to make a quick start in circuit simulation with SPICE OPUS should read this chapter first. It provides an overview of the features SPICE OPUS offers and also contains brief installation instructions.

The third, fourth, and fifth chapters constitute a manual for SPICE OPUS. They describe all features of SPICE OPUS, starting with the input file syntax, followed by circuit analysis methods offered by SPICE OPUS, and finally the built-in scripting language (NUTMEG). These chapters are intended for users of SPICE OPUS primarily as a reference.

The sixth chapter returns to the theory that makes SPICE tick. It is less general than the first chapter and focuses mostly on the simulation techniques used by SPICE. The material is described in a more gradual way, starting with linear resistive circuits and ending with dynamic circuits and numerical integration algorithms. The target audience is undergraduate students and simulator developers. For the latter, the second half of the chapter will be more interesting because it describes the inner workings of SPICE, ranging from tricks that make operating point analysis possible to algorithms for time-step control.

The seventh and final chapter is a collection of examples. The examples cover the simulation of various circuits and systems, ranging from simple serially connected transmission lines to a complete phase-locked loop. Some of the most common uses of SPICE OPUS are presented, with an emphasis on the built-in scripting language (NUTMEG). Chapter 7 is intended primarily for SPICE OPUS users.

Ljubljana, Slovenia   *Tadej Tuma*
June 2009   *Árpád Bűrmen*

Chapter 1
Introduction to Circuit Simulation

This chapter introduces the reader to the basic notions in the field of circuit simulation. It is highly recommended that you read this chapter before reading any other part of the book simply to become familiar with the terminology used in the chapters that follow.

We will start with an introduction to circuit operating point calculation. Then we will continue with operating point sweeps and definitions of differential gain, resistance, and conductance. Next we move on to time-domain modeling and analysis.

A large part of this chapter is dedicated to the frequency-domain analysis of circuits with small signal excitations. We will give a brief introduction to Fourier transformation and small signal circuit modeling followed by a short tutorial on small signal noise modeling and analysis. At the end of the chapter, we move to the Laplace transformation, s-domain modeling, and pole-zero analysis.

The order in which analyses are described in this chapter is somewhat different than in subsequent chapters. We have done this because the basic principles of the Fourier and Laplace transformations are harder to understand than circuit analysis in the time domain. On the other hand, the implementation of the two transforms is much more related to operating point analysis than it is to time-domain analysis.

Be aware that Y in this chapter denotes the Fourier transform of the circuit solution vector. In all the remaining chapters, Y denotes the complex admittance, as used everywhere else in the literature on circuit theory.

1.1 Signals and Linear Systems

A signal y is a map $y : \mathbb{R} \to \mathbb{R}$. The domain of the map is also referred to as time (t). We denote a signal by $y(t)$.

A system S maps an input signal $y_{in}(t)$, also referred to as excitation, to an output signal $y_{out}(t) = S(y_{in}(t))$ (response) (Fig. 1.1). Suppose we have two input signals $y_{in1}(t)$ and $y_{in2}(t)$. Let $y_{out1}(t)$ and $y_{out2}(t)$ denote the corresponding responses of system S. If $y_{in1}(t) = y_{in2}(t)$ for all $t < T$ implies $y_{out1}(t) = y_{out2}(t)$ for all $t < T$ regardless of the choice of T, we say that the system is causal. Circuits are causal systems.

T. Tuma and Á. Bűrmen, *Circuit Simulation with SPICE OPUS*, Modeling and
Simulation in Science, Engineering and Technology, DOI 10.1007/978-0-8176-4867-1_1,
© Birkhäuser Boston, a part of Springer Science+Business Media, LLC 2009

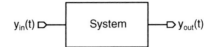

$y_{in}(t)$ ▷─| System |─▷ $y_{out}(t)$

Fig. 1.1 A system maps the input signal $y_{in}(t)$ (excitation) to the output signal $y_{out}(t)$ (response)

A system $\mathcal{S}$ is linear if it satisfies the following two requirements:

$$\mathcal{S}(Ay_{in}(t)) = A\mathcal{S}(y_{in}(t)) \tag{1.1}$$

$$\mathcal{S}(y_{in1}(t) + y_{in2}(t)) = \mathcal{S}(y_{in1}(t)) + \mathcal{S}(y_{in2}(t)) \tag{1.2}$$

Property (1.2) is also referred to as the superposition property. It means that the response of a linear system to multiple superpositioned excitations is simply the superposition of responses to individual excitations. This makes it possible to obtain the response of a system by adding up the responses to individual components of the excitation.

Linear systems possess many other useful properties. The interested reader may refer to [47]. Despite the fact that we are discussing nonlinear circuit simulation, much useful information can be obtained from linearized circuit models, which in turn can be treated as linear systems.

A system is time invariant if $\mathcal{S}(y_{in}(t + T)) = y_{out}(t + T)$ for all values of T where $y_{out} = \mathcal{S}(y_{in}(t))$. A linear time-invariant (LTI) system is characterized by its impulse response $h(t)$. The impulse response is the response of the system to a particular type of input signal – the Dirac impulse $\delta(t)$. $\delta(t)$ is zero for all values of t, except for 0. Nothing can be said about the value of $\delta(0)$, except that it is infinite. In fact, the Dirac impulse is defined by its integral $\int_{-\infty}^{\infty} \delta(t)dt = 1$.

The response to an arbitrary excitation can be expressed with the impulse response in the following manner:

$$y_{out}(t) = h(t) * y_{in}(t) = \int_{-\infty}^{\infty} h(t - \tau)y_{in}(\tau)d\tau. \tag{1.3}$$

The operation denoted by $*$ is called convolution. An LTI system is causal if its impulse response satisfies $h(t) = 0$ for all $t < 0$. For causal LTI systems the convolution simplifies to

$$y_{out}(t) = h(t) * y_{in}(t) = \int_{-\infty}^{t} h(t - \tau)y_{in}(\tau)d\tau. \tag{1.4}$$

If an LTI system is excited with a sinusoidal signal of frequency f (e.g., $A_{in}\cos(2\pi ft + \varphi_0)$), it responds with a superposition of a transient signal and a phase-shifted sinusoidal signal $A_{out}\cos(2\pi ft + \varphi_1)$.

The unit of the impulse response is the unit of the output signal divided by the unit of the input signal and the unit of time. If the input is a voltage (V) and the output is a current (A), the unit of the impulse response is $\frac{A}{Vs}$. For more on linear systems theory see [47].

1.2 Lumped Circuits

Suppose the circuits we are considering are small. By small we mean that they are significantly smaller than the wavelength of electromagnetic radiation that corresponds to the highest frequency at which the circuit operates. Such circuits belong to the class of lumped circuits.

If we take, for instance, an audio amplifier, we can be pretty sure that the frequencies of signals in such a circuit are below 100 kHz. The wavelengths that correspond to these frequencies are greater than 3 km (assuming that the speed of light is $3 \cdot 10^8$ m/s). So any reasonably sized audio amplifier can be treated as a lumped circuit. On the other hand, for a computer with its bus operating at 800 MHz the corresponding wavelength is 37.5 cm. With typical distances of 10 cm between the processor and the memory, such circuits cannot be classified as lumped circuits. The design techniques that apply to lumped circuits cannot be used for motherboards of such computers.

Lumped circuits can be described as a set of branches. Every branch is connected to a pair of nodes (P and N). The state of a node A is described by a voltage (v_A). This voltage is often referred to as the node potential or node voltage. Every branch B is associated with a branch current (i_B) and a branch voltage (v_B). The orientation of the branch voltage is indicated by the $+$ and $-$ signs and is associated with the branch current which flows from the $+$ sign to the $-$ sign (see Fig. 1.2).

A large variety of different branches can be defined. They differ in the way the branch current (or voltage) is expressed using other node voltages, branch voltages, and branch currents in the circuit. For instance, a simple linear time-independent resistive branch can be defined with the following branch equation:

$$i_B - \frac{v_B}{R} = 0, \tag{1.5}$$

where R is a property of the branch and as such a constant. Similarly, a linear time-independent capacitive branch is defined with the following branch equation:

$$i_B - C \frac{dv_B}{dt} = 0, \tag{1.6}$$

where t denotes the time.

Note that time independence does not mean that the voltage and the current are time independent. On the contrary, they depend on time in general. Time independence means that the resistance (R) and the capacitance C are constant. Similarly,

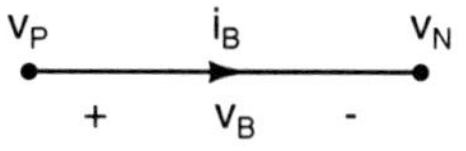

Fig. 1.2 A branch with its associated branch voltage (v_B) and branch current (i_B)

branches for nonlinear resistors and capacitors are obtained by replacing relations (1.5) and (1.6) with $i_B - f_i(v_B) = 0$ and $i_B - \frac{d}{dt} f_q(v_B) = 0$, where $f_i(x)$ and $f_q(x)$ are nonlinear functions.

An element model consists of one or more branches and zero or more internal nodes. Take, for instance, a diode (Fig. 1.3, left). Its model comprises one linear resistive branch (R_s), one nonlinear resistive branch ($f_g(v_i)$), one nonlinear capacitive branch ($f_q(v_i)$), and one internal node (N).

Until now we have not mentioned anything about how the connection of multiple branches to a node is modeled mathematically. In lumped circuits nodes cannot accumulate charge, so the sum of all branch currents flowing into a node must be equal to the sum of all branch currents flowing out of the node. In other words the sum of all incident branch currents must be zero for every node. This requirement is also referred to as the Kirchhoff current law (KCL) and is illustrated in Fig. 1.4.

Any branch voltage can be expressed as the difference of the node voltages that belong to the two nodes to which the branch is connected ($v_B = v_P - v_N$, see Fig. 1.2). This is just a special form of the Kirchhoff voltage law (KVL).

The KVL states that the sum of all branch voltages in any closed loop consisting of branches must be 0. Branch voltages are taken as negative if the branch direction does not agree with the direction of the loop (see Fig. 1.5). We can quickly see that by defining the branch voltage as the difference of two node voltages, the KVL is satisfied by definition.

A circuit consists of one or more elements connected to nodes. By replacing the elements with their respective models, we can write a set of equations that uniquely

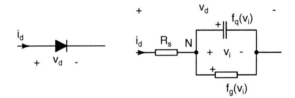

Fig. 1.3 A diode (*left*) and its associated model (*right*) comprising a linear resistive branch (R_s), a nonlinear resistive branch ($f_g(v_i)$), a nonlinear capacitive branch ($f_q(v_i)$), and an internal node (N). The + sign beside the capacitance and resistance symbols indicates the polarity of the corresponding nonlinear branch

Fig. 1.4 Kirchhoff's current law. The sum of all incident branch currents at a node must be 0. For node N this requirement can be written as $i_1 + i_2 - i_3 - i_4 = 0$. Branch currents that by orientation flow into the node are taken as negative

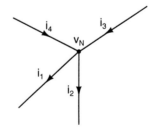

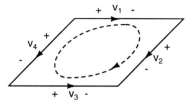

Fig. 1.5 Kirchhoff's voltage law. The sum of all branch voltages in a closed loop must be 0. For the loop above, this requirement can be written as $v_1 + v_2 - v_3 - v_4 = 0$. Branch voltages that by orientation do not agree with the direction of the loops are taken as negative

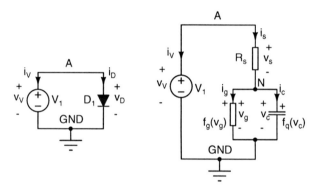

Fig. 1.6 A simple circuit (*left*) and its model (*right*)

determine the circuit's behavior. There are several different ways to obtain such a system. Furthermore, a circuit can be described by different sets of equations. For a particular circuit all such sets of equations are equivalent. One way to obtain a system of equations is to

1. Select one node (denote it by GND) and set its node voltage to $v_{GND} = 0$
2. Write down KCL equations for all nodes except for node GND
3. Write down KVL equations of the form $v_B = v_P - v_N$ for all branches
4. Write down all branch equations

If we have b branches and $n + 1$ nodes, we end up with $n + 2b$ equations. Such a set of equations is also referred to as the circuit tableau.

Let us illustrate the circuit tableau on an example in Fig. 1.6. The circuit has two nodes (A and GND), whereas its model has an additional node N. The model is obtained by replacing the diode with its equivalent circuit (see Fig. 1.3, right). The model consists of four branches. Now let us write the tableau for the circuit in Fig. 1.6. We set the GND node voltage to 0. The KCL equations for nodes A and N are

$$i_V + i_S = 0, \tag{1.7}$$
$$-i_s + i_g + i_c = 0. \tag{1.8}$$

We have four branches which result in four KVL equations,

$$v_V = v_A, \tag{1.9}$$

$$v_s = v_A - v_N, \tag{1.10}$$

$$v_g = v_N, \tag{1.11}$$

$$v_c = v_N. \tag{1.12}$$

Finally the four branches contribute four branch equations,

$$v_V - V_1 = 0, \tag{1.13}$$

$$i_s - \frac{v_s}{R_s} = 0, \tag{1.14}$$

$$i_g - f_g(v_g) = 0, \tag{1.15}$$

$$i_c - \frac{d}{dt} f_q(v_c) = 0. \tag{1.16}$$

Equations (1.7)–(1.16) are the circuit tableau. Note that all node voltages, branch voltages, and branch currents are functions of time. The tableau equations can be simplified by substituting (1.9)–(1.12) into the branch equations (1.13)–(1.16). This way we eliminate branch voltages and obtain the following set of equations:

$$i_V + i_S = 0, \tag{1.17}$$

$$-i_s + i_g + i_c = 0, \tag{1.18}$$

$$v_A - V_1 = 0, \tag{1.19}$$

$$i_s - \frac{v_A - v_N}{R_s} = 0, \tag{1.20}$$

$$i_g - f_g(v_N) = 0, \tag{1.21}$$

$$i_c - \frac{d}{dt} f_q(v_N) = 0. \tag{1.22}$$

We see that some branch currents can be expressed explicitly. If we substitute them into the KCL equations (1.17)–(1.18) we end up with the following three equations:

$$i_V + \frac{v_A - v_N}{R_s} = 0, \tag{1.23}$$

$$-\frac{v_A - v_N}{R_s} + f_g(v_N) + \frac{d}{dt} f_q(v_N) = 0, \tag{1.24}$$

$$v_A - V_1 = 0. \tag{1.25}$$

Equations (1.23)–(1.25) are the modified nodal (MN) equations. The unknowns in the system are the two nodal voltages (v_A and v_N) and one branch current (i_V). This is the current that we were not able to express explicitly from its branch equation (which in turn is $v_V - V_1 = 0$). The definitions of functions f_g (diode DC characteristic) and f_q (diode charge storage) are

$$f_g(v) = \text{IS}(e^{-\frac{vq}{NkT}} - 1), \tag{1.26}$$

$$f_q(v) = \begin{cases} \frac{C0 \cdot VJ}{1-MJ}\left(1 - (1 - \frac{v}{VJ})^{1-MJ}\right) & v < \text{FC} \cdot \text{VJ} \\ C' + \frac{C0 \cdot v'}{(1-FC)^{MJ}}\left(1 + \frac{MJ \cdot v'}{2VJ(1-FC)}\right) & v \geq \text{FC} \cdot \text{VJ}, \end{cases} \tag{1.27}$$

where $C' = \frac{C0 \cdot VJ}{1-MJ}\left(1 - (1 - FC)^{1-MJ}\right)$ and $v' = v - \text{FC} \cdot \text{VJ}$. Parameters IS, N, C0, VJ, M, and FC are diode model parameters, k is the Boltzmann constant, q is the elementary charge, and T is the diode temperature. See Figs. 1.7 and 1.8 for the characteristics of a typical silicon diode.

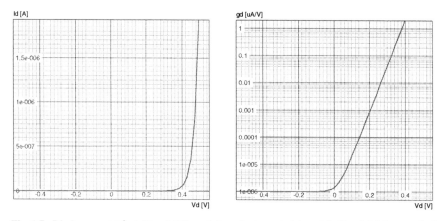

Fig. 1.7 Diode current (f_g, *left*) and differential conductance ($g(v) = \mathrm{d}f_g/\mathrm{d}v$, *right*) for a typical silicon p–n diode with IS $= 0.01$ pA

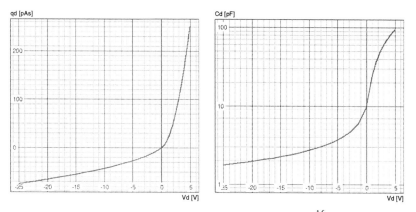

Fig. 1.8 Diode charge (f_q, *left*) and differential capacitance ($c(v) = \frac{\mathrm{d}f_q}{\mathrm{d}v}$, *right*) for a typical silicon p–n diode with C0 $= 10$ pF

In general the process of writing down circuit equations results in the following system of equations:

$$f_{g1}(\mathbf{y}(t),t) + \frac{\mathrm{d}}{\mathrm{d}t} f_{q1}(\mathbf{y}(t),t) + f_{e1}(t) = 0, \qquad (1.28)$$

$$f_{g2}(\mathbf{y}(t),t) + \frac{\mathrm{d}}{\mathrm{d}t} f_{q2}(\mathbf{y}(t),t) + f_{e2}(t) = 0, \qquad (1.29)$$

...

$$f_{g(n+m)}(\mathbf{y}(t),t) + \frac{\mathrm{d}}{\mathrm{d}t} f_{q(n+m)}(\mathbf{y}(t),t) + f_{e(n+m)}(t) = 0, \qquad (1.30)$$

where $\mathbf{y}(t) = [v_1,\ldots,v_n,i_1,\ldots,i_m] = [y_1,\ldots,y_{n+m}]$ is a real-valued vector of unknown node voltages and branch currents. $m < b$ is the number of branches for which the branch current cannot be explicitly expressed from the branch equation. MN equations are generally nonlinear ordinary differential equations (ODE). In the formulation (1.28)–(1.30) only the first derivative with respect to time can appear.

The system of MN equations can be written more concisely in the vector form

$$\mathbf{f}_g(\mathbf{y}(t),t) + \frac{\mathrm{d}}{\mathrm{d}t}\mathbf{f}_q(\mathbf{y}(t),t) + \mathbf{f}_e(t) = \mathbf{0}, \qquad (1.31)$$

where $\mathbf{f}_g$ and $\mathbf{f}_q$ are vector-valued functions of $\mathbf{y}$, and $\mathbf{f}_e$ is a vector-valued function of time. Equation (1.31) represents the time-domain model of the circuit.

In the beginning of this section we mentioned that lumped circuit analysis techniques cannot be applied to circuits where the wavelength of the electromagnetic radiation is in the same order of magnitude or even smaller than the size of the physical circuit. This is only partially true. The parts of the circuit that cannot be modeled as lumped can be described as black boxes (Fig. 1.9). These black boxes are connected to the nodes of the lumped circuit. Such black boxes can often be described in a way that is compatible with lumped circuit modeling.

A typical example is that of coupled linear lossy transmission lines. The response (r currents flowing into the inputs of the black box) can be expressed

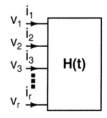

Fig. 1.9 A linear black-box circuit

as a convolution of the black-box's impulse response matrix and the vector of r node voltages at nodes to which the black-box inputs are connected,

$$\mathbf{i}(t) = \mathbf{H}(t) * \mathbf{v}(t) \tag{1.32}$$

The impulse response matrix is an $r \times r$ symmetrical matrix. The entries in the matrix are real-valued functions of time. The relation becomes particularly simple in the frequency domain, where it reduces to simple (in our case matrix) multiplication (see Sect. 1.6). Although the branch currents in (1.32) are explicitly expressed, we often do not substitute them into KCL equations because convolution in the time domain is calculated numerically and therefore cannot be handled by most numerical equation solvers.

If we formally extend the vector $\mathbf{v}(t)$ to $\mathbf{y}(t)$, the matrix $\mathbf{H}(t)$ expands to an $(n + m) \times (n + m)$ matrix,

$$\mathbf{H}(t) = \begin{bmatrix} h_{11}(t) & h_{12}(t) & \dots & h_{1(n+m)}(t) \\ h_{21}(t) & h_{22}(t) & \dots & h_{2(n+m)}(t) \\ \dots & \dots & \dots & \dots \\ h_{(n+m)1}(t) & h_{(n+m)2}(t) & \dots & h_{(n+m)(n+m)}(t) \end{bmatrix}. \tag{1.33}$$

Now we can rewrite the general formulation of MN equations from (1.31) as

$$\mathbf{f}_g(\mathbf{y}(t), t) + \frac{\mathrm{d}}{\mathrm{d}t} \mathbf{f}_q(\mathbf{y}(t), t) + \mathbf{f}_e(t) + \mathbf{H}(t) * \mathbf{y}(t) = \mathbf{0}. \tag{1.34}$$

$\mathbf{H}(t) * \mathbf{y}(t)$ is a vector containing $n + m$ functions of time. Assuming that the circuit is a causal system, we can express one component of this vector as

$$(\mathbf{H}(t) * \mathbf{y}(t))_k = \sum_{l=1}^{n+m} \int_{-\infty}^{t} h_{kl}(t - \tau) y_l(\tau) \mathrm{d}\tau, \tag{1.35}$$

where $y_l(t)$ is the l-th component of $\mathbf{y}(t)$.

Equation (1.32) can be used to model the effects introduced by interconnects in the circuit (e.g., on printed circuit boards). This extends the range of circuits to which the lumped circuit approach can be applied.

Finally, a note on the time dependence of $\mathbf{f}_g(\mathbf{x}, t)$ and $\mathbf{f}_q(\mathbf{x}, t)$. Suppose a circuit responds to excitation $\mathbf{f}_e(t)$ with response $\mathbf{y}(t)$. It is fairly easy to show that the circuit is time invariant if the following two conditions are satisfied:

$$\frac{\partial \mathbf{f}_g(\mathbf{y}, t)}{\partial t} = 0, \tag{1.36}$$

$$\frac{\partial \mathbf{f}_q(\mathbf{y}, t)}{\partial t} = 0. \tag{1.37}$$

More on lumped circuits and the formulation of circuit equations can be found in [6, 23].

1.3   DC Solutions and the Operating Point of a Circuit

Often we are interested in the response of a time-invariant circuit under the following assumptions:

$$\frac{d}{dt}\mathbf{y}(t) = \mathbf{0}, \tag{1.38}$$

$$\frac{d}{dt}\mathbf{f}_e(t) = \mathbf{0}. \tag{1.39}$$

This corresponds to the situation when all signals and all excitations in the circuit, be they voltages or currents, are constant. Equation (1.34) changes into a system of nonlinear algebraic equations,

$$\mathbf{f}_g(\mathbf{y}) + \mathbf{f}_e + \mathbf{H}_0\mathbf{y} = \mathbf{0}, \tag{1.40}$$

where $\mathbf{f}_e$ is a constant vector and $\mathbf{H}_0 = \int_{-\infty}^{\infty}\mathbf{H}(\tau)d\tau$. Equation (1.40) represents the DC model of the circuit. The obtained solutions of the circuit equations are called DC solutions. These solutions remain unchanged if we remove all capacitors and short all inductors in the circuit (set $\mathbf{f}_q$ to $\mathbf{0}$).

Some DC solutions correspond to stable operating points of real-world circuits. However, some solutions are unstable. A real-world circuit transitions from an unstable to a stable operating point due to noise which is always present, either as disturbances coming from the environment or as inherent noise exhibited by every physical circuit component. Sometimes a circuit oscillates around unstable DC operating points. Such circuits are called oscillators.

A method for determining the operating point stability will be described in Sect. 1.12.

An example of a circuit with multiple DC solutions is a ring of inverters. A complementary metal oxide semiconductor (CMOS) inverter is depicted in Fig. 1.10 (left). Note that when using the symbol we do not draw connections for the power supply nodes V_{dd} and V_{ss} on inverter symbols. We implicitly assume that V_{dd} is connected to 1.8 V and V_{ss} is grounded. The response of a CMOS inverter is depicted in Fig. 1.10 (right).

Now consider a single inverter with its output connected to its input (Fig. 1.11, left). The DC solution for this circuit can be obtained graphically by finding the intersection of the inverter characteristic with the line $V_{out} = V_{in}$. The solution is marked with a square in Fig. 1.11 (right). This is the only DC solution. Because it is stable, small disturbances coming from the environment cannot make the circuit oscillate.

Another example is depicted in Fig. 1.12 (left). It is a ring with two inverters. The DC solution can again be obtained graphically by overlaying an inverter characteristic mirrored across $V_{out} = V_{in}$ over another inverter characteristic (Fig. 1.12, right). The intersections of the two curves represent the DC solutions of the circuit. Now

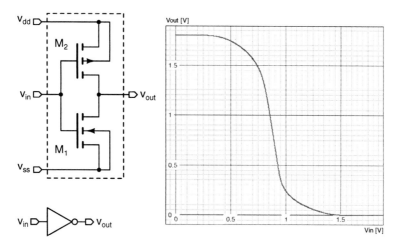

Fig. 1.10 CMOS inverter schematic (*top left*), symbol (*bottom left*), and characteristic (*right*)

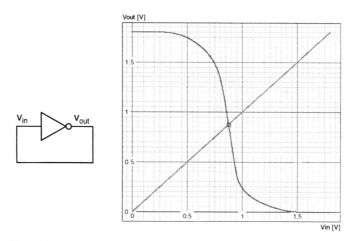

Fig. 1.11 CMOS inverter with its output connected to its input (*left*) and the corresponding DC solution denoted by a square (*right*)

we have three DC solutions. The left one ($V_{in1} = 0$ V and $V_{out1} = 1.8$ V) and the right one ($V_{in1} = 1.8$ V and $V_{out1} = 0$ V) are stable, whereas the one in the middle ($V_{in1} = V_{out1} = 0.8756748$ V) is unstable.

The circuit in Fig. 1.12 (left) persists in an unstable state until some disturbance makes it transition to one of the two stable DC solutions. The disturbance can be modeled using a pulsed voltage source. The pulse causes a transition to a stable DC solution (see Fig. 1.13, right). The response in the figure was obtained with a pulse that appeared at $t = 100$ ps. The height of the pulse was 1 nV and its width was 11 ps. The transition ended at $t = 800$ ps.

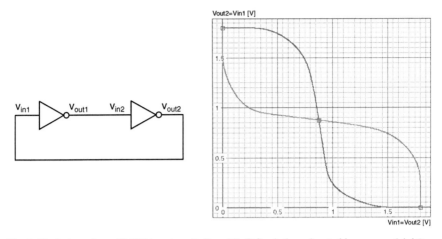

Fig. 1.12 A ring of two CMOS inverters (*left*) and its DC solutions denoted by squares (*right*)

Just to illustrate how easy it is for a circuit to transition from an unstable state to a stable DC solution, consider that the root mean square (RMS) value of thermal noise voltage on a 0.1 Ω resistor at 300 K across a 1 MHz bandwidth is 40 nV. The inverters in Fig. 1.13 (left) have more than 1 GHz bandwidth, so even the thermal noise of the interconnects is sufficient for starting the transition (remember that the transition in Fig. 1.13 was initiated by a 1 nV pulse).

A ring of two inverters can be used as a memory element capable of storing 1 bit. One stable DC solution corresponds to logic "1" and the other one to logic "0." Such circuits with two stable DC solutions are often referred to as bistable multivibrators and are the basic building blocks of static memory. The transition from one stable DC solution to the other can be initiated with appropriately polarized current pulses at inverter outputs. See Fig. 1.14.

The DC solutions of a ring with three inverters can be obtained in a similar way as for the solution of the ring with two inverters. The mirrored characteristic of a single inverter is laid over the characteristic of a two-inverter chain (Fig. 1.15, right). Only one DC solution exists and even that one is unstable (Fig. 1.16, left).

As the only DC solution is unstable, we expect this circuit to oscillate. This is confirmed in Fig. 1.16 (right) where the time-domain response of a ring with inverters is depicted. Larger rings with an odd number of inverters behave similarly to a ring with three inverters. Rings with an even number of inverters behave similarly to a ring with two inverters.

Inverter rings are used as ring oscillators. The oscillation frequency can be adjusted by changing the MOS transistor's capacitance. Larger transistors have a larger gate capacitance and result in lower frequencies. The frequency can also be adjusted by changing the on resistance of the transistor. This can be achieved by connecting a resistor in series with the drain. If a MOS transistor is used instead of a resistor, we can even adjust the resistance with the MOS gate voltage. Such oscillators are

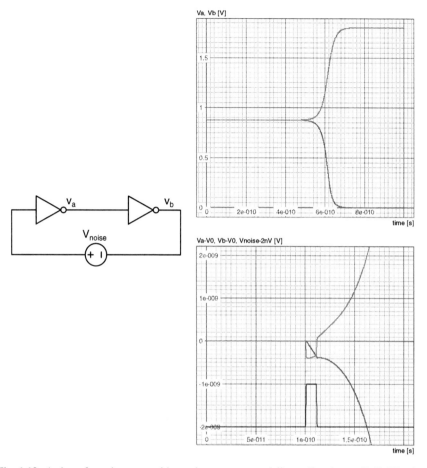

Fig. 1.13 A ring of two inverters with a voltage source modeling a disturbance (*left*). The ring is initially in an unstable state. Due to the pulse generated by V_{noise} it transitions to a stable DC solution (*top right*). A detail shows the pulse and the first moments of the transition (*bottom right*). The unstable DC solution is $V_a = V_b = V_0$

called current-starved ring oscillators. Because their frequency can be adjusted by changing the voltage, they are called voltage-controlled oscillators (VCOs).

1.4 Incremental DC Circuit Model

For a moment put aside the fact that a circuit can have multiple DC solutions. In this section we are going to consider only a single DC solution. Often we are interested in how this solution changes if some circuit parameter varies. Such a dependence is called the DC characteristic of a circuit.

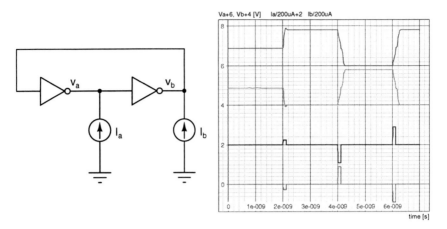

Fig. 1.14 Switching a ring of two inverters from one DC solution to the other with two current sources (*left*). The response shows the two output voltages and switch currents (*right*)

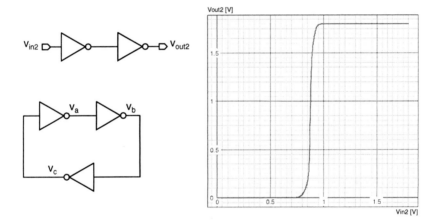

Fig. 1.15 A chain with two inverters (*top left*), a ring with three inverters (*bottom left*), and the characteristic of the two-inverter chain (*right*)

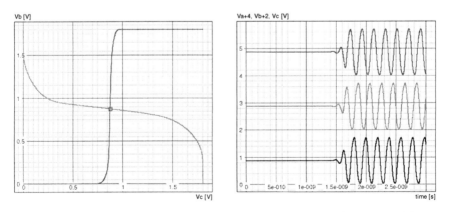

Fig. 1.16 The DC solution of a three-inverter ring (*left*) and its time-domain response (*right*)

Fig. 1.17 A simple CMOS
amplifier. The v_{bias} voltage
sets the drain current at the
DC solution. We assume that
v_{bias} and V_{dd} are constant. The
input signal is generated by
V_{in}. The V_{out} node voltage
is the output signal. The
amplifier load is represented
by the I_{load} current source

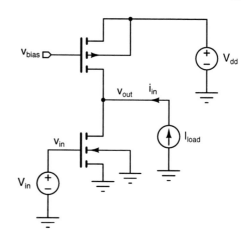

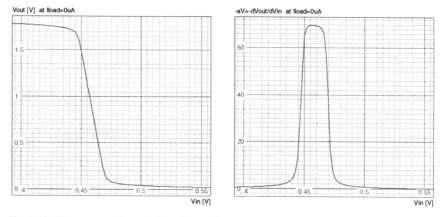

Fig. 1.18 The output voltage (*left*) and the differential voltage gain (*right*) vs. the input voltage
for the amplifier in Fig. 1.17. The curves were obtained for $I_{load} = 0\,A$

Let p denote a circuit parameter. Then the DC characteristic of some voltage (or
current) is expressed as $v = v(p)$ (or $i = i(p)$). Sometimes we use a current or a
voltage in the circuit as the parameter. Take, for instance, the circuit in Fig. 1.17. If
we set a fixed value for I_{load} (for instance $0\,A$) and plot the output voltage vs. the
input voltage, we obtain a curve $v_{out} = v_{out}(v_{in})$.

The derivative $A_V = dv_{out}(dv_{in})$ curve is referred to as the differential voltage
gain. We can see that the voltage gain depends on the operating point (which in turn
depends on the V_{in} voltage and I_{load} current). The $v_{out}(v_{in})$ and the $dv_{out}(dv_{in})$ curves
are plotted against v_{in} in Fig. 1.18.

On the other hand, if we set V_{in} to a fixed value (for instance $0.46\,V$) and plot
the output voltage vs. the load current we obtain a curve $v_{out}(I_{load})$. The derivative $R_{out} = dv_{out}/dI_{load}$ is referred to as the differential output resistance. Just like
the differential voltage gain, it depends on the operating point. The $v_{out}(I_{load})$ and
the dv_{out}/dI_{load} curves are depicted in Fig. 1.19.

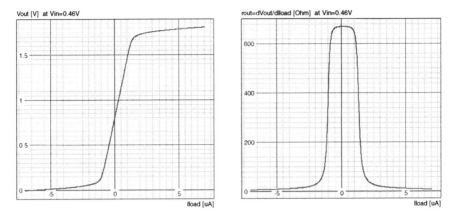

Fig. 1.19 The output voltage (*left*) and the differential output resistance (*right*) vs. the load current for the amplifier in Fig. 1.17. The curves were obtained for $V_{in} = 0.46$ V

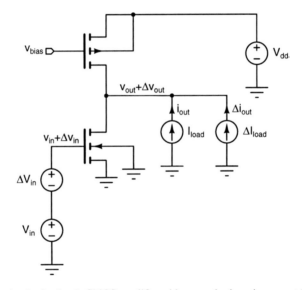

Fig. 1.20 The circuit of a simple CMOS amplifier with an emphasis on incremental signals. Small changes of V_{in} are modeled with the independent voltage source ΔV_{in}. Small changes of I_{load} are modeled with the independent current source ΔI_{load}. The small changes in V_{in} or I_{load} result in a small change of the output voltage (Δv_{out})

The differential gain and resistance can be explained with the help of the incremental circuit in Fig. 1.20. Small changes of V_{in} or I_{load} are modeled as independent sources dV_{in} and dI_{load}. These small changes cause a small change of the operating point from v_{out} to $v_{out} + dv_{out}$. By small we usually mean small enough so that dv_{out} is proportionate to the changes in V_{in} or I_{load}. This means that the circuit (although it is nonlinear by nature) behaves as linear for the incremental quantities dV_{in}, dI_{load},

dv_{in}, dv_{out}, and di_{out}. Linear circuits [6] are much easier to analyze and possess many useful properties.

The linearization of the circuit can be explained with the DC circuit equations (1.40). Let $\mathbf{y}_0$ denote a DC solution of the circuit for $\mathbf{f}_e = \mathbf{f}_{e0}$. Suppose the circuit excitation vector changes to

$$\mathbf{f}_e = \mathbf{f}_{e0} + \Delta\mathbf{f}_e. \tag{1.41}$$

The circuit responds to the changed excitation with

$$\mathbf{y} = \mathbf{y}_0 + \Delta\mathbf{y}. \tag{1.42}$$

If the incremental change of the excitation ($\Delta\mathbf{f}_e$) is small, the incremental change of the response ($\Delta\mathbf{y}$) is also small. Therefore, we can use the first two terms of Taylor's expansion [53] and replace $\mathbf{f}_g(\mathbf{y}_0 + \Delta\mathbf{y})$ with

$$\mathbf{f}_g(\mathbf{y}_0 + \Delta\mathbf{y}) \rightarrow \mathbf{f}_g(\mathbf{y}_0) + \mathbf{G}_0\Delta\mathbf{y}, \tag{1.43}$$

where the matrix $\mathbf{G}_0$ (the Jacobi matrix [27] of $\mathbf{f}_g$ at $\mathbf{y} = \mathbf{y}_0$) is defined as

$$\mathbf{G}_0 = \mathbf{G}(\mathbf{y}_0) = \begin{bmatrix} \dfrac{df_{g1}}{dy_1} & \dfrac{df_{g1}}{dy_2} & \cdots & \dfrac{df_{g1}}{dy_{n+m}} \\ \dfrac{df_{g2}}{dy_1} & \dfrac{df_{g2}}{dy_2} & \cdots & \dfrac{df_{g2}}{dy_{n+m}} \\ \cdots & \cdots & \cdots & \cdots \\ \dfrac{df_{g(n+m)}}{dy_1} & \dfrac{df_{g(n+m)}}{dy_2} & \cdots & \dfrac{df_{g(n+m)}}{dy_{n+m}} \end{bmatrix}_{\mathbf{y}=\mathbf{y}_0}. \tag{1.44}$$

Equation (1.40) can now be written as

$$\mathbf{f}_g(\mathbf{y}_0) + \mathbf{f}_{e0} + \mathbf{H}_0\mathbf{y}_0 + \mathbf{G}_0\Delta\mathbf{y} + \Delta\mathbf{f}_e + \mathbf{H}_0\Delta\mathbf{y} = \mathbf{0}. \tag{1.45}$$

Because the first three terms add up to $\mathbf{0}$ (remember that $\mathbf{y}_0$ is a DC solution for $\mathbf{f}_e = \mathbf{f}_{e0}$), we end up with

$$\mathbf{G}_0\Delta\mathbf{y} + \Delta\mathbf{f}_e + \mathbf{H}_0\Delta\mathbf{y} = \mathbf{0}. \tag{1.46}$$

Equation (1.46) is an equation of a linear circuit excited by incremental excitations $\Delta\mathbf{f}_e$ resulting in incremental response $\Delta\mathbf{y}$. Quotients of incremental signals often have a special meaning. For instance,

$$a_V = \frac{\Delta v_2}{\Delta v_1} \tag{1.47}$$

is the differential voltage gain. The differential current gain is defined as

$$a_I = \frac{\Delta i_2}{\Delta i_1}. \tag{1.48}$$

The differential gain is equal to the derivative of the respective DC characteristic $v_2(v_1)$ (or $i_2(i_1)$) with respect to v_1 (or i_1).

Quotients of incremental voltage and current are also referred to as differential conductance ($g = \frac{\Delta i}{\Delta v}$) and differential resistance ($r = \frac{\Delta v}{\Delta i}$). Differential conductance is equal to the derivative of the respective DC characteristic $i(v)$ with respect to voltage v. Similarly, differential resistance is equal to the derivative of the respective DC characteristic $v(i)$ with respect to current i. If the voltage refers to a different branch than the current, then the terms transconductance or transresistance are used.

1.5   Circuit's Response in the Time Domain

When we are talking about the circuit's response in the time domain we mean the voltages and currents (members of vector $\mathbf{y}$) as functions of time. Usually we observe these functions for times greater than 0. The time-domain response at time 0 (denoted by $\mathbf{y}(0)$) is also referred to as the initial condition.

The time-domain response can be obtained by solving (1.34) using $\mathbf{y}(0)$ as the initial condition. There exist many approaches for solving the DC circuit equations (1.40). Unfortunately, they cannot be applied directly to (1.34) due to the $\mathbf{f}_q$ term and the convolution term ($\mathbf{H}(t) * y(t)$).

One approach for solving (1.34) involves time discretization. This means that $\mathbf{y}(t)$ is obtained for times from a finite set $\{t_0, t_1, t_2, \ldots, t_{end}\}$. Values t_i are also referred to as the time points. Let $h_k = t_{k+1} - t_k$ denote the time step between consecutive time points.

When we are solving the circuit at time point $t = t_{k+1}$ we take our knowledge of the circuit solution at past time points $t_k, t_{k-1}, \ldots$. Numerical techniques for time-domain circuit simulation attempt to reformulate the circuit equations (1.34) in the form

$$\mathbf{F}(\mathbf{y}(t_{k+1})) = \mathbf{0}, \tag{1.49}$$

which can be solved in the same manner as DC circuit equations (1.40) are solved. Note that function $\mathbf{F}$ depends on the circuit's response at past time points.

Using numerical quadrature techniques, the third term in (1.34) (convolution term) can be simplified to

$$\mathbf{H}(t) * \mathbf{y}(t) \big|_{t=t_{k+1}} = \int_{-\infty}^{t_{k+1}} \mathbf{H}(t_{k+1} - \tau)\mathbf{y}(\tau)d\tau$$

$$\approx \sum_{i=0}^{\infty} w_i \mathbf{H}(t_{k+1} - t_{k-i+1})\mathbf{y}(t_{k-i+1})$$

$$\approx \Phi_0 \mathbf{y}(t_{k+1}) + \Phi_1. \tag{1.50}$$

The coefficients w_i are defined by the numerical quadrature formula [1] (e.g., Simpson's rule). If the time step is not constant, they depend on the ratios of time steps. Matrix Φ_0 and vector Φ_1 are defined as

$$\Phi_0 = w_0 \mathbf{H}(0), \tag{1.51}$$

$$\Phi_1 = \sum_{i=1}^{\infty} w_i \, \mathbf{H}(t_{k+1} - t_{k-i+1}) \mathbf{y}(t_{k-i+1}). \tag{1.52}$$

The value of Φ_1 depends on the values of $\mathbf{y}(t)$ at past time points, whereas Φ_0 does not.

Often we use a finite lower bound for the integral in (1.50). This is justified if members of matrix $\mathbf{H}(t)$ approach zero fast enough when time approaches infinity. In this case the upper bound for the sum is N, meaning that we take into account only N past time steps. The result is the following expression for Φ_1:

$$\Phi_1 = \sum_{i=1}^{N} w_i - \mathbf{H}(t_{k+1} - t_{k-i+1}) \mathbf{y}(t_{k-i+1}). \tag{1.53}$$

Let $\mathbf{f}_q(t)$ and $\dot{\mathbf{f}}_q(t)$ denote $\mathbf{f}_q(\mathbf{y}(t), t)$ and $\frac{d}{dt}\mathbf{f}_q(\mathbf{y}(t), t)$. We can eliminate the derivative with respect to time if we express $\dot{\mathbf{f}}_q(t_{k+1})$ with past values of $\dot{\mathbf{f}}_q(t)$ and $\mathbf{f}_q(t)$ including the unknown value $\mathbf{f}_q(t_{k+1})$. This approach is referred to as numerical integration [24] (do not confuse it with numerical quadrature, which is used for evaluating a definite integral). A very simple numerical integration algorithm is

$$\dot{\mathbf{f}}_q(t_{k+1}) \approx \frac{\mathbf{f}_q(t_{k+1}) - \mathbf{f}_q(t_k)}{h_k}, \tag{1.54}$$

which is also known as the backward Euler formula. The formula is exact if $\mathbf{f}_q(t)$ is a linear function of time. The backward Euler formula is an implicit integration method because it expresses $\dot{\mathbf{f}}_q(t)$ with $\mathbf{f}_q(t_{k+1})$, which in turn is unknown until the circuit equations are solved for $t = t_{k+1}$.

A general implicit integration formula can be written as

$$\mathbf{f}_q(t_{k+1}) \approx \sum_{i=0}^{p} a_i \mathbf{f}_q(t_{k-i}) + \sum_{i=-1}^{r} b_i \dot{\mathbf{f}}_q(t_{k-i}). \tag{1.55}$$

The coefficients a_i and b_i depend on the ratios of past time steps. See Chap. 6 for a more detailed explanation of numerical integration algorithms.

From (1.55) we can express $\dot{\mathbf{f}}_q(t_{k+1})$.

$$\dot{\mathbf{f}}_q(t_{k+1}) \approx \frac{1}{h_k b_{-1}} \mathbf{f}_q(t_{k+1}) - \sum_{i=0}^{p} \frac{a_i}{h_k b_i} \mathbf{f}_q(t_{k-i}) - \sum_{i=0}^{r} \frac{b_i}{b_{-1}} \dot{\mathbf{f}}_q(t_{k-i})$$
$$\approx \psi_0 \mathbf{f}_q(t_{k+1}) + \Psi_1, \tag{1.56}$$

where scalar ψ_0 and vector Ψ_1 can be expressed as

$$\psi_0 = \frac{1}{h_k b_{-1}} \tag{1.57}$$

$$\Psi_1 = -\sum_{i=0}^{p} \frac{a_i}{h_k b_i} \mathbf{f}_q(t_{k-i}) - \sum_{i=0}^{r} \frac{b_i}{b_{-1}} \dot{\mathbf{f}}_q(t_{k-i}). \tag{1.58}$$

Only $\mathbf{f}_q(t_{k+1})$ is unknown. $\mathbf{f}_q(t_{k-i})$ and $\dot{\mathbf{f}}_q(t_{k-i})$ are known for all past time steps. Let $\mathbf{y}_{k+1}$ denote $\mathbf{y}(t_{k+1})$. Now we can insert (1.50) and (1.56) into (1.34),

$$\mathbf{f}_g(\mathbf{y}_{k+1}, t_{k+1}) + \psi_0 \mathbf{f}_q(\mathbf{y}_{k+1}, t_{k+1}) + \Phi_0 \mathbf{y}_{k+1} + \Psi_1 + \mathbf{f}_e(t_{k+1}) + \Phi_1 \approx 0. \tag{1.59}$$

Equation (1.59) can be solved for $\mathbf{y}(t_{k+1})$ using the same techniques used in (1.40) to solve for the circuit's DC solutions. In the beginning (for the first time point after the initial conditions at $t = 0$) simple numerical integration algorithms that require only a single past time point are used. Later, as the number of past time points accumulates, more complex formulae can be employed.

The approximate equality is caused by the use of approximation. Remember that we approximated the convolution integral (numerical quadrature). We also approximated the first derivative of $\mathbf{f}_q(\mathbf{y}(t), t)$ with respect to time (numerical integration). The latter approximation introduces the local truncation error (LTE). This error grows as time step h_k increases. The main means for keeping the LTE bounded is by controlling the time step. More about time-step control and the LTE can be found in Chap. 6.

In this section we have shown that the problem of solving the circuit's equations for one time step can be reduced to solving a system of nonlinear algebraic equations.

1.6 Fourier Series and Fourier Transformation

Any periodic complex-valued signal $y(t)$ with period $T > 0$ (meaning that $y(t + nT) = y(t)$ for every integer n) for which the following holds:

$$\int_{t=t_0}^{t_0+T} |f(t)|^2 dt < \infty, \tag{1.60}$$

can be expressed in the form of a Fourier series,

$$y(t) = \sum_{n=-\infty}^{\infty} c_{yn} e^{jn2\pi f_1 t}, \tag{1.61}$$

where j is the imaginary unit ($j^2 = -1$) and $f_1 = 1/T$ is the fundamental harmonic frequency. Coefficients c_{yn} are complex and can be obtained as

$$c_{yn} = \frac{1}{T} \int_{t_0}^{t_0+T} y(t)e^{-jn2\pi f_1 t}\,dt. \tag{1.62}$$

Coefficient c_{yn} corresponds to the n-th harmonic frequency nf_1 and can be written in the polar form

$$c_{yn} = |c_{yn}|e^{j\varphi_{yn}}, \tag{1.63}$$

where $|c_{yn}|$ and φ_{yn} are the magnitude and the phase of the n-th harmonic. The set of coefficients $\{c_{yn}\}$ is also referred to as the spectrum of a periodic signal. Similarly, set $\{|c_{yn}|^2\}$ represents the power spectrum of a periodic signal.

Parseval's theorem relates the power spectrum to the power of a signal,

$$\sum_{n=-\infty}^{\infty} |c_{yn}|^2 = \frac{1}{T} \int_{t_0}^{t_0+T} |y(t)|^2\,dt. \tag{1.64}$$

For real-valued signals the relation $c_{y(-n)} = c_{yn}^*$ holds where $*$ denotes complex conjugation. In the chapters that follow we will be considering only real-valued signals. By collecting the n-th and the $-n$-th coefficients we obtain a form of the Fourier series which can be applied to real-valued signals,

$$y(t) = c_{y0} + \sum_{n=1}^{\infty} 2|c_{yn}| \cos(n2\pi f_1 t + \varphi_{yn}). \tag{1.65}$$

What the Fourier series does for periodic signals the Fourier transform does for aperiodic signals. We obtain the Fourier transform of a signal in the limit when $T \to \infty$. The increase of the period towards infinity causes the set of harmonic frequencies to become more dense, until in the limit it becomes continuous. If we substitute $nf_1 \to f$ and $Tc_{yn} \to Y(f)$ the expression for the Fourier coefficient (1.62) becomes the Fourier transformation of aperiodic signal $y(t)$,

$$\mathcal{F}(y(t)) = Y(f) = \int_{-\infty}^{\infty} y(t)e^{-j2\pi f t}\,dt. \tag{1.66}$$

$Y(f)$ is complex and is referred to as the spectrum of aperiodic signal $y(t)$. The inverse Fourier transformation reconstructs the signal from its spectrum,

$$\mathcal{F}^{-1}(Y(f)) = y(t) = \int_{-\infty}^{\infty} Y(f)e^{j2\pi f t}\,df. \tag{1.67}$$

The Fourier transform of a signal $y(t)$ exists if the following relation holds:

$$\int_{-\infty}^{\infty} |y(t)|\,dt < \infty. \tag{1.68}$$

Similarly as for periodic signals, we can define the magnitude ($|Y(f)|$), phase ($\phi_y(f)$), and the energy spectrum density ($|Y(f)|^2$). Parseval's theorem relates the energy spectrum density of a signal to the energy of a signal,

$$\int_{-\infty}^{\infty} |Y(f)|^2 \mathrm{d}f = \int_{-\infty}^{\infty} |y(t)|^2 \mathrm{d}t. \tag{1.69}$$

The spectrum of a periodic signal can be treated within the framework of the Fourier transform if the spectrum is expressed as

$$Y(f) = \sum_{n=-\infty}^{\infty} c_{yn}\delta(f - nf_1), \tag{1.70}$$

where $\delta(t)$ is the Dirac impulse. Note that (1.70) cannot be used with Parseval's theorem (1.69) for aperiodic signals. This is due to the fact that the energy of a periodic signal is infinite.

Due to (1.70) and (1.73), an LTI system affects the spectrum of a periodic signal in the same way as it affects the spectrum of an aperiodic signal. Let $c_{in(n)}$ denote the Fourier coefficients of the input signal $y_{in}(t)$ with period T. Then the Fourier coefficients of the output signal $y_{out}(t)$ (denoted by $c_{out(n)}$) can be expressed as

$$c_{out(n)} = H(nf_1)c_{in(n)}. \tag{1.71}$$

The Fourier transformation can be used for solving linear ODEs because it converts the derivative with respect to time in simple multiplication,

$$\mathcal{F}(\frac{\mathrm{d}}{\mathrm{d}t}y(t)) = j2\pi f \mathcal{F}(y(t)) = j2\pi f Y(f) = j\omega Y(f), \tag{1.72}$$

where j is the imaginary unit ($j^2 = -1$). The term $2\pi f$ is often denoted by ω and referred to as the angular frequency. The Fourier transformation converts convolution into multiplication,

$$\mathcal{F}(h(t) * y(t)) = \mathcal{F}(h(t)) \cdot \mathcal{F}(y(t)) = H(f)Y(f), \tag{1.73}$$

where $H(f)$ is the Fourier transform of the impulse response. The Fourier series and the Fourier transform have many other properties. The interested reader can refer to [7] for details.

1.7 Incremental Circuit Model in the Time Domain

A similar simplification to that in Sect. 1.4 can be applied to the circuit in the time domain, provided that

- The incremental part of the excitation is small enough so that the circuit can be treated as a linear circuit
- The circuit is time invariant for the duration of the observation

The first assumption implies that the excitation and the response can be written in the form

$$\mathbf{f}_e(t) = \mathbf{f}_0 + \Delta\mathbf{f}_e(t) \tag{1.74}$$

$$\mathbf{y}(t) = \mathbf{y}_0 + \Delta\mathbf{y}(t). \tag{1.75}$$

The assumption of time invariance simplifies $\mathbf{f}_g(\mathbf{y}(t), t)$ and $\mathbf{f}_q(\mathbf{y}(t), t)$ to $\mathbf{f}_g(\mathbf{y}(t))$ and $\mathbf{f}_q(\mathbf{y}(t))$. Let $\mathbf{y}_0$ denote a DC solution of the circuit for $\mathbf{f}_e = \mathbf{f}_{e0}$. Then we can linearize $\mathbf{f}_g$ and $\mathbf{f}_q$.

$$\mathbf{f}_g(\mathbf{y}_0 + \Delta\mathbf{y}(t)) \rightarrow \mathbf{f}_g(\mathbf{y}_0) + \mathbf{G}\Delta\mathbf{y}(t), \tag{1.76}$$

$$\mathbf{f}_q(\mathbf{y}_0 + \Delta\mathbf{y}(t)) \rightarrow \mathbf{f}_q(\mathbf{y}_0) + \mathbf{C}\Delta\mathbf{y}(t), \tag{1.77}$$

where $\mathbf{G}$ and $\mathbf{C}$ are the Jacobi matrices of $\mathbf{f}_g(\mathbf{y})$ and $\mathbf{f}_q(\mathbf{y})$ at $\mathbf{y} = \mathbf{y}_0$, respectively. The linearity of convolution implies

$$\mathbf{H}(t) * (\mathbf{y}_0 + \Delta\mathbf{y}(t)) = \mathbf{H}_0\mathbf{y}_0 + \mathbf{H}(t) * \Delta\mathbf{y}(t). \tag{1.78}$$

With time invariance in mind we can rewrite the circuit equations (1.34) as

$$\mathbf{f}_g(\mathbf{y}_0) + \mathbf{f}_{e0} + \mathbf{H}_0\mathbf{y}_0$$

$$+ \mathbf{G}\Delta\mathbf{y}(t) + \mathbf{C}\frac{\mathrm{d}}{\mathrm{d}t}\Delta\mathbf{y}(t) + \Delta\mathbf{f}_e(t) + \mathbf{H}(t) * \Delta\mathbf{y}(t) = 0. \tag{1.79}$$

Because $\mathbf{y}_0$ is a DC solution for $\mathbf{f}_e = \mathbf{f}_{e0}$ the first three terms add up to 0 and we end up with

$$\mathbf{G}\Delta\mathbf{y}(t) + \mathbf{C}\frac{\mathrm{d}}{\mathrm{d}t}\Delta\mathbf{y}(t) + \Delta\mathbf{f}_e(t) + \mathbf{H}(t) * \Delta\mathbf{y}(t) = 0. \tag{1.80}$$

Equation (1.80) is an equation of a linear circuit. It is excited with incremental excitation $\Delta\mathbf{f}_e$ and responds with incremental response Δbfy. To summarize, we first split the excitation into its DC part $\mathbf{f}_{e0}$ and incremental part $\Delta\mathbf{f}_e(t)$, which is small enough so that the linearization of the circuit is justified. Due to the superposition property of linear systems, we split the complete response of the nonlinear circuit into two parts:

- DC operating point solution $\mathbf{y}_0$ for excitation $\mathbf{f}_{e0}$ (obtained from (1.40)
- Incremental time domain response $\Delta\mathbf{y}(t)$ of a linear circuit (obtained from (1.80))

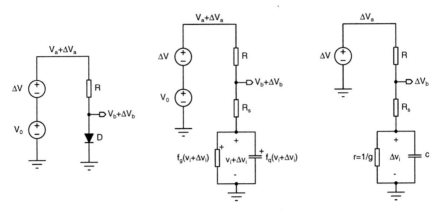

Fig. 1.21 A nonlinear voltage divider (*left*). The nonlinear circuit model (using a diode model from Sect. 1.2) is based on (1.34) (*middle*). The time-domain incremental model based on (1.80) consists of linear elements only (*right*). The incremental model describes changes in the circuit's response (V_a and V_b) accurately if the changes of the excitation (ΔV) are sufficiently small so that linearization (1.76)–(1.77) can be applied to the circuit

We will illustrate the circuit linearization with a simple example. The operating point (V_a, V_b, and v_i) of the circuit in Fig. 1.21 (left) is obtained as the solution of the circuit model (Fig. 1.21, middle) when $\Delta V = 0$. The latter implies $\Delta V_a = 0$, $\Delta V_b = 0$, and $\Delta v_i = 0$.

The incremental model of a diode (Fig. 1.21, right) consists of a linear series resistance (R_s), differential junction resistance ($r = 1/g$), and differential junction capacitance (c). r and g originate from the Jacobi matrices **G** and **C** and can be expressed as the derivatives of diode current ($f_g(v)$) and diode charge ($f_q(v)$) (see (1.26)–(1.27)) with respect to v at $v = v_i$.

$$g(v) = \frac{d}{dv} f_g(v) = \frac{IS \cdot q}{NkT} e^{-\frac{vq}{NkT}}, \tag{1.81}$$

$$c(v) = \frac{d}{dv} f_q(v) = \begin{cases} C0(1 - \frac{v}{VJ})^{-MJ} & v < FC \cdot VJ \\ \frac{C0}{(1-FC)^{MJ}} \left(1 + \frac{MJ(v-FC\cdot VJ)}{VJ(1-FC)}\right) & v \geq FC \cdot VJ \end{cases}. \tag{1.82}$$

Suppose the excitation that is applied to the circuit in Fig. 1.21 consists of a constant V_0 and $\Delta V = V_{mag} \sin(2\pi ft)$. The DC solution of the circuit (we assume $V_0 = 0.8$ V and $\Delta V = 0$ V) is at $V_a = 0.8$ V and $V_b = 0.663$ V.

The time-domain response of the circuit for $V_0 = 0.8$ V, $V_{mag} = 50$ mV, and $f = 500$ MHz is depicted in Fig. 1.22. We can see that the steady state solution is a superposition of the DC circuit solution and incremental circuit response (which is a 500 MHz sinusoidal waveform). As long as V_{mag} is small enough, linearization can be applied to nonlinear circuit elements and the superposition is justified.

The steady state response of the circuit in Fig. 1.21 (left) for two different V_0 values (0.8 and 1.2 V), $V_{mag} = 50$ mV, and $f = 1$ kHz is depicted in Fig. 1.23.

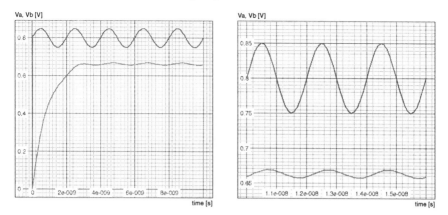

Fig. 1.22 The response of the circuit in Fig. 1.21 (*left*) for $V_0 = 0.8\,$V, $V_a = 50\,$mV, and $f = 500\,$MHz. The complete transient assuming that the initial charge in the nonlinear capacitor is 0As (*left*). A detail of the steady state (*right*)

Fig. 1.23 The steady state response of the circuit in Fig. 1.21 (*left*) for two different V_0 values (0.8 and 1.2 V), $V_{mag} = 50\,$mV, and $f = 1\,$kHz

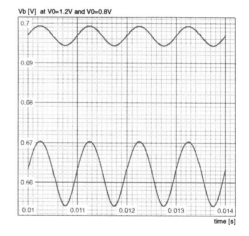

For higher values of V_0 we get a smaller amplitude of the output sinusoidal signal because a higher V_0 means a higher diode current and thus a lower differential resistance. Lower differential resistance results in a lower ratio of the output incremental signal (ΔV_b) with respect to the input incremental signal (ΔV_a) (see Fig. 1.21, right).

1.8 Incremental Circuit Model in the Frequency Domain

Because the incremental circuit model in the time domain is linear, we can apply the Fourier transformation to it. This changes (1.80) into

$$\mathbf{G}\Delta\mathbf{Y}(f) + j\omega\mathbf{C}\Delta\mathbf{Y}(f) + \Delta\mathbf{F}_e(f) + \mathbf{H}(f)\Delta\mathbf{Y}(f) = 0, \qquad (1.83)$$

where $\Delta\mathbf{F}_e(f)$ and $\mathbf{H}(f)$ are the Fourier transform of the excitation vector and the black-box impulse response, respectively. From (1.83) we can easily express the Fourier transform of the incremental circuit's response,

$$\Delta\mathbf{Y}(f) = -(\mathbf{G} + j\omega\mathbf{C} + \mathbf{H}(f))^{-1}\Delta\mathbf{F}_e(f). \tag{1.84}$$

The term

$$T(f) = -(\mathbf{G} + j\omega\mathbf{C} + \mathbf{H}(f))^{-1} = -(\mathbf{G} + j2\pi f\mathbf{C} + \mathbf{H}(f))^{-1} \tag{1.85}$$

is the circuit's transfer function. Multiplying the spectrum of the excitation by the transfer function results in the spectrum of the response. If the incremental excitation consists of sinusoidal signals with a common frequency f, the response signals are also sinusoidal with frequency f.

For the circuit in Fig. 1.21 (left) the incremental frequency-domain model is the same as the one in Fig. 1.21 (right). The only difference is that now we can consider the capacitor as an element with complex impedance $1/(j\omega C)$. The frequency-domain transfer function

$$T_{ba}(f) = \frac{\Delta V_b(f)}{\Delta V_a(f)} \tag{1.86}$$

is depicted in Fig. 1.24. The magnitude in decibels is obtained from

$$|T_{ba}(f)|_{dB} = 20\log|T_{ba}(f)|, \tag{1.87}$$

where $\log(x)$ denotes the base-10 logarithm of x. If the DC component of the excitation (V_0) is increased, the differential conductance of the diode increases and the transfer function magnitude decreases.

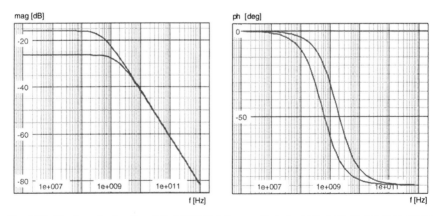

Fig. 1.24 The transfer function $\Delta V_b(f)/\Delta V_a(f)$ magnitude in decibels (*left*) and phase in degrees (*right*) for $V_0 = 0.8\,V$ (higher magnitude, phase curve on the *left*) and $V_0 = 1.2\,V$ (lower magnitude, phase curve on the *right*)

1.9   Noise Signals in LTI Systems

Until now we have encountered two classes of signals:

- Periodic signals with finite power ($\frac{1}{T}\int_{t_0}^{t_0+T}|y(t)|^2\mathrm{d}t < \infty$, where T is the period of the signal)
- Aperiodic signals with finite energy ($\int_{-\infty}^{\infty}|y(t)|^2\mathrm{d}t < \infty$)

The first category was modeled using the Fourier series and the second one using the Fourier transform. In this section we introduce another class of signals – noise signals with finite power. This means that

$$\lim_{T\to\infty}\frac{1}{T}\int_{-T/2}^{T/2}|y(t)|^2\mathrm{d}t < \infty. \tag{1.88}$$

Such signals are aperiodic ($T = \infty$) and yet their energy ($\int_{-\infty}^{\infty}|y(t)|^2\mathrm{d}t$) is infinite (that is, infinite if the whole range of time from $-\infty$ to ∞ is considered). Noise signals are generated by random processes [37]. A realization of a noise signal ($y(t)$) is just one among infinitely many possible realizations. All realizations of a particular noise signal have some common properties. To mathematically formulate this we will need the notion of a correlation function.

The correlation function $c_{xy}(\tau)$ of two signals $x(t)$ and $y(t)$ is defined as

$$c_{xy}(t, \tau) = E[x(t)y(t + \tau)], \tag{1.89}$$

where $E[\cdot]$ denotes expectation. If the random processes that generate $x(t)$ and $y(t)$ are stationary, c_{xy} depends only on τ. Furthermore, if the processes are ergodic, we can replace expectation by the limit of a time average,

$$c_{xy}(\tau) = \lim_{T\to\infty}\frac{1}{T}\int_{-T/2}^{T/2}x(t)y(t + \tau)\mathrm{d}t. \tag{1.90}$$

If $y(t) = x(t)$ the term autocorrelation function is used. Two signals are said to be uncorrelated if $c_{xy}(\tau) = 0$. The value of the autocorrelation function at $\tau = 0$ is equal to the signal's time average square value,

$$c_{yy}(0) = \lim_{T\to\infty}\frac{1}{T}\int_{-T/2}^{T/2}|y(t)|^2\mathrm{d}t. \tag{1.91}$$

In the remainder of this section we will restrict ourselves to stationary ergodic processes. Such processes are used for circuit component noise modeling. For signals generated by stationary processes, the time average square value can be treated as the signal's power.

The power spectrum density of a noise signal $y(t)$ is denoted by $S_{yy}(f)$ and is obtained as the Fourier transform of the signal's autocorrelation function, $c_{yy}(\tau)$,

$$S_{yy}(f) = \mathcal{F}(c_{yy}(\tau)). \tag{1.92}$$

Parseval's theorem relates the signal's power spectrum density to the signal's power,

$$\int_{-\infty}^{\infty} S_{yy}(f)df = c_{yy}(0) = \lim_{T \to \infty} \frac{1}{T} \int_{-T/2}^{T/2} |y(t)|^2 dt. \tag{1.93}$$

When an LTI system is excited by a noise signal $y_{in}(t)$, it responds with an output noise signal $y_{out}(t)$. The power spectrum density of the output signal is defined as

$$S_{out,out}(f) = |H(f)|^2 S_{in,in}(f), \tag{1.94}$$

where $H(f)$ is the Fourier transform of the LTI system's impulse response $h(t)$. This is no surprise because the power spectrum density of deterministic signals is also shaped by $|H(f)|^2$ as the signal pass through the LTI system.

The autocorrelation function of a superposition of two uncorrelated noise signals $x(t)$ and $y(t)$ is the sum of individual autocorrelation functions,

$$c_{x+y,x+y}(\tau) = c_{xx}(\tau) + c_{yy}(\tau). \tag{1.95}$$

As the Fourier transform is linear, this also holds for the noise power spectra,

$$S_{x+y,x+y}(f) = S_{xx}(f) + S_{yy}(f). \tag{1.96}$$

Very often signals are characterized by their RMS value,

$$y_{RMS} = \left(\lim_{T \to \infty} \frac{1}{T} \int_{-T/2}^{T/2} |y(t)|^2 dt \right)^{\frac{1}{2}}. \tag{1.97}$$

The RMS values of uncorrelated signals are added similarly as orthogonal vectors,

$$(x + y)_{RMS} = (x_{RMS}^2 + y_{RMS}^2)^{\frac{1}{2}}. \tag{1.98}$$

The latter is a direct consequence of (1.95). Suppose we obtain signal $y_{out}(t)$ by sending the noise signal $y_{in}(t)$ through an ideal passband filter ($f_1 \le f \le f_2$). Then the RMS value of the output signal is

$$y_{outRMS} = \left(\int_{-f_2}^{-f_1} S_{in,in}(f)df + \int_{f_1}^{f_2} S_{in,in}(f)df \right)^{\frac{1}{2}}. \tag{1.99}$$

Because autocorrelation is symmetric ($c_{yy}(-t) = c_{yy}(t)$), the power spectrum density $S_{yy}(f)$ is also symmetric ($S_{yy}(-f) = S_{yy}(f)$). Due to this property, the power spectrum density is often given in one-sided form,

$$S_{yy}^{+}(f) = 2S_{yy}(f), \tag{1.100}$$

which is defined only for positive frequencies ($f \geq 0$). This changes Parseval's theorem into

$$\int_{0}^{\infty} S_{yy}^{+}(f)\mathrm{d}f = c_{yy}(0) = \lim_{T \to \infty} \frac{1}{T} \int_{-T/2}^{T/2} |y(t)|^2 \mathrm{d}t. \tag{1.101}$$

The RMS value of a noise signal $y_{out}(t)$ obtained from $y_{in}(t)$ by filtering with a bandpass filter is now

$$y_{outRMS} = \left(\int_{f_1}^{f_2} S_{in,in}^{+}(f)\mathrm{d}f \right)^{\frac{1}{2}}. \tag{1.102}$$

1.10 Circuit Noise Modeling and Analysis

Circuit elements generate many different types of noise [54]. The most simple and common is thermal (Johnson–Nyquist) noise. Such noise is caused by the chaotic movement of electrons in the conductor and it is generated by every resistor in the circuit.

The resistor noise model consists of a noiseless resistor and a current source representing thermal noise (see Fig. 1.25). The power spectrum density of the current source is

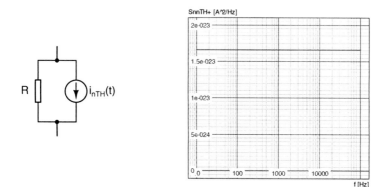

Fig. 1.25 Resistor noise model (*left*) and the power spectrum density of the noise current at $T = 300\,\mathrm{K}$ for a $1\,\mathrm{k\Omega}$ resistor

$$S_{nn\text{TH}}^+(f) = \frac{4Rhf}{R(e^{\frac{hf}{kT}} - 1)},\qquad(1.103)$$

where R is the resistance, k is the Boltzmann constant, and h is the Planck constant. Thermal noise is often considered "white" meaning that its power spectrum density is constant for all frequencies. This holds quite well for frequencies up to kT/h, where the exponential can be approximated by the first two terms of the Taylor series, resulting in

$$S_{nn\text{TH}}^+(f) = \frac{4kT}{R}.\qquad(1.104)$$

For $T = 300\,K$ the frequency above which the thermal noise power spectrum density cannot be considered white is $6\,THz$. The bandwidth of almost all circuits is lower than this frequency. Note that the polarity of the noise source is not important because the power spectrum density remains the same if the current is reversed.

The noise model of a diode is more complicated. Two additional noise signals appear at the p–n junction: shot noise and flicker noise ($1/f$ noise). Shot noise is caused by the random fluctuations of the diode current. The fluctuations occur because the current consists of a flow of discrete charges (electrons). The power spectrum density of a shot noise current $(i_{nS}(t))$ is

$$S_{nnS}^+(f) = 2qI,\qquad(1.105)$$

where q is the elementary (electron) charge and I is the average of the diode junction current $(i(t))$. The power spectrum density of the flicker noise current $(i_{nF}(t))$ is

$$S_{nnF}^+(f) = \frac{K_f I^{A_f}}{f},\qquad(1.106)$$

where K_f is the flicker noise coefficient and A_f is the flicker noise exponent.

The frequency at which the shot noise is equal to the flicker noise is called the corner frequency f_a,

$$f_a = \frac{K_f I^{A_f-1}}{2q}.\qquad(1.107)$$

Noise models of other circuit elements are similar to the diode noise model, where the noise is modeled as a noise current source with a given power spectrum density (see Fig. 1.26).

As noise signals are small by nature, we can assume that the circuit behaves linearly within the magnitude of the noise signals. Therefore, noise signals can be treated as incremental signals, and the incremental time-domain circuit model can be used to obtain the response of the circuit to its inherent noise sources. This response is superimposed to the circuit's operating point.

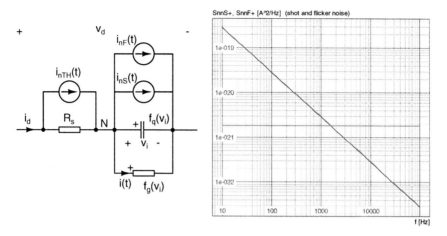

Fig. 1.26 Diode noise model (*left*) and the power spectrum density of the noise currents at $T = 300$ K for a diode with operating point current $I = 5.67$ mA, $K_f = 0.5$ fA, and $A_f = 1$ (*right*). The corner frequency (f_a) is 1.56 kHz

The noise sources are given in the frequency domain (through their power spectrum densities), thus the most natural way to analyze the incremental circuit is in the frequency domain (Sect. 1.8). Suppose a circuit has N noise sources. The processes that are modeled by the noise sources are usually uncorrelated, so we can assume that the noise sources themselves are also uncorrelated.

Due to the circuit's linearity, the analysis can be split into N parts, where every partial analysis results in the incremental circuit's response to one of the noise sources. Usually we are interested in the noise at some output (i.e., branch or node voltage, or branch current). Let $\Delta y_{out}(t)$ denote the incremental output signal for which we want to obtain the noise power spectrum density $\Delta S_{out,out}(f)$.

Suppose that in the i-th partial analysis the contribution of noise source $\Delta i_{n,i}(t)$ to the total noise signal $\Delta y_{out}(t)$ is evaluated. The contribution is obtained by analyzing the incremental circuit with all signal sources set to 0, except for the noise source $\Delta i_{n,i}(t)$. Let $\Delta y_{out,i}(t)$ denote the resulting noise signal contribution and $\Delta S_{out,out(i)}(f)$ its power spectrum density. Then the power spectrum density of the total noise can be obtained using the following equation (based on (1.96)):

$$\Delta S_{out,out}(f) = \sum_{i=1}^{N} \Delta S_{out,out(i)}(f). \qquad (1.108)$$

Note that every one of the N partial analyses is conducted in the frequency domain using the incremental circuit model (1.83). A partial analysis evaluates the circuit's transfer function ($T(f)$) from the noise source in question to the output signal $\Delta Y_{out}(f)$. The power spectrum density of the noise source is multiplied by $|T(f)|^2$ to obtain the power spectrum density $\Delta S_{out,out(i)}(f)$ of contribution $\Delta y_{out,i}(t)$ (see (1.94)).

1.11 Laplace Transformation and the s-Domain

The Laplace transform of a function $f(t)$ defined for $t \geq 0$ is obtained as

$$\mathcal{L}(f(t)) = F(s) = \lim_{\epsilon \uparrow 0} \int_{\epsilon}^{\infty} e^{-st} f(t) dt. \tag{1.109}$$

The lower limit of the integral is such that the whole Dirac impulse $\delta(t)$ is included (ϵ approaches 0 from below). The parameter s is complex and is often referred to as the complex frequency. The Laplace transformation transforms a function from the time domain onto the complex frequency domain (s-domain). The inverse Laplace transformation is defined as

$$\mathcal{L}^{-1}(F(s)) = \frac{1}{2\pi j} \int_{\gamma - j \cdot \infty}^{\gamma + j \cdot \infty} e^{st} F(s) ds. \tag{1.110}$$

The value of γ must be such that $\text{Re}(s_p) < \gamma$ holds for all singularities s_p of the transform $F(s_p)$. The Laplace transform $F_L(s)$ of $f(t)$ is related to the Fourier transform $F_F(f)$ through

$$F_F(f) = F_L(j2\pi f). \tag{1.111}$$

The Laplace transform of a periodic function $f(t)$ with period $T > 0$ can be expressed as

$$\mathcal{L}(f(t)) = \frac{1}{1 - e^{-Ts}} \int_{0}^{T} e^{-st} f(t) dt. \tag{1.112}$$

Suppose $Y(s)$ is given. We are interested in $y(t) = \mathcal{L}^{-1}(Y(s))$, in particular, its stability (i.e., whether $y(t)$ dies out as t approaches infinity). The stability depends on the singularities of $Y(s)$. A singularity s_p is a point where $|Y(s)|$ goes to infinity when s approaches s_p. If all of the singularities lie to the left of the imaginary axis (i.e., $\text{Re}(s_p) < 0$), then the response $y(t)$ dies out as t approaches infinity and $y(t)$ is stable.

The Laplace transformation handles differentiation and convolution in a similar manner to that of the Fourier transformation,

$$\mathcal{L}(\frac{d}{dt} f(t)) = s\mathcal{L}(f(t)) - \lim_{\epsilon \uparrow 0} f(\epsilon) = sF(s) - \lim_{\epsilon \uparrow 0} f(\epsilon) \tag{1.113}$$

$$\mathcal{L}(f(t) * g(t)) = \mathcal{L}(f(t)) \cdot \mathcal{L}(g(t)) = F(s) \cdot G(s). \tag{1.114}$$

More details on the Laplace transformation can be found in [47].

1.12   Circuit Analysis in the s-Domain, DC Solution Stability

Let us recall the linear equation of the incremental circuit in the time domain (1.80),

$$\mathbf{G}\Delta\mathbf{y}(t) + \mathbf{C}\frac{d}{dt}\Delta\mathbf{y}(t) + \Delta\mathbf{f}_e(t) + \mathbf{H}(t) * \Delta\mathbf{y}(t) = 0. \qquad (1.115)$$

Let $\Delta\mathbf{y}(0)$ denote the initial conditions (incremental response at $t = 0$) of the incremental circuit. If we apply the Laplace transformation to the circuit equation, we get

$$\mathbf{G}\Delta\mathbf{Y}(s) + \mathbf{C}(s\Delta\mathbf{Y}(s) + \Delta\mathbf{y}(0)) + \Delta\mathbf{F}_e(s) + \mathbf{H}(s)\cdot\Delta\mathbf{Y}(s) = 0, \qquad (1.116)$$

where $\Delta\mathbf{F}_e(s)$ and $\mathbf{H}(s)$ denote the Laplace transforms of the incremental excitation vector and black-box impulse response, respectively. The circuit's incremental response can now be expressed as

$$\Delta\mathbf{Y}(s) = -(\mathbf{G} + s\mathbf{C} + \mathbf{H}(s))^{-1}(\Delta\mathbf{F}_e(s) + \mathbf{C}\Delta\mathbf{y}(0)). \qquad (1.117)$$

When we are analyzing the circuit's response to small perturbations in the circuit's excitation, we assume that the initial conditions (the initial value of the incremental circuit's response) are zero. This simplifies (1.117) to

$$\Delta\mathbf{Y}(s) = -(\mathbf{G} + s\mathbf{C} + \mathbf{H}(s))^{-1}\Delta\mathbf{F}_e(s) = \mathbf{T}(s)\Delta\mathbf{F}_e(s). \qquad (1.118)$$

The elements of the incremental circuit depend on the DC solution around which the original circuit was linearized. The stability of the DC solution can be judged based on the incremental circuit's response. If the incremental circuit responds to incremental excitations that die out as t approaches infinity ($\lim_{t\to\infty}\Delta\mathbf{f}_e(t) = \mathbf{0}$) with an incremental response that also dies out ($\lim_{t\to\infty}\Delta\mathbf{y}(t) = \mathbf{0}$), we conclude that the operating point is stable.

Because (1.118) is linear, we can express components of the incremental response vector as

$$\Delta Y_k(s) = \sum_{l=1}^{n+m} T_{kl}(s)\Delta F_{e(l)}(s) \quad k = 1, 2, \ldots, n+m, \qquad (1.119)$$

where $T_{kl}(s)$ are functions of the complex frequency s and represent entries in the circuit's transfer function matrix $\mathbf{T}(s)$. The DC solution is stable if the singularities s_p of all $T_{kl}(s)$ lie to the left of the imaginary axis ($\mathrm{Re}(s_p) < 0$).

If the circuit contains no black-box elements, then $\mathbf{H}(s) = 0$ and the circuit's response becomes

$$\Delta\mathbf{Y}(s) = -(\mathbf{G} + s\mathbf{C})^{-1}\Delta\mathbf{F}_e(s). \qquad (1.120)$$

In this case the members of the circuit's transfer function matrix $\mathbf{T}(s)$ are rational functions of s,

$$T_{kl}(s) = \frac{P_{kl}(s)}{\prod_{l=1}^{N}(s - s_{p(l)})}, \qquad (1.121)$$

where $P_{kl}(s)$ are polynomials and $s_{p(l)}$ are the poles of the rational functions. Note that all $T_{kl}(s)$ share the same set of poles. The DC solution $\mathbf{y}_0$ is stable if all of the poles $s_{p(l)}$ corresponding to the incremental circuit linearized at $\mathbf{y}_0$ are located to the left of the imaginary axis ($\mathrm{Re}(s_{pi}) < 0$).

The poles of the CMOS inverter rings of different lengths are depicted in Fig. 1.27. The stable DC solution of the single-inverter ring and the unstable DC solution of the two-inverter and three-inverter rings is at $V_{in} = V_{in1} = V_{in2} = V_a =$

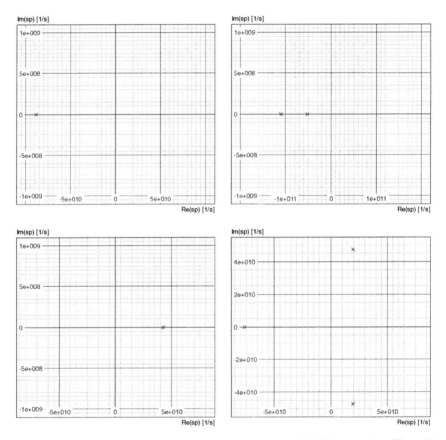

Fig. 1.27 The poles of the circuit's transfer function for various CMOS inverter rings. The stable DC solution's poles for the single-inverter ring from Fig. 1.11 (*top left*). The stable (*top right*) and the unstable (*bottom left*) DC solution's poles for the two-inverter ring from Fig. 1.12. The unstable DC solution's poles for the three-inverter ring (*bottom right*) from Fig. 1.15

$V_b = V_c = 0.8756748$ V. The stable solution of the two-inverter ring for which the poles are depicted in Fig. 1.27 is at $V_{in1} = 0$ V and $V_{in2} = 1.8$ V.

We can see that the stable DC solutions have poles only to the left of the imaginary axis. Unstable DC solutions have at least one pole to the right of the imaginary axis. The unstable solution of the three-inverter ring has a complex conjugate pair of poles to the right of the imaginary axis. Such DC operating points lead to oscillations. The oscillation frequency can be estimated from

$$ f = \frac{|\mathrm{Im}(s_p)|}{2\pi}, \tag{1.122} $$

where s_p is one of the two complex conjugate poles. In our case the frequency according to (1.122) is approximately 7.5 GHz. If we measure the oscillation frequency in Fig. 1.16 we get 5.1 GHz, which is quite close. The discrepancy is due to the nonlinearity of the circuit (the oscillations are not sinusoidal). If the magnitude of the oscillations is small enough so that the circuit's nonlinearities can be neglected, the two frequencies match.

Chapter 2
Short Tutorial

In this chapter we first discuss the installation of SPICE OPUS (for both Linux and Windows). We will go through the startup sequence and the command line options. After that, we will embark on a journey through the most important features of the SPICE OPUS simulator. They include the outline of the input file format, interactive interpreter, all types of analysis, and the built-in scripting language NUT-MEG. These features will be illustrated through a simple example (a bipolar junction transistor amplifier). For readers with a circuit analysis background (the enigmatic matrix and circuit model business), we provide circuit models for all kinds of analysis in SPICE.

2.1 Installation

Windows

Start the installer. Follow the instructions to install SPICE OPUS. The usual location of the installation is c:\SpiceOpus. The installed directory structure is as follows (PREFIX is usually c:\SpiceOpus):

PREFIX\bin ... the binaries (SPICE OPUS bin directory),

PREFIX\documentation ... html documentation,

PREFIX\lib ... SPICE OPUS lib directory,

PREFIX\lib\cm ... cm modules,

PREFIX\lib\helpdir ... help files,

PREFIX\lib\scripts ... scripts (e.g., spinit),

PREFIX\source ... cm module source files,

PREFIX\source\include ... include files for building cm modules,

PREFIX\source\cmlib ... C source files of cm modules,

PREFIX\build ... Microsoft Visual Studio project and workspace files for building cm modules.

After the installation is completed, the OPUSHOME environmental variable must be set to the directory where SPICE OPUS is installed (usually c:\SpiceOpus).

T. Tuma and Á. Bűrmen, *Circuit Simulation with SPICE OPUS*, Modeling and
Simulation in Science, Engineering and Technology, DOI 10.1007/978-0-8176-4867-1_2,
© Birkhäuser Boston, a part of Springer Science+Business Media, LLC 2009

Linux

After unpacking the tar gzipped archive, go to the newly created directory and run
the installation script (`install`) by typing
`./install PREFIX`
   Usually you will want to install SPICE OPUS in the `/usr/local` tree, so the
correct command is
`./spice_install /usr/local`
   The installation script looks for an older version of SPICE OPUS in `PREFIX`
and removes it before the latest version is installed. The following directories are
created or updated (`PREFIX` is usually `/usr/local`):
`PREFIX/bin` ... the binaries (SPICE OPUS bin directory),
`PREFIX/lib/spiceopus` ... SPICE OPUS lib directory,
`PREFIX/lib/spiceopus/documentation` ... html documentation,
`PREFIX/lib/spiceopus/cm` ... cm modules,
`PREFIX/lib/spiceopus/helpdir` ... help files,
`PREFIX/lib/spiceopus/scripts` ... scripts (e.g., `spinit`),
`PREFIX/src/spiceopus` ... cm module source files and makefiles,
`PREFIX/src/spiceopus/include` ... include files for building cm modules,
`PREFIX/src/spiceopus/cmlib` ... C source files of cm modules.

   The installer also removes the older (pre-2006) version of SPICE OPUS from
`PREFIX`. This affects the `PREFIX/lib/spice` directory and files `multidec`,
`multidec.bin`, `proc2mod`, `proc2mod.bin`, `spice3`, `spice3.bin`, and
`spice-config` in the `PREFIX/bin` directory.
   There is no need to set the `OPUSHOME` environmental variable as it is incorporated
into the `PREFIX/bin/spiceopus-config` shell script at installation. This script is
used for starting all binaries in the SPICE OPUS suite.

2.2 Starting SPICE OPUS

In Windows you can start SPICE OPUS by double-clicking on the corresponding
icon in the Start menu or on the desktop. The usual location of the installation is
`c:\SpiceOpus`. We will assume this directory throughout the book. The binary is
named `spiceopus.exe` and is located in the `c:\SpiceOpus\bin` directory.
   Under Linux the directory where the SPICE OPUS binary is installed is usually
also listed in the `PATH` environmental variable, so SPICE can be started by simply
typing `spiceopus`. We assume throughout this book that SPICE OPUS is installed
in the `/usr/local` tree.
   The complete syntax of the `spiceopus` command (Linux and Windows) is
`spiceopus [option1] [option2] ... [input_file]`

Table 2.1 SPICE OPUS command line options

Option	Meaning
-b	Batch mode, exit after the simulations are finished
-i	Interactive mode (default)
-c	Console mode
-g	GUI (graphical user interface) mode (default)
-pl *name*	Set the lib directory
-pe *name*	Set the bin directory
-pw *name*	Set the working directory
-n	Don't process the spinit file
-o, -o *name*	Send output to a file, default is spice.out
-r, -r *name*	Set the default raw file name, default is rawfile.raw

Table 2.1 lists the available command line options. In batch mode SPICE OPUS exits as soon as the input file is processed. In console mode graphics (e.g., plotting) is not available. The lib directory contains the help files and the spice startup script spinit. The working directory is by default the directory where the spiceopus binary was started from (Linux) or the directory set under "Start in" in the shortcut (Windows). The -o option redirects the output to the specified file (spice.out by default). In console mode it causes the output to be no longer visible in the console (as it is directed to a file).

After startup SPICE OPUS looks for the OPUSHOME environmental variable. If it exists, the spice bin directory is set to OPUSHOME/bin and the spice lib directory is set to OPUSHOME/lib (Windows) or OPUSHOME/lib/spiceopus (Linux). If the OPUSHOME environmental variable is not set, the following environmental variables are used:

SPICE_EXEC_DIR ... spice bin directory, and
SPICE_LIB_DIR ... spice lib directory.

If the -pl or -pe command line option is specified, it takes precedence over the settings provided by the environmental variables.

The installation paths c:\SpiceOpus and /usr/local are hardcoded into the Windows and Linux binaries. If SPICE OPUS is installed in a different directory under Windows, the environmental variable (OPUSHOME or SPICE_LIB_DIR and SPICE_EXEC_DIR) must be set for SPICE to work correctly. Under Linux no variables need to be set as the settings are hardcoded into the shell script spiceopus-config. This script is used for invoking the binaries from the SPICE OPUS suite.

A failure to provide the information on the location of SPICE OPUS installation to the SPICE OPUS binary (either through command line options or environmental variables) will prevent SPICE OPUS from running the startup script spinit, loading the supplied set of cm modules (extensions to SPICE OPUS based on XSPICE [11]), and disable the online help.

When the spiceopus binary is started (provided that the -n command line option is not specified), the spinit file is read from the scripts subdirectory of the

lib directory. spinit contains NUTMEG commands that are executed at startup. A
sample spinit file is provided with the installation.

Next the input file (if specified on the command line) is read and processed. After
the processing is finished, SPICE OPUS awaits further commands (if it was started
in interactive mode) or exits (if it was started in batch mode).

2.3 The Command Window

If SPICE OPUS is started in graphical user interface (GUI) mode, the command
window (Fig. 2.1) pops up right after startup. The central part of the window is
occupied by the SPICE terminal. The main menu offers the following options:

- **File/Print** (Ctrl+P) prints the contents of the terminal window.
- **File/Exit** (Alt+F4) closes the command window and exits SPICE OPUS.
- **Edit/Copy** (Ctrl+C) copies the marked text to the clipboard.
- **Edit/Paste** (Ctrl+V) pastes the text from the clipboard to the terminal window.
- **Edit/Select All** (Ctrl+A) selects the entire contents of the terminal window.
- **Edit/Loging** (Ctrl+L) enables/disables output to the output file specified by the
  -o command line option.
- **Edit/Clear Terminal History** (Ctrl+T) clears the text from the terminal
  window.
- **Control/Stop Execution** (Ctrl+Alt+S) stops the running simulation.
- **Window/Close All Plot Windows** closes all plot windows.
- **Help/About** displays information about the simulator.

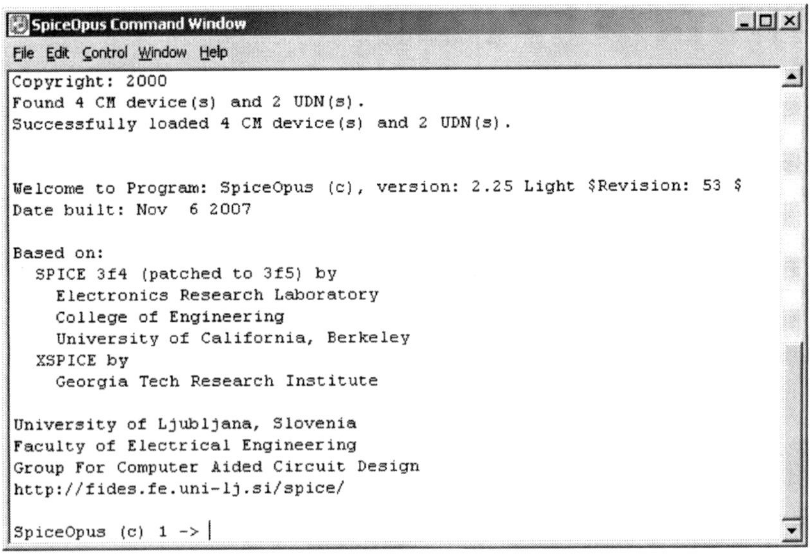

```
SpiceOpus Command Window                                      _|□|×|
File  Edit  Control  Window  Help
Copyright: 2000                                                    ▲
Found 4 CM device(s) and 2 UDN(s).
Successfully loaded 4 CM device(s) and 2 UDN(s).


Welcome to Program: SpiceOpus (c), version: 2.25 Light $Revision: 53 $
Date built: Nov  6 2007

Based on:
  SPICE 3f4 (patched to 3f5) by
    Electronics Research Laboratory
    College of Engineering
    University of California, Berkeley
  XSPICE by
    Georgia Tech Research Institute

University of Ljubljana, Slovenia
Faculty of Electrical Engineering
Group For Computer Aided Circuit Design
http://fides.fe.uni-lj.si/spice/

SpiceOpus (c) 1 -> |                                               ▼
```

Fig. 2.1 The SPICE OPUS command window

In GUI mode you can scroll through the terminal window by using the scrollbar or with the Page Up and Page Down keys. The command line editor allows you to use the Backspace, Delete, Home, End, Cursor Left, and Cursor Right keys for moving the cursor and editing the input. To discard the input, press the Esc key. Command history is available through the Cursor Up and Cursor Down keys. Ctrl+Cursor Up moves to the oldest command in the history, and Ctrl+Cursor Down moves to the newest command in the history.

An interactive session can be closed by entering the quit command or (if SPICE OPUS is running in the GUI mode) by selecting the **File/Exit** option in the main menu.

2.4 Describing a Circuit

2.4.1 Input File Structure

One describes circuits that are to be simulated by means of files. A file with a SPICE OPUS circuit description is called a netlist. Netlists can also include sequences of NUTMEG commands that are executed after the netlist is loaded and the circuit is parsed by the simulator. The general structure of a netlist is

```
Circuit name
circuit description
.control
NUTMEG commands
.endc
.end
```

The first line of a netlist is always interpreted as the circuit name. The control block (.control, NUTMEG commands, and .endc part of the netlist can be omitted). The commands from the control block are executed after the circuit description is parsed.

The parts of the netlist that describe the circuit, .control, .endc, and .end are interpreted as case insensitive. It is recommended that you type these lines in lowercase. The contents of the control block are case sensitive. Although some parts of it may be written in uppercase (e.g., strings that will be printed in the command window), lowercase is recommended here, too.

Leading spaces and tab characters at the beginning of the line are ignored. Comments in the netlist are denoted by an asterisk (*). Such a line is entirely interpreted as a comment and does not affect the behavior of the SPICE OPUS environment.

```
* This is a comment.
```

Comments can also comprise a part of the line. This can be achieved by using the $ character. The $ character and all the characters that follow up to the end of the line are interpreted as a comment. This type of comment is available only in the circuit description part of the netlist.

```
r1 (9 5) r=1k $ Resistor r1 is connected to nodes 9 and 5
```

You can split very long lines by using the line joining + character in the following manner:

```
Thisisanextremelylonglinewithnospacesinbetween.
Thisisanextremely
+longlinewithno
+spacesinbetween.
```

The first line of the above example and the last three lines both produce a single logical line with the same contents. It is recommended to put a space or a tab after the line joining + character. The following example illustrates why.

```
m1 (2 5 0 0) w=1u l=1u
+m=1u
```

If there is no trailing space or tab characters after l=1u the result will be

```
m1 (2 5 0 0) w=1u l=1um=1u
```

which is clearly incorrect as there is no space between l=1u and m=1u. Adding a space or a tab after the line joining + character automatically avoids such problems.

Take special care when you are writing formulas.

```
b1 (2 5) v=v(10)*v(20)
+ v(11)
```

The above two lines will produce

```
b1 (2 5) v=v(10)*v(20) v(11)
```

which obviously is not the correct formula. The + character was interpreted as a line joining + character.

You must also use caution regarding comments in conjunction with the line joining + character.

```
This is the first
* comment
+ part of the line.
```

The above example will result in a single line with the comment ignored.

```
This is the first part of the line.
```

2.4.2 Numbers in SPICE

You can write numbers in SPICE in the usual C style. So, for instance, you could write −0.00025 and 2500 as any of the following:

```
-0.00025    2500
-0.25e-3    2.5e3
-0.25m      2.5k
```

The last form is specific to SPICE. It uses the SI prefix milli- (10^{-3}) for the first number and kilo- (10^3) for the second one. The complete list of SI prefixes that are recognized by SPICE OPUS is given in Table 2.2.

Table 2.2 SI prefixes in SPICE

SI prefix	Notation in SPICE	Value	Example
femto-	f	10^{-15}	5f or 5e-15
pico-	p	10^{-12}	8p or 8e-12
nano-	n	10^{-9}	1.2n or 1.2e-9
micro-	u	10^{-6}	4.7u or 4.7e-6
milli-	m	10^{-3}	10m or 0.01
kilo-	k	10^{3}	100k or 100e3
mega-	meg	10^{6}	10meg or 1e7
giga-	g	10^{9}	1g or 1e9
tera-	t	10^{12}	5t or 5e12

Fig. 2.2 Simple circuit

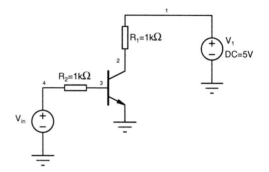

Be careful with milli- and mega-. A common mistake is to use an uppercase M as mega-. Because the circuit description is case insensitive, it will be interpreted as milli-. Therefore, 5M will actually be $5 \cdot 10^{-3}$ instead of $5 \cdot 10^{6}$. The correct way to obtain 5 mega would be to use 5meg.

In many different netlists you will often encounter not only SI prefixes, but also units. For instance, if you type 5uV it will be interpreted as $5 \cdot 10^{-6}$. The unit (volts) is ignored by SPICE. Writing units can also be a source of mistakes. Say you want to write 8 farads in the netlist and type 8F. SPICE circuit descriptions are case independent so it will be interpreted as 5 femto- $(5 \cdot 10^{-15})$. Generally, it is not recommended to write SI units in SPICE. Use SI prefixes only.

2.4.3 A Simple Circuit

Now let us take a look at a sample circuit and its description. Figure 2.2 shows a simple bipolar junction transistor (BJT) amplifier. The nodes are labeled with small numbers. Most elements (in SPICE also referred to as instances) require a single line in the circuit description. Take, for instance, resistor r1.

```
r1 (1 2) r=1k
```

The above line describes a resistor as the first letter on the line is an r. The name of the resistor is r1. It is connected between nodes 1 and 2 and its resistance is 1 kΩ. Note that the use of parentheses around the list of the instance's connections is optional.

Generally, instance names can be arbitrary strings that start with a letter (a..z) and contain only letters, numbers (0..9), and underscores (_). The same applies to node names, except that node names can also start with a number (in which case they can contain only numbers) or an underscore.

Valid node names would be for instance 10, _a, and in10 whereas 10a and #b are not valid node names. Similarly, rfb10 is a valid element name, but _rfb10 is not.

Now let us describe an independent voltage source. Take for instance v1.

```
v1 (1 0) dc=5
```

This tells SPICE that v1 is connected to nodes 1 (positive node) and 0 (negative node) and behaves as a 5 V DC source. You probably guessed by now that the names of independent voltage sources start with v.

Note that the ground node is named (0). This name is reserved for the global ground node. Every node in the circuit must have a DC path to the global ground node. If this is not the case, the set of modified nodal (MN) circuit equations [23] is not solvable (the matrix is singular). Problems occur when you try to analyze such circuits. SPICE OPUS has a mechanism that warns you when there are such floating nodes in the circuit. The warning is issued when you try to analyze the circuit. Of course, the analysis is skipped, if floating nodes are detected.

Finally, let us take a look at something we have not considered so far – the BJT. SPICE knows that it is dealing with a BJT from its name, which must start with q.

```
q1 (2 3 0) t2n2222
```

This line tells SPICE that the collector is connected to node 2, the base to node 3, and the emitter to node 0. The transistor is of type t2n2222. The type is also called "model name" in SPICE. Now we have said everything about the connectivity of the transistor, but nothing about its polarity (PNP or NPN) or its parameters (like the forward current gain). The latter two are provided to SPICE through a "model line."

```
.model t2n2222 npn
+ is=19f  bf=150 vaf=100 ikf=0.18  ise=50p
+ ne=2.5  br=7.5  var=6.4 ikr=12m   isc=8.7p
+ nc=1.2  rb=50   re=0.4  rc=0.3     cje=26p tf=0.5n
+ cjc=11p tr=7n   xtb=1.5 kf=0.032f af=1
```

This line tells SPICE that all t2n2222 transistors in the circuit are NPN transistors with the parameters listed in the four lines that follow. The meaning of the parameters can be found in Sect. 3.6.18. Note that the above excerpt represents a single line that was split using a line joining + character. For an exercise, write the description of a Darlington pair consisting of t2n2222 transistors q1a and q1b that is connected to the same three nodes as q1.

Finally, the complete circuit description would be

```
A simple circuit
v1 (1 0) dc=5
r1 (1 2) r=1k
q1 (2 3 0) t2n2222
r2 (3 4) r=1k
vin (4 0) dc=0.68
.model t2n2222 npn
+ is=19f   bf=150 vaf=100 ikf=0.18  ise=50p
+ ne=2.5   br=7.5  var=6.4 ikr=12m  isc=8.7p
+ nc=1.2   rb=50   re=0.4  rc=0.3   cje=26p  tf=0.5n
+ cjc=11p  tr=7n   xtb=1.5 kf=0.032f af=1
.end
```

Independent voltage source `vin` is a DC source set to 0.68 V. This establishes the correct bias for the amplifier and puts it in the class A operating region.

To learn more about circuit description syntax read Chap. 3.

2.5 Operating Point (OP) Analysis

Regarding operating regions, let us check if the amplifier in Fig. 2.2 is truly a class A amplifier. Save the netlist to a file named `sample1` in your working directory and start SPICE OPUS. Now you must load the circuit into the simulator. But first change the working directory to the place where the `sample1` netlist is stored (we'll assume the directory is named `c:\samples`). Enter the following command in the SPICE OPUS command window:

```
cd c:/samples
```

SPICE responds with

```
current directory: C:/samples
```

and you are back at the prompt. Now load the circuit by typing

```
source sample1
```

SPICE outputs the name of the circuit

```
Circuit: A simple circuit
```

and awaits further commands. Now we will solve for the operating point of the circuit. Issue the following command:

```
op
```

It takes only a moment and you are back at the prompt. For small circuits the operating point analysis is very fast. Although it seems that nothing happened, the results of analysis are awaiting your inspection in the computer's memory. Type

```
display
```

and you will be flooded with a list of so-called vectors. Vectors are groups of numbers that represent results of the simulation. In our case every vector contains only one number, because the operating point is described by a single numeric value per every circuit quantity like voltage or current.

```
Here are the vectors currently active:

Title: A simple circuit
Name: op1 (Operating Point)
Date: Tue Nov 23 11:51:12  2004

    V(1)                : voltage, real, 1 long [default scale]
    V(2)                : voltage, real, 1 long
    V(3)                : voltage, real, 1 long
    V(4)                : voltage, real, 1 long
    q1#base             : voltage, real, 1 long
    q1#collector        : voltage, real, 1 long
    q1#emitter          : voltage, real, 1 long
    v1#branch           : current, real, 1 long
    vin#branch          : current, real, 1 long
```

The output starts with the circuit name followed by the name of the group of vectors that are being listed (op1). Suppose you performed another operating point analysis; then the group would be named op2. In SPICE such groups are called plots. Next comes the date when the plot was created, followed by the list of vectors. For every vector SPICE displays its type and length. One vector is also marked to be the default scale (default vector) of the plot. We will get back to default scales when we work with graphs, but for now just ignore it. Let's take a look at the voltage at node 2.

```
print v(2)
```

SPICE promptly responds with

```
v(2) = 2.837273e+000
```

This is the potential of node 2 in volts (2.8 V). The potential of node 0 is assumed to be 0 V. As we can see, the amplifier is in the class A operating region because the operating point of the output is between the highest and the lowest voltage in the circuit.

When SPICE calculates the operating point of the circuit, it actually solves a set of nonlinear equations that describe the circuit in Fig. 2.3. This circuit is the DC model of the circuit in Fig. 2.2. Note that the two controlled current sources are nonlinear functions of node voltages. There are three internal nodes in the BJT model. These nodes are called q1#base, q1#collector, and q1#emitter in SPICE.

You probably guessed that node voltages are accessed through expressions of the form v(*node_name*). Differential node voltages can be accessed by using the syntax v(*node1*,*node2*). Such notation is equivalent to typing v(*node1*)-v(*node2*). Besides node voltages, SPICE also evaluates the currents flowing through independent voltage sources (v1#branch and vin#branch). You can access these currents by typing i(*source_name*). The currents are considered positive if they flow into the positive node of the voltage source.

Let's print the value of the current supplied by v1 to the transistor.

```
print i(v1)
```

SPICE responds with

```
i(v1) = -2.16273e-003
```

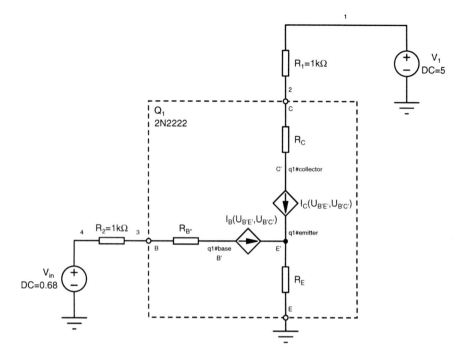

Fig. 2.3 Model of circuit in Fig. 2.2 evaluated in the operating point analysis

The current is $2.16\,\text{mA}$. Look at the polarity of the current. It is negative because positive currents flow into the positive node of the voltage source, whereas in our case the current flows out of the voltage source's positive node.

As you can see, SPICE returns only the node voltages and currents flowing through voltage sources. If you are interested in some current that is not available, you should put a 0 V voltage source in series with the branch where the current flows. Suppose you want to measure the current that flows from node 2 into capacitor c1. You should change the description of capacitor c1 and add a 0 V voltage source to the netlist.

```
vmeas (2 measure_2) dc=0
c1 (measure_2 0) c=1n
```

There are other ways for obtaining device currents (using the save command), but we will defer these to later chapters.

One of the main advantages of interactive operation is that you can change the circuit without altering the netlist and then resimulate it. Suppose you would like to see what effect a lower operating voltage has on the output node's operating point. Type

```
let @v1[dc]=3.3
```

Repeat the operating point analysis and print the output node's voltage. You should get 1.18 V. Note that the netlist remained unchanged, yet we still simulated a different circuit. In order to restore the circuit that was loaded from the netlist, type

```
reset
```

All commands you issued in the SPICE OPUS command window can be embedded into the netlist. Add the following before the .end line in the netlist.

```
.control
echo 5V power source
op
print v(2) i(v1)
echo 3.3V power source
let @v1[dc]=3.3
op
print v(2) i(v1)
echo Restoring original circuit
reset
.endc
```

The echo command displays a string in the command window. After saving the netlist and reloading it (source sample1), you will notice that the commands between .control and .endc that constitute the control block are executed. The following output appears:

```
Circuit: A simple circuit

5V power source
v(2) = 2.837273e+000
i(v1) = -2.16273e-003
3.3V power source
v(2) = 1.175656e+000
i(v1) = -2.12434e-003
Restoring original circuit
```

This way you can automate analyses so that you don't have to type all the commands in the command window. After the execution of the control block is finished, you are presented with the SPICE OPUS prompt and may continue your interactive SPICE session.

See Sect. 4.1 to find out more about operating point analysis.

2.6 Operating Point Sweep (DC), Plotting of Simulation Results

Suppose you would like to know how the operating point changes as the DC voltage of the input source vin changes. In such cases you can use the dc analysis. Let us look at an example:

```
dc vin 0 1.4 0.01
```

The DC analysis sweeps the input voltage source from 0 to 1.4 V in 10 mV steps. For every vin value the operating point of the circuit is evaluated. The operating points for a single node are collected in a vector which has as many points as there were operating point evaluations. Look at the list of vectors by typing display.

```
Title: A simple circuit
Name: dc1 (DC transfer characteristic)
Date: Tue Nov 23 14:50:16  2004

    V(1)                : voltage, real, 140 long
    V(2)                : voltage, real, 140 long
    V(3)                : voltage, real, 140 long
    V(4)                : voltage, real, 140 long
    q1#base             : voltage, real, 140 long
    q1#collector        : voltage, real, 140 long
    q1#emitter          : voltage, real, 140 long
    sweep               : voltage, real, 140 long [default scale]
    v1#branch           : current, real, 140 long
    vin#branch          : current, real, 140 long
```

The list contains the same vectors as those produced by the operating point analysis, except that the vectors now comprise 140 points each. Vector `sweep` was not present in the operating point analysis results. This vector contains the values of the swept source. Take a look at the output node's voltage with respect to the sweep vector.

```
print sweep v(2)
```

The following output is produced:

```
A simple circuit
DC transfer characteristic  Tue Nov 23 14:50:16  2004
-----------------------------------------------------------
Index   sweep             v(2)
-----------------------------------------------------------
0       0.000000e+000     5.000000e+000
1       1.000000e-002     5.000000e+000
2       2.000000e-002     5.000000e+000
3       3.000000e-002     5.000000e+000
4       4.000000e-002     5.000000e+000
5       5.000000e-002     5.000000e+000
...        ...               ...
138     1.380000e+000     1.299366e-001
139     1.390000e+000     1.295236e-001
```

Because every vector now contains more than one point, the output has the form of a table. The first column represents the index that runs from 0 to 139 as we go through the 140 points, the second column contains the swept source's dc parameter values, and the last column the corresponding output node voltage.

Of course, this tabular output is not as interesting as a graph. The simplest possible way to plot a graph is to use the `plot` command and list the vectors that we want to plot against the default scale of the current plot.

```
plot v(2)
```

A plot window shown in Fig. 2.4 appears. Do not be confused by the axis labels (`x`, `Real` and `y`, `Imag`). They are just the default labels provided by SPICE OPUS. There is no legend in the graph, so if there were multiple curves plotted in a single plot window it would be hard to recognize the individual curves. In such cases automatic curve identification comes in handy. Move the mouse cursor in the neighborhood of the curve and press SPACE. A green line connects the cursor with the curve, and the name of the curve appears in the bottom of the plot window (Fig. 2.5 left).

Fig. 2.4 Output voltage with
respect to default vector

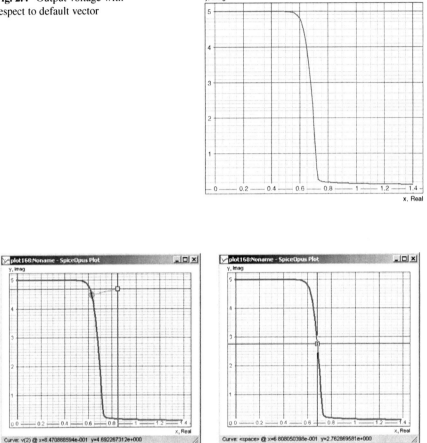

Fig. 2.5 *Left*: identifying a curve. *Right*: measuring with the mouse cursor

The position of the cursor is displayed in the bottom of the plot window (Fig. 2.5 right). It changes as you move the cursor.

Now we will zoom the curve so we can measure the gain. Make the plot window active and press Z. The cursor changes shape, indicating that the box zoom mode is on. If you want to turn the zoom mode off, press Z again. If you turned the zoom mode off, turn it back on, move the cursor to the upper left part of the rectangle you wish to zoom, and press the left mouse button. Hold the button down, drag the mouse, and in this way select the rectangle you want to zoom in on. After you release the left mouse button, the selected part of the graph is zoomed and the cursor is no longer in the box zoom mode. By pressing Ctrl+A the graph is autoscaled back to its full size.

Click with the right mouse button on the graph in the plot window. A menu appears (Fig. 2.6). The items in the menu are

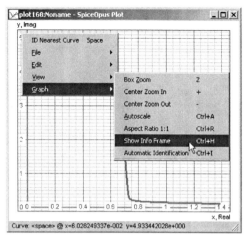

Fig. 2.6 The context menu of a graph obtained by clicking with the right mouse button on the graph

- **ID Nearest Curve** (SPACE) identifies the curve that is nearest to the cursor.
- **File/Print** (Ctrl+P) prints the contents of the plot window.
- **File/Close** (Alt+F4) closes the plot window.
- **File/Close All Plots** closes all plot windows.
- **Edit/Copy** (Ctrl+C) copies the selected text from the info frame to the clipboard.
- **Edit/Freeze Messages** (Ctrl+F) freezes the output in the info frame, so that you can select it and copy it to the clipboard.
- **Edit/Clear Marker** (Esc) clears the marker that was positioned by clicking with the left mouse button on the graph.
- **View/Box Zoom** (Z) turns box zoom mode on/off.
- **View/Center Zoom In** (+) zooms in on the graph around the cursor.
- **View/Center Zoom Out** (−) zooms out on the graph around the cursor.
- **View/Autoscale** (Ctrl+A) autoscales the graph.
- **View/Aspect Ratio 1:1** (Ctrl+R) corrects the aspect ratio of the graph to 1:1 (with aspect ratio 1:1 circles look like true circles).
- **View/Show Info Frame** (Ctrl+H) turns the info frame on/off.
- **View/Automatic Identification** (Ctrl+I) turns automatic curve identification on/off (with automatic identification turned on the nearest curve is automatically identified after the cursor does not move for some period of time).
- **Graph/Curve Representation/Default** displays the curves on the graph in their original style.
- **Graph/Curve Representation/Points** displays the curves on the graph in point mode (Fig. 2.7 left).
- **Graph/Curve Representation/Line** displays the curves on the graph in line mode (this is the default).

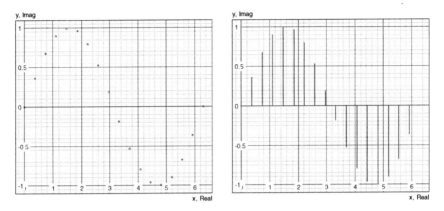

Fig. 2.7 *Left*: curve in point mode. *Right*: curve in comb mode

- **Graph/Curve Representation/Comb** displays the curves on the graph in comb mode (Fig. 2.7 right).
- **Graph/Curve Representation/Width: Default** selects the line thickness for the curves that was set as the graph was plotted.
- **Graph/Curve Representation/Width: x** forces the desired line thickness for the curves in the graph.
- **Graph/Curve Representation/Point: Default** selects the point size that was set as the curves were plotted. Applies only to curves displayed in point mode.
- **Graph/Curve Representation/Point: x** forces the desired point size for the curves displayed in point mode.
- **Graph/Grid/X Log Scale (for x–y grid)** turns logarithmic x-axis on/off.
- **Graph/Grid/Y Log Scale (for x–y grid)** turns logarithmic y-axis on/off.
- **Graph/Grid/x–y Grid** switches to x–y grid mode.
- **Graph/Grid/Polar Grid** switches to polar grid mode.
- **Graph/Grid/Smith Z Grid** switches to Smith impedance grid.
- **Graph/Grid/Smith Y Grid** switches to Smith admittance grid.

We defer the explanation of the remaining items from the **Graph** and **Graph/ Units** menus to later sections in this chapter, after the user becomes familiar with complex numbers and AC analysis in SPICE.

Select **View/Show Info Frame**. A frame appears in the bottom of the plot window (Fig. 2.8 left) displaying cursor position and nearest curve identification. Zoom in on the part of the curve shown in the left part of Fig. 2.8. Place the cursor on the curve where the input voltage is 0.658 V. Press the left mouse button. A marker appears. At the same time the info frame starts to show the marker position (in the Marker row). As you move the mouse cursor, the info frame displays the absolute cursor position (Cursor row) and its relative position with respect to the marker (Delta row). Move the cursor so that the line connecting the cursor and the marker is parallel with the curve (see Fig. 2.8 right). The gain can be obtained by looking

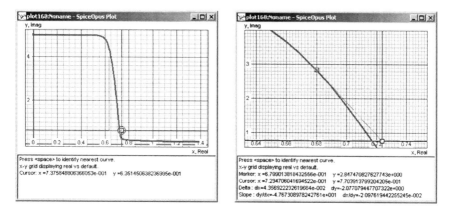

Fig. 2.8 *Left*: zooming the curve for gain measurement. *Right*: measuring the gain

at the dy/dx value in the Slope row of the info frame where the quotient of the two
values from the Delta row is displayed. The gain should be somewhere around −48.

To find out more about operating point sweep analysis see Sect. 4.2. More details
about the `plot` command can be found in Sect. 5.8.

2.7 Small Signal Analysis (AC and TF)

In small signal analysis SPICE first evaluates the operating point of the circuit. Then
the circuit is linearized around the obtained operating point. This means that the
capacitances become linear with their capacitance equal to corresponding dq/du
derivatives at the circuit's operating point. All nonlinear resistances and controlled
sources become linear. They are now represented by conductance matrices with
matrix elements equal to corresponding di/du derivatives at the circuit's operating
point. Figure 2.9 shows the small signal model of the circuit in Fig. 2.2.

All independent sources that have no ac parameter specified are turned off (i.e.,
voltage sources represent short circuits, and current sources represent open circuits).
The remaining independent sources are represented by complex voltages and currents. Parameter `acmag` sets the complex magnitude of a source in AC analysis. To
set the phase of an independent source, one can use the `acphase` parameter. By
default, the phase is set to 0.

Next the circuit is analyzed across a set of frequencies. Because the circuit is
linear, it can be represented by a complex admittance matrix that depends only on
the operating point of the circuit and the frequency. Effectively, AC analysis evaluates the response of the linearized circuit to a set of sinusoidal inputs (defined
by independent sources that have the `acmag` parameter set to some nonzero value).
The magnitude of the sinusoidal input is determined by the `ac` parameter of every
independent source. By default, the value of the `acmag` parameter is 0.

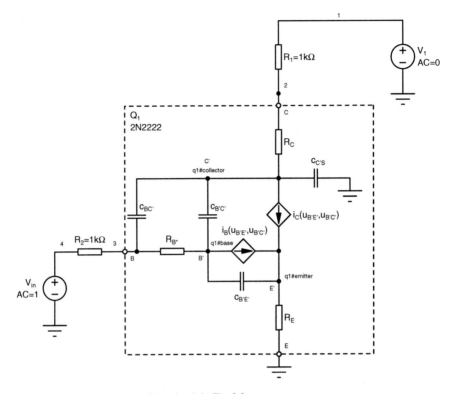

Fig. 2.9 Small signal model of the circuit in Fig. 2.2

Note that you do not have to set the `acmag` parameter to some small value in order to remain in the linear region around the operating point. The circuit is already linearized before the small signal response is evaluated. So if you get an output magnitude of, say, 5 V at input excitation magnitude set to 1 mV, you will get 5,000 V at input excitation magnitude 1 V. This does not mean that your circuit is capable of producing a 5 kV output from 1 V input, but merely that the small signal gain of the circuit is 1,000 (or in engineering lingo, 60 dB). Probably your circuit is capable of producing only a few volts of output magnitude. In the real world (where circuits are not linearized) an excitation in excess of a couple of millivolts will cause saturation effects. Always keep in mind that in AC analysis you are dealing with a linearized circuit.

The resulting vectors contain as many components as there were frequencies at which the circuit was analyzed. Individual vector components are complex and represent the magnitude and the phase of the response. The default scale of the result is the frequency, which is also a complex quantity, but with the imaginary part set to 0.

Before we start an AC analysis, we must set the `acmag` parameter of the excitation source `vin` to 1. Therefore, we replace

```
vin (4 0) dc=0.68
```

with

```
vin (4 0) dc=0.68 acmag=1
```

and reload the netlist (`source sample1`). Then we can start an AC analysis by typing

```
ac dec 10 100 1g
```

The command starts an AC sweep across the frequency range from 100 Hz to 1 GHz. The sweep is logarithmic with 10 points per decade (keyword `dec`). Let us take a look at the contents of the plot created by the AC analysis by typing `display`.

```
Title: A simple circuit
Name: ac1 (AC Analysis)
Date: Fri Nov 26 11:10:39  2004

    V(1)                   : voltage, complex, 71 long
    V(2)                   : voltage, complex, 71 long
    V(3)                   : voltage, complex, 71 long
    V(4)                   : voltage, complex, 71 long
    frequency              : frequency, complex, 71 long, grid = xlog [default scale]
    q1#base                : voltage, complex, 71 long
    q1#collector           : voltage, complex, 71 long
    q1#emitter             : voltage, complex, 71 long
    v1#branch              : current, complex, 71 long
    vin#branch             : current, complex, 71 long
```

We see that the plot is named `ac1` and that it contains the same set of vectors as the plot that was the result of an OP analysis. Vector `frequency` is new and contains the frequencies at which the small signal circuit model (Fig. 2.9) was analyzed. We can also see that all vectors are complex.

To learn some more, let us print the frequency and the output voltage.

```
print frequency v(2)
```

We get

```
A simple circuit
AC Analysis  Fri Nov 26 11:15:05  2004
--------------------------------------------------------------------------
Index    frequency                       v(2)
--------------------------------------------------------------------------
0     1.000000e+002, 0.000000e+000   -4.720902e+001,  1.214921e-002
1     1.258925e+002, 0.000000e+000   -4.720901e+001,  1.529495e-002
2     1.584893e+002, 0.000000e+000   -4.720901e+001,  1.925520e-002
3     1.995262e+002, 0.000000e+000   -4.720901e+001,  2.424085e-002
4     2.511886e+002, 0.000000e+000   -4.720900e+001,  3.051742e-002
5     3.162278e+002, 0.000000e+000   -4.720899e+001,  3.841915e-002
...
69    7.943282e+008, 0.000000e+000    6.061657e-003, -9.713294e-004
70    1.000000e+009, 0.000000e+000    4.036111e-003, -1.151247e-003
```

The second and fourth columns represent the real part of the frequency and the output voltage vector. The third and the fifth columns represent the imaginary part of these two vectors. As an exercise you can print the input node voltage (at node 4). It should be real with magnitude 1 V, as it is dictated by the `ac` parameter of the input voltage source `vin`.

Pictures say more than words, so we will take a look at the output voltage magnitude in the logarithmic y-axis scale.

```
plot abs(v(2)) ylog
```

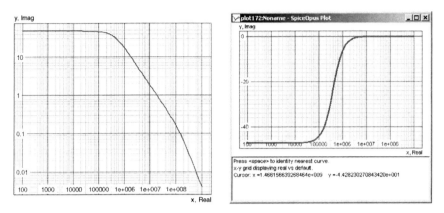

Fig. 2.10 *Left*: output voltage magnitude. *Right*: output voltage, real part

The plot window depicted in Fig. 2.10 (left) pops up.

The x-axis is logarithmic, too. This is due to the frequency vector forcing a logarithmic axis. Let us see what happens if we plot just the output voltage without the abs function.

```
plot v(2)
```

After the plot window pops up, turn the info frame on (Ctrl+H which results in the plot window from Fig. 2.10, right). The graph represents the real part of the output voltage with respect to frequency (info frame says x-y grid displaying real vs. default). If you want to get the imaginary part of the vector, just right-click the graph and select **Graph/Imaginary**. There is another way to display the imaginary part of the vector. Type

```
plot imag(v(2))
```

This will result in a real vector with components equal to the imaginary part of the output voltage. This way the plot window will show x-y grid displaying real vs. default, but the real part it displays will in fact be the imaginary part of the output voltage.

The plot window by default displays the real part of a vector. If you select **Graph/Imaginary (imag)** in the plot window menu, you will get a horizontal line. This happens because you are now displaying the imaginary part of the plotted vector (which in turn is 0 because the vector was real).

For the x-axis value the real part of the default vector is used (in our case frequency) unless otherwise specified with the following syntax.

```
plot v(2) vs frequency
```

This will result in the same graph as the previous plot command, except that we now have explicitly specified to use the frequency vector for the x-axis.

Let us now take a brief look at the remaining options in the plot window menu.

- **Graph/Real (real)** displays the real part of the plotted vectors.
- **Graph/Imaginary (imag)** displays the imaginary part of the plotted vectors.
- **Graph/Magnitude (mag)** displays the magnitude of the plotted vectors. See the **Graph/Units** submenu for additional settings.
- **Graph/Phase (phase)** displays the phase of the plotted vectors. See the **Graph/Units** submenu for additional settings.
- **Graph/R (realz)** interprets the complex vectors as reflectances[1] and plots the real part of the corresponding normalized impedance.
- **Graph/X (imagz)** interprets the complex vectors as reflectances and plots the imaginary part of the corresponding normalized impedance.
- **Graph/G (realy)** interprets the complex vectors as reflectances and plots the real part of the corresponding normalized admittance.
- **Graph/B (imagy)** interprets the complex vectors as reflectances and plots the imaginary part of the corresponding normalized admittance.
- **Graph/Complex (cx), x–y grid** plots the real part of the complex vector on the x-axis and the imaginary part on the y-axis.
- **Graph/Complex (cx), polar grid** same as previous, except that it turns on polar grid.
- **Graph/Complex (cx), Smith Z grid** turns on Smith impedance grid. Vectors are interpreted as reflectances and are plotted as in the **Complex (cx), x–y grid** mode.
- **Graph/Complex (cx), Smith Y grid** turns on Smith admittance grid. Vectors are interpreted as reflectances and are plotted as in the **Complex (cx), x–y grid** mode.

Where it is not explicitly stated, the x–y grid is used.

To choose the type of the grid you can use the **Graph/Grid** menu. The **Graph/Units** menu enables you to select the units for the angle and magnitude while displaying the magnitude (**Graph/Magnitude**) or phase (**Graph/Phase**) of the complex vectors.

Now, say you want to plot the polar chart for the output voltage. You could plot the complex output voltage vector and then switch to complex polar grid mode (**Graph/Complex (cx), polar grid**) or you could do this from the command line by typing

```
plot mode cx polar v(2)
```

The plot window in Fig. 2.11 (left) pops up.

What usually interests us when dealing with amplifiers is the gain. The gain of a linearized circuit is the output voltage divided by the input voltage. The gain is complex. Take a look at the magnitude of the gain in dB ($20\log_{10}(\cdot)$ of gain magnitude).

```
plot db(v(2)/v(4))
```

---

[1] A reflectance is the quotient of the reflected signal and the incoming signal calculated in the frequency domain. See [48] for details.

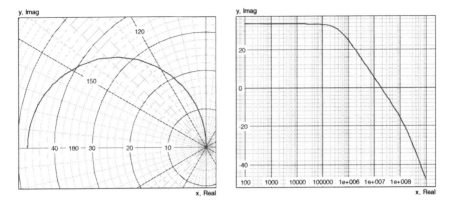

Fig. 2.11 *Left*: polar plot of output voltage. *Right*: voltage gain in dB

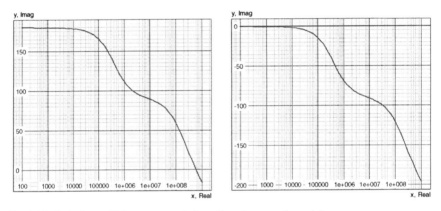

Fig. 2.12 *Left*: phase. *Right*: unwrapped phase (based on negative gain)

The plot window in Fig. 2.11 (right) pops up. We could leave out the division by output input voltage as it is 1 V for every frequency (dictated by the ac parameter of the input voltage source `vin`). Nothing would change if we wrote `plot db(v(2))`. Of course, this holds only when the input voltage source's ac parameter is set to 1.

Now we will take a look at the phase. If we type

```
plot ph(v(2)/v(4))
```

we get the phase plot, but unfortunately the phase is in radians. To set degrees as the angle unit in SPICE, we must first type

```
set units=degrees
```

and then plot the phase again. This time, the phase is in degrees (Fig. 2.12, left).

Now suppose you want the phase to start at 0°. Because our amplifier is an inverting amplifier, we must plot the phase of the negative gain. This can be done by entering

```
plot ph(-v(2)/v(4))
```

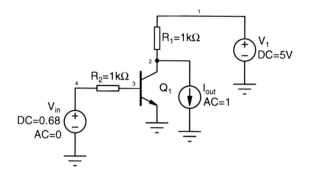

Fig. 2.13 Measuring the output impedance of the circuit in Fig. 2.2

From the resulting graph we can see that the phase wraps around at ±180°. This effect can be quite annoying. Fortunately, it can be avoided by using the unwrap function.

```
plot unwrap(ph(-v(2)/v(4)))
```

The result is shown in Fig. 2.12 (right). Another interesting item is the input impedance. You can obtain the absolute value of the input impedance by typing

```
plot abs(-v(4)/i(vin))
```

The result is in Fig. 2.14 (left). This is the impedance felt by the input voltage source vin. To measure the output impedance, we must first add a current source with its ac parameter set to 1 to the output (see Fig. 2.13). At the same time the ac parameter of the input source must be set to 0. This way the excitation is now at the output (output current source) and not at the input. As a response we can observe the output impedance as the negative voltage of the output node. To summarize the changes, we should replace

```
vin (4 0) dc=0.68 acmag=1
```

with

```
vin (4 0) dc=0.68 acmag=0
iout (2 0) dc=0 acmag=1
```

In AC analysis the iout independent current source pulls current from node 2. It has no effect on the operating point because its dc parameter is set to 0 meaning that the source actually is an open circuit.

Now we can type

```
ac dec 10 100 1g
plot abs(-v(2))
```

A plot window (Fig. 2.14, right) pops up, showing the output impedance magnitude with respect to frequency. Note that this output impedance is felt by any load connected between node 2 and ground (where the output current source is connected) when the input is shorted.

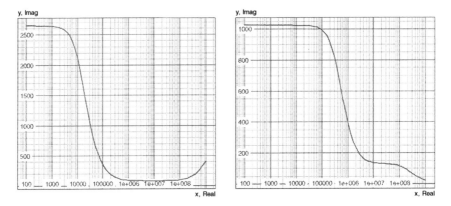

Fig. 2.14 *Left*: input impedance magnitude. *Right*: output impedance magnitude

So much for the AC analysis. Let us revert back to the original circuit in Fig. 2.2. Make sure that you remove `iout` from the circuit and that the `ac` parameter of the input voltage source `vin` is set to 1. Reload the circuit in SPICE (`source sample1`).

We are often interested in the low frequency small signal response. For that we need not do a complete AC analysis. We can simply say

```
tf v(2) vin
```

This starts a transfer function (TF) analysis. The circuit model is the same as for an AC analysis, except that all reactances (like capacitors and inductors) are removed from the circuit. What remains is a pure DC linearized circuit. Of course, the linearization depends on the operating point of the circuit. This circuit model is analyzed by the TF analysis.

The first parameter of the `tf` command defines the output of the circuit. It can also be a differential voltage (e.g., `v(1,2)`). The second parameter defines the input excitation source. Note that the input excitation source does not have to be set to `acmag=1`. The `acmag` parameter affects AC analysis and NOISE analysis only.

To see what kind of results TF analysis produces, type `display`. You should get the following:

```
Title: A simple circuit
Name: tf1 (Transfer Function)
Date: Fri Nov 26 13:42:33 2004

    input_impedance      : voltage, real, 1 long
    output_impedance     : voltage, real, 1 long
    transfer_function    : voltage, real, 1 long [default scale]
```

Now print the three vectors by typing

```
print input_impedance output_impedance transfer_function
```

You should get the following result:

```
input_impedance = 2.672775e+003
output_impedance = 9.774287e+002
transfer_function = -4.72090e+001
```

If we compare these numbers to the gain (Fig. 2.11, right), input impedance, and output impedance (Fig. 2.14) at low frequencies, we can see that they are identical. No wonder – the TF analysis worked with the same linearized circuit as the AC analysis, except that the reactances were missing. Therefore, the result should be the same as that for the AC analysis at low frequencies.

More details about small signal analysis can be found in Sects. 4.4 (AC analysis) and 4.3 (TF analysis).

2.8 Pole-Zero Analysis (PZ)

The pole-zero (PZ) analysis calculates poles and zeros of a transfer function. Any transfer function of a lumped circuit can be written in the form

$$H(s) = A \frac{(s - z_1)(s - z_2) \cdot \ldots \cdot (s - z_m)}{(s - p_1)(s - p_2) \cdot \ldots \cdot (s - p_n)}. \tag{2.1}$$

The m complex numbers $z_1, z_2, \ldots, z_m$ represent the zeros, and the n complex numbers $p_1, p_2, \ldots, p_n$ represent the poles of the transfer function.

PZ analysis takes place in the frequency domain. The circuit is first linearized at its operating point. In PZ analysis independent voltage sources are handled a bit differently than in AC analysis. An independent voltage source that has no `acmag` parameter in the netlist represents a short circuit in PZ analysis. If, however, the `acmag` parameter is provided, the source is removed from the circuit, thus representing an open circuit. This is different than in AC analysis, where such a source is not removed, as it represents a complex signal source.

The reasoning is the following. Independent voltage sources with no `acmag` parameter provide the power to the analyzed circuit. Such sources should be represented by a short circuit when the circuit is linearized. Independent voltage sources that have the `acmag` parameter specified in the netlist are signal sources. Such a source needs to be removed from the circuit before the transfer function is calculated, as it is (almost always) located at the input of the circuit. If such a source is not removed (and replaced by an open circuit), but merely disabled (replaced by a short circuit), it will cause a short circuit at the input, and prevent SPICE from evaluating the transfer function.

Now you could remove such a source (in our case `vin`) from the netlist by hand and then reload the circuit before the PZ analysis takes place. Unfortunately, this would change the operating point of the circuit. The DC operating point is defined by the `dc` parameter of `vin`, so removing the source changes the DC operating point, which in turn changes the linearized circuit. But because SPICE takes care of removing the source in PZ analysis as long as the source has an `acmag` parameter, everything is fine.

Now we will take a look at the PZ analysis in practice. Type

```
pz 4 0 2 0 vol pz
```

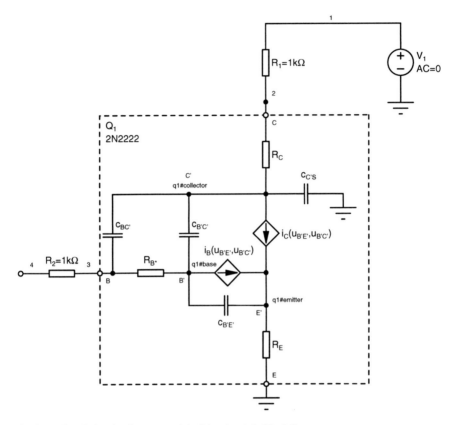

Fig. 2.15 Small signal pole-zero model of the circuit in Fig. 2.2

The first two parameters of the `pz` command (4 and 0) represent the nodes of the input. The next two (2 and 0) represent the nodes of the output. `vol` specifies that the transfer function is of type output voltage/input voltage, and finally `pz` causes SPICE to evaluate poles and zeros of the transfer function. If you look at Fig. 2.15, you can see that the linearized circuit lacks the input source `vin`, which was removed because its `acmag` parameter is specified in the netlist. The voltage source `v1`, which provides power to the circuit, is effectively replaced by a short circuit (its `acmag` parameter is not specified).

After the analysis is finished, look at the list of vectors by typing `display`. The following output is produced:

```
Title: A simple circuit
Name: pz1 (Pole-Zero Analysis)
Date: Mon Dec 06 10:43:59  2004

    pole              : voltage, complex, 2 long
    polevalid         : voltage, complex, 2 long [default scale]
    zero              : voltage, complex, 2 long
    zerovalid         : voltage, complex, 2 long
```

The `polevalid` (`zerovalid`) vector contains the pole (zero) multiplicity for the poles (zeros) in vector `pole` (`zero`). If some pole (zero) multiplicity is 0, the respective pole (zero) is not valid and must be ignored. Let us print the `polevalid` and `pole` vectors (`print polevalid pole`).

```
A simple circuit
Pole-Zero Analysis  Mon Dec 06 10:43:59  2004
-----------------------------------------------------------------
Index   polevalid                       pole
-----------------------------------------------------------------
0    1.000000e+000,  0.000000e+000  -1.174277e+009,  0.000000e+000
1    1.000000e+000,  0.000000e+000  -2.675330e+006,  0.000000e+000
```

Both poles are on the left (negative real) side of the complex plane. This means that the circuit is stable. The dominant pole is $p = -2\pi f = -2.675 \cdot 10^6$. From this the frequency corresponding to the dominant pole can be obtained (425.792 kHz). If we look at the right part of Fig. 2.11, we can see that the point where the gain drops by 3 dB is somewhere near this frequency.

Now we will display the poles and zeros on a graph. First we will create a plot window and give it a tag.

```
plot create pz
```

An empty plot window pops up (Fig. 2.16, left). In the left part of the titlebar you can see the tag (pz). Now we set the plot style for poles (one cross for every pole).

```
set plottype=point
set pointtype=x
set linewidth=1
set pointsize=3
```

The first line sets the plot type to be points that are not interconnected by lines. The second line sets "x" as the point style. The last two lines set the line thickness and point symbol size. Now, everything we plot will be plotted with the style we just specified. Let's plot the poles.

```
plot append pz pole[0,(sum(abs(polevalid) gt 0)-1)]
```

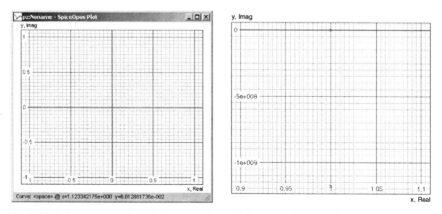

Fig. 2.16 *Left*: an empty plot window tagged pz. *Right*: the real part of the pole vector vs. the real part of the `polevalid` vector

The `append` keyword is used to add the listed vectors to the plot tagged with pz. The part enclosed in brackets selects only the valid poles. Suppose there was only one valid pole. SPICE would put it in the beginning of the `pole` vector. The first element of the `polevalid` vector would then be 1 (or more) and the second one would be 0. The expression in the brackets would select only the first element of the `pole` vector. But as both of our poles are valid, we could do just fine even without the part in the brackets.

For the more curious readers, the brackets select a subvector from the `pole` vector starting at index 0 and ending at index `sum(abs(polevalid) gt 0)-1`. The `abs` function converts the `polevalid` vector to a real vector. Then the nonzero entries are counted by simply adding up all members of the resulting vector that are greater than zero (function `sum`). Finally, 1 is subtracted as indices in SPICE start with 0 and not 1 as is usually the case in mathematics.

The plot window will be updated with two crosses (Fig. 2.16, right). But the crosses are not in the correct place. If you turn on the info frame, you will find out why. The info window says `x-y grid displaying real vs. default` meaning that the real part of the `pole` vector is plotted against the real part of the default scale (vector `polevalid`). If you switch the plot window display mode to **View/Graph/Complex, Polar** the poles will jump to the correct position (Fig. 2.17, left). Now add the zeros. They will be represented by circles.

```
set pointtype=o
plot append pz zero[0,(sum(abs(zerovalid) gt 0)-1)]
```

The plot window will be updated with the zeros (Fig. 2.17, right), albeit everything will be displayed in the wrong mode (real vs. default). Finally, we will switch to the correct display mode, but this time using the command line.

```
plot append pz mode cx polar
```

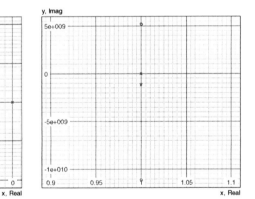

Fig. 2.17 *Left*: complex polar mode displaying the poles. *Right*: poles and zeros in the real vs. default mode

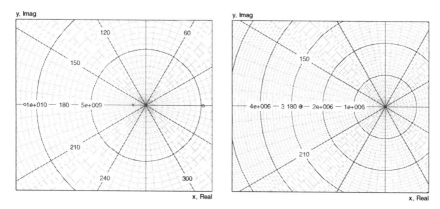

Fig. 2.18 *Left*: complex polar mode displaying poles and zeros. *Right*: dominant pole, zoomed

The plot window will change to complex polar mode (Fig. 2.18, left). It may seem that there is one pole in the origin of the coordinate system, but if you zoom in, you will see that this is not the case (Fig. 2.18, right).

Finally, let us change the plot type for all subsequent plots back to the usual style (points connected with lines).

```
set plottype=line
```

As a final note, keep in mind that PZ analysis can handle only lumped circuits. Therefore circuits containing transmission lines cannot be analyzed with the PZ analysis, unless the transmission line is represented by a series of lumped elements (like, for instance, the case with the uniformly distributed RC transmission line model (URC)).

Interested readers may refer to Sect. 4.5 for details on pole-zero analysis.

2.9 Small Signal Noise Analysis (NOISE)

We are often interested in the way noise is generated and propagated through a circuit. The magnitude of noise is usually small enough so that the circuit can be treated as linear. Therefore, it is safe to assume a linearized circuit for noise evaluation. Every resistor and semiconductor element is a source of noise. Noise is modeled by means of current sources connected in parallel with elements that are sources of noise. A dual approach with voltage sources is also possible, but as it introduces an additional node for every voltage source (and by that an additional unknown in the system of equations) it is not used in SPICE. In contrast to signal current sources that have a constant spectrum determined by their `ac`, `acmag`, and `acphase` parameters, noise current sources have a frequency-dependent spectrum in general.

A voltage (current) noise signal can be described by its root mean square (RMS) value V_{RMS} in volts (I_{RMS} in amperes). This figure, however, does not describe the frequency content of the signal, which is represented by a voltage (current) noise spectrum density. The voltage (current) noise spectrum density is a nonnegative real function of frequency with V^2/Hz (A^2/Hz) for units. The integral of a voltage (current) noise spectrum density across all frequencies from 0 to ∞ is the mean square value of the noise signal V^2_{RMS} (I^2_{RMS}). If the integral is taken across a narrower frequency band, its value represents the mean square value of the noise signal after it passes through an ideal bandpass filter with a corresponding passband.

The first step in analyzing the noise in a circuit is to perform an operating point analysis and construct a linearized circuit. Because the circuit is linear, contributions of individual noise sources to the output noise spectrum can be evaluated independently. In Fig. 2.19 we have seven noise current sources, an independent voltage source v1 that provides the power to the circuit, and an independent voltage source vin, which is the source of a small signal in AC analysis.

Small signal noise analysis of the circuit in Fig. 2.19 would require seven partial noise evaluations. For every noise evaluation all independent sources are disabled (current sources are replaced by open circuits and voltage sources by short circuits). Of all the noise sources, only the one whose contribution is being evaluated remains enabled; all others are disabled. The output noise power spectrum density

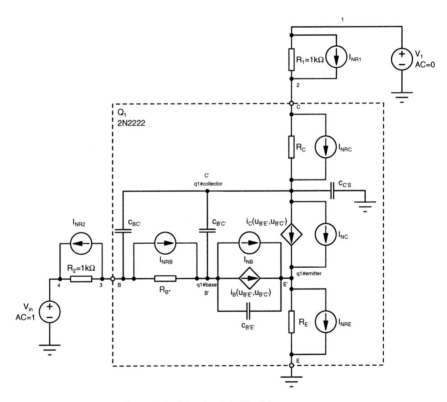

Fig. 2.19 Small signal noise model of the circuit in Fig. 2.2

contribution is obtained by multiplying the noise power spectrum density of the source by the squared magnitude of the transfer function from the noise source to the output of the circuit.

Contributions of individual noise sources to the output noise are not correlated. For uncorrelated noise signals the mean square (V_{RMS}^2) of the sum of noise signals is the sum of mean square values of individual noise signals. The same is valid for noise power spectrum densities. Therefore, SPICE obtains the total output noise power spectrum density by adding up individual output noise power spectrum density contributions.

Another important characteristic is the equivalent input noise power spectrum density. It represents the spectrum of the noise signal that must be injected at the input of the circuit without any noise sources in order to produce the given output noise power spectrum density. It can be obtained by dividing the output noise power spectrum density by the squared magnitude of the transfer function from the input to the output of the circuit (needless to say, this magnitude is frequency dependent). Now make sure the ac parameter of the input voltage source vin is set to 1. Reload the circuit and then issue the following command:

```
noise v(2) vin dec 10 0.5 100meg 1
```

The above statement performs a small signal noise analysis. The output is between node 2 and ground. The input is located at the independent voltage source vin (this information is needed solely for evaluating the equivalent input noise power spectrum density). The frequency range between 0.5 Hz and 100 MHz is divided in a logarithmic scale with 10 points per decade. The last number represents the number of calculated noise power spectrum density points for which one point is put into the resulting vectors.

Let's take a look at the results by typing display. We get

```
Title: A simple circuit
Name: noise2 (Integrated Noise - V^2 or A^2)
Date: Tue Dec 07 11:24:17  2004

    inoise_total          : voltage, real, 1 long
    inoise_total_q1       : voltage, real, 1 long
    inoise_total_q1_1overf: voltage, real, 1 long
    inoise_total_q1_ib    : voltage, real, 1 long
    inoise_total_q1_ic    : voltage, real, 1 long
    inoise_total_q1_rb    : voltage, real, 1 long
    inoise_total_q1_rc    : voltage, real, 1 long
    inoise_total_q1_re    : voltage, real, 1 long
    inoise_total_r1       : voltage, real, 1 long
    inoise_total_r2       : voltage, real, 1 long
    onoise_total          : voltage, real, 1 long
    onoise_total_q1       : voltage, real, 1 long
    onoise_total_q1_1overf: voltage, real, 1 long
    onoise_total_q1_ib    : voltage, real, 1 long
    onoise_total_q1_ic    : voltage, real, 1 long
    onoise_total_q1_rb    : voltage, real, 1 long
    onoise_total_q1_rc    : voltage, real, 1 long [default scale]
    onoise_total_q1_re    : voltage, real, 1 long
    onoise_total_r1       : voltage, real, 1 long
    onoise_total_r2       : voltage, real, 1 long
```

As we can see, the current plot (noise2) contains integrated noise representing the mean square value of noise on the frequency range given with the noise

Table 2.3 Noise contributions from the circuit in Fig. 2.19

Name postfix	Contribution of
_q1	Total noise of q1
_q1_1overf	$1/f$ noise of q1
_q1_ib	Base junction shot noise of q1
_q1_ic	Collector junction shot noise of q1
_q1_rb	Base thermal noise of q1
_q1_rc	Collector thermal noise of q1
_q1_re	Emitter thermal noise of q1
_r1	Thermal noise of r1
_r2	Thermal noise of r2

statement. Vectors `onoise_total` and `inoise_total` represent integrated output and equivalent input noise. The remaining vectors represent integrated contributions to input (output) noise. Table 2.3 lists the noise contributions from the circuit in Fig. 2.19.

We print the total output noise RMS value by typing

```
print sqrt(onoise_total)
```

We get

```
sqrt(onoise_total) = 1.828401e-004
```

If you add up contributions `onoise_total_q1_1overf`, `onoise_total_q1_ib`, `onoise_total_q1_ic`, `onoise_total_q1_rg`, `onoise_total_q1_rc`, and `onoise_total_q1_re` you should get `onoise_total_q1`. Similarly, if you add up `onoise_total_q1`, `onoise_total_r1`, and `onoise_total_rin` you should get `onoise_total`. Equivalent relationships are valid for input noise contributions, too.

NOISE analysis creates two plots (groups of results). The one we just examined is the noise summary plot. Besides the noise summary, SPICE also produces noise power spectrum densities. To switch to the plot containing power spectrum densities, type `setplot previous`. For switching back to the noise summary plot, type `setplot next`. But let us skip switching back for now and stay in the plot where the noise power spectrum densities are stored.

Let us take a look at the vectors in the plot by typing `display`. We get

```
Title: A simple circuit
Name: noise1 (Noise Spectral Density Curves - (V^2 or A^2)/Hz)
Date: Tue Dec 07 11:29:41  2004

    frequency           : frequency, real, 84 long, grid = xlog [default scale]
    inoise_spectrum     : voltage, real, 84 long
    onoise_q1           : voltage, real, 84 long
    onoise_q1_1overf    : voltage, real, 84 long
    onoise_q1_ib        : voltage, real, 84 long
    onoise_q1_ic        : voltage, real, 84 long
    onoise_q1_rb        : voltage, real, 84 long
    onoise_q1_rc        : voltage, real, 84 long
    onoise_q1_re        : voltage, real, 84 long
    onoise_r1           : voltage, real, 84 long
    onoise_rin          : voltage, real, 84 long
    onoise_spectrum     : voltage, real, 84 long
```

As you can see, we have more than one point per vector in this plot. The default scale is the `frequency` vector. This time there are no equivalent input noise spectrum contributions available. We only have output noise spectrum contributions and, of course, the `inoise_spectrum` and the `onoise_spectrum`. We plot the output noise spectrum contributions of q1 with logarithmic scale on the y-axis.

```
plot ylog
+    onoise_q1_1overf
+    onoise_q1_ib onoise_q1_ic
+    onoise_q1_rb onoise_q1_rc onoise_q1_re
+    onoise_q1
```

From the plot window in Fig. 2.20 (left) we can see that the individual noise contributions of q1 result in the top curve which represents the contribution of q1 to the output noise power spectrum density. Similarly, we can plot the noise contributions of individual devices and the complete output noise power spectrum density by typing

```
plot ylog onoise_q1 onoise_r1 onoise_r2 onoise_spectrum
```

A plot window pops up, displaying the graph in Fig. 2.20 (right). The topmost curve (output noise power spectrum density) is the sum of all three partial noise contributions.

Now we will look at the equivalent input noise power spectrum density and the output noise power spectrum density on the same graph.

```
plot ylog onoise_spectrum inoise_spectrum
```

A plot window pops up (see Fig. 2.21, left) showing both equivalent input noise power spectrum density (lower curve) and output noise power spectrum density. The right part of Fig. 2.21 shows the square root of the ratio of the output and equivalent input noise, plotted in dB. If you compare it to Fig. 2.11 (right) you will see that it is identical to the gain magnitude curve.

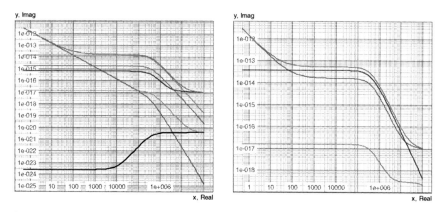

Fig. 2.20 *Left*: partial noise contributions and the complete output noise contribution of q1. *Right*: partial contributions of the devices q1, r1, and r2 to output noise and the complete output noise power spectrum density

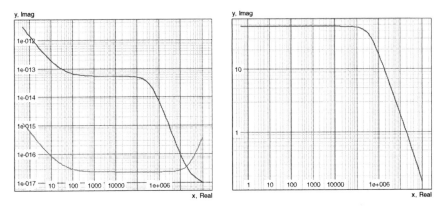

Fig. 2.21 *Left*: equivalent input and output noise power spectrum density. *Right*: square root of the ratio of output and equivalent input noise power spectrum densities

Now all these results were obtained with the `acmag` parameter of the input voltage source `vin` set to 1. If you set it to 10, you will notice that the equivalent input noise is two decades lower than before and that the square root of the ratio of the output noise and the equivalent input noise is 10 times (20 dB) higher. This is caused by the fact that SPICE does not divide the output differential voltage with the input sources's `acmag` parameter value when calculating the gain of the circuit. So make sure to set the input voltage source's `acmag` parameter to a nonzero value before the NOISE analysis. If you do not specify the `acmag` parameter for the input source, the output noise will still be correct, but the equivalent input noise power spectrum densities and the corresponding integrated values will be incorrect.

Now we will look at the results if the noise analysis is started with

```
noise v(2) vin dec 10 0.5 100MHz 5
```

This will output one point into the noise power spectrum densities for every fifth point of the frequency scale. If you look at the vectors in the noise spectrum plot, you will see that they contain only 17 points instead of 84.

If you completely leave out the last parameter, like in the following command:

```
noise v(2) vin dec 10 0.5 100MHz
```

you will get only the input and output integrated noise and noise power spectrum densities. No partial contributions will be saved. However, the number of points in the noise spectra will be the same as if 1 were set for the last parameter.

Let us summarize. NOISE analysis produces two plots. The one you see immediately after the analysis is the integrated noise plot. It contains the integrated noise contributions to output and equivalent input noise for all devices. You also get the integrated output and equivalent input noise. If you switch to the other plot with `setplot previous` you will have access to noise power spectrum densities. In this plot you have output and equivalent input noise power spectrum densities along with the contributions of all devices to the output noise power spectrum density. If the last parameter of the `noise` statement is omitted, the device contributions are not saved. If the last parameter is set to something other than 1, the noise power spectrum densities contain fewer points. To switch back to the integrated noise plot, type `setplot next`.

Interested readers may refer to Sect. 4.6 for details on small signal noise analysis.

2.10   Time-Domain Analysis (TRAN)

Time-domain analysis evaluates the time-domain response of the circuit to an input stimulus also specified in the time domain. As an example, suppose that the input voltage source (`vin`) produces a voltage step from 0.68 to 0.78 V. First we must specify the stimulus. Change the line in the netlist describing `vin` to

```
vin (4 0) dc=0.68 acmag=1 pulse=(0.68 0.78 1u 1n)
```

This means that `vin` produces the specified pulse that starts at 1 µs with 1 ns rise time. Reload the netlist by typing `source sample1`. Of course, you can change the settings for source `vin` without making changes to the netlist and then reloading it. In the command window you could type

```
let @vin[pulse]=(0.68;0.78;1u;1n)
```

Note how individual values describing the pulse source are now separated with semicolons. The `dc` parameter value specifies the value of the source in the OP and DC analysis (if the source is not swept by the DC analysis). It also specifies the source's value when the circuit is linearized in the beginning of AC, PZ, TF, and NOISE analyses. The `acmag` parameter specifies the source's value in the AC and NOISE analysis (when the gain for evaluating the equivalent input noise is calculated). Omitting the `acmag` parameter is equivalent to setting it to 0. If you do not specify the `dc` parameter, its value is assumed to be 0, unless some transient source (like `pulse`) is specified. In the latter case the assumed `dc` value is the value of the transient source at time 0.

In order to turn off the transient source specified with the `let` statement (just as if it were not specified in the netlist), type

```
let @vin[function]=0
```

In case you typed any of the above `let` statements reload the netlist. Now let's start a TRAN analysis.

```
tran 1n 2u
```

This means that the initial time step will be 1 ns and that the analysis will finish at 2 µs. Take a look at the results (`display`). We get

```
Title: A simple circuit
Name: tran1 (Transient Analysis)
Date: Tue Dec 21 09:50:59  2004
```

```
V(1)                 : voltage, real, 410 long
V(2)                 : voltage, real, 410 long
V(3)                 : voltage, real, 410 long
V(4)                 : voltage, real, 410 long
q1#base              : voltage, real, 410 long
q1#collector         : voltage, real, 410 long
q1#emitter           : voltage, real, 410 long
time                 : time, real, 410 long [default scale]
v1#branch            : current, real, 410 long
vin#branch           : current, real, 410 long
```

Again, we get the same set of vectors as in the operating point analysis, except that this time the default scale is the `time` vector. Plot the output and input voltages by typing

```
plot v(2) v(4)
```

The plot window in Fig. 2.22 (left) pops up. The time step is not constant. Let us plot the output voltage in comb mode around the point when the transition in the input pulse occurs. After the `plot` command is finished, switch back to line mode.

```
set plottype=comb
plot v(2) xl 0.99u 1.01u
set plottype=line
```

The `xl 0.99u 1.10u` part of the `plot` command specifies the x-axis range to display in the plot window. The window in Fig. 2.22 (right) pops up. From the figure it is obvious that the time step adapts to the waveform dynamics.

Of course, a variable time step is not always desirable. In order to interpolate the waveforms to the uniform time step of 1 ns, type

```
linearize 1n
```

At first sight nothing happens, but if you look at the list of plots (`setplot`) you will see two TRAN plots. The first one is the one created by the TRAN analysis,

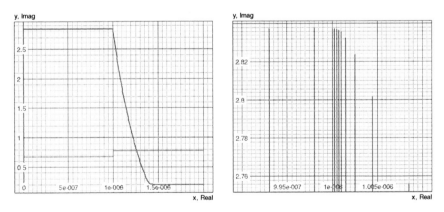

Fig. 2.22 *Left*: input pulse signal and the corresponding output response. *Right*: variable time step at pulse transition

and the second one is created by the `linearize` statement. This is also the active plot after the `linearize` command finishes.

If you repeat the commands producing the comb plot, you will see that the time step is now uniform.

Now let us look at the circuit's response to a sinusoidal signal. Note that this is not a sinusoidal signal like in the AC analysis where it was specified with magnitude and phase. In AC analysis the analyzed circuit was linearized. In the TRAN analysis the simulator takes into account all nonlinear effects caused by nonlinear circuit components. First set up a sinusoidal source. You could change the corresponding line in the netlist to

```
vin (4 0) dc=0.68 acmag=1 sin=(0.68 80m 1k)
```

After reloading the netlist the `vin` source becomes a sinusoidal source in TRAN analysis. Its offset is 0.68 V, magnitude 80 mV, and frequency 1 kHz. Of course, this can be achieved without changing the netlist and reloading the circuit, if you type

```
let @vin[sin]=(0.68;80m;1k)
```

in the command window. Now start a TRAN analysis by typing

```
tran 1u 5m
```

and plot the resulting output and input waveforms

```
plot v(4) v(2)
```

The window in Fig. 2.23 (left) pops up. Note that the simulator did not start the simulation from the 0 state. Instead, it first calculated the operating point of the circuit with the sources set to their respective 0 s transient waveform values. The result was then used as the initial state for the TRAN analysis. If the circuit is a simple amplifier, this takes care of the initial transient that occurs before the amplifier reaches the steady state. To see how to set up your own initial conditions for the circuit, take a look at the `ic` and `tran` commands in the command reference and the `ic` parameters of devices containing reactances.

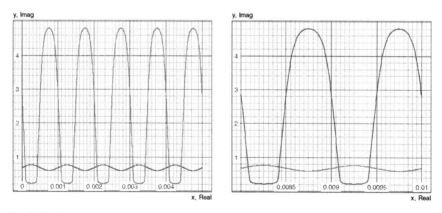

Fig. 2.23 *Left*: response to a 80 mV, 1 kHz input sinusoidal signal. *Right*: response between 8 and 10 ms

Often we are interested in the transient response starting with some time greater than zero. Say you are interested in the response between 8 and 10 ms. You can use the following form of the `tran` command:

```
tran 1u 10m 8m
```

Note that the simulator still simulates the circuit from 0 to 8 ms, except that now the resulting points for these times are not saved in the vectors. Plot the output and input voltage.

```
plot v(2) v(4)
```

The result is shown in Fig. 2.23 (right). As you can see, the output voltage is distorted significantly.

Now look at the harmonics in the output signal. Type

```
fourier 1k v(2)
```

The fundamental frequency is set to 1 kHz by the `fourier` command. You will get the following output:

```
Fourier analysis for v(2):
  No. Harmonics: 10, THD: 17.8501 %, Gridsize: 200, Interpolation Degree: 1

Harmonic Frequency    Magnitude    Phase       Norm. Mag    Norm. Phase
-------- ---------    ---------    -----       ---------    -----------
0        0           2.52906      0           0            0
1        1000        2.66094      179.835     1            0
2        2000        0.108587     88.618      0.0408078    -91.217
3        3000        0.426804     179.457     0.160396     -0.37856
4        4000        0.149295     88.5968     0.0561061    -91.239
5        5000        0.038619     179.926     0.0145133    0.0905974
6        6000        0.0731063    88.2524     0.0274738    -91.583
7        7000        0.0425601    -2.5498     0.0159944    -182.39
8        8000        0.00173449   95.134      0.000651833  -84.701
9        9000        0.0267194    -2.8358     0.0100413    -182.67
```

From the output you can see that the total harmonic distortion (THD) is 17.9%. The `fourier` command uses the last period of the signal (1 ms) between 8 and 10 ms. The signal is then interpolated to a uniformly spaced grid with 200 points per period (gridsize). The interpolation polynomial is linear by default (interpolation degree). Ten harmonics are evaluated. The normalized magnitude and phase for the dc component are both 0. For all other components the normalized magnitude is the ratio between the magnitude of the harmonic and the magnitude of the fundamental harmonic. Similarly, the normalized phase is the difference between the phase of a harmonic and the phase of the fundamental harmonic. The THD is calculated as the root summed square of normalized magnitudes starting with the second harmonic.

Not only a few harmonics, but also complete spectra can be evaluated by SPICE. The command that enables you to do that is `spec`. First we must produce a linearized plot so that the time step between neighboring points is uniform.

```
linearize 1u
```

Set up the windowing function. We do not want a windowing function in this case, so

```
set specwindow=none
```

Now evaluate the spectra of input and output signals.

```
spec 0 10k 1k v(2) v(4)
```

This will calculate the discrete Fourier transform of the output and input voltages. The analyzed signal is between 8 and 10 ms. The spectrum is calculated for the frequencies from 0 Hz to 10 kHz in 1 kHz steps.

Now take a look at the resulting plot. It is called `sp1` and the list of vectors obtained by typing `display` is

```
Title: A simple circuit
Name: sp1 (Spectrum)
Date: Tue Dec 21 11:05:52  2004

   frequency           : frequency, real, 11 long [default scale]
   v(2)                : voltage, complex, 11 long
   v(4)                : voltage, complex, 11 long
```

As we can see, the resulting spectra are complex. We plot the input signal's spectrum in comb style with the logarithmic y-axis.

```
set plottype=comb
plot ylog abs(v(4))
```

The plot window in Fig. 2.24 (left) pops up.

The input signal contains only the DC component and the fundamental harmonic. Everything else in the spectrum is numerical noise. If we plot the output spectrum, we get Fig. 2.24 (right).

```
plot ylog abs(v(2))
```

Finally, let's change the plot style back to normal.

```
set plottype=line
```

More details on transient analysis can be found in Sect. 4.7.

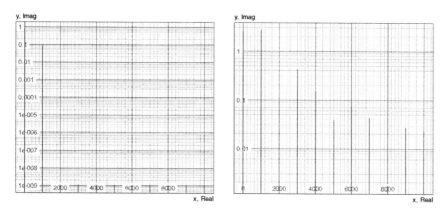

Fig. 2.24 *Left*: input voltage spectrum. *Right*: output voltage spectrum

2.11   Interactive Interpreter and Advanced Features

In order to illustrate the capabilities of the interactive interpreter, we will look at two examples. First we look at automated measurements. Chapter 5 explains the built-in interpreter in more detail.

2.11.1   Expressions, Vectors, Plots, and Automated Measurements

Often you want to measure some quantity from the plots created by various analyses. Making several such measurements by hand can be tedious. Therefore, SPICE offers automated measurements through its cursor facility. First we will illustrate such measurements through a simple example.

Create a new plot.

```
setplot new
```

The newly created plot with no vectors becomes the current plot. It is called unnamed1. Now rename it myplot.

```
nameplot myplot
```

Let us create a default scale vector named x with 1001 values from 0 to 2. The function vector(1001) creates a vector with 1001 components ranging from 1 to 1000. If we divide it by 1001 and multiply it by 2, we get an equidistant scale across a period of length 2. The last point (2) is not present in the vector because it represents the beginning of the second period.

```
let x=vector(1001)/1001*2
```

The first vector created in a plot is the default scale vector. Function vector(n) returns values from 0 to $n-1$. These values are then scaled in order to produce the desired scale.

On the given scale create two vectors containing the function values corresponding to points in x. Let the first function be $\sin(2\pi x)$ and the second one $\cos(2\pi x)$. Remember to switch the angle units to radians.

```
set units=radians
let s=sin(2*pi*x)
let c=cos(2*pi*x)
```

For the sake of checking, plot the two vectors (Fig. 2.25, left).

```
plot s c
```

Now create a cursor named cur.

```
let cur=0
```

A cursor is a vector with a single component whose value represents an index into other vectors. Now move the cursor to the index where x equals 0.25.

```
cursor cur right x 0.25
```

This moves cur towards higher index (right) values until it reaches the value 0.25 in vector x. Now print the value of cur.

```
print cur
```

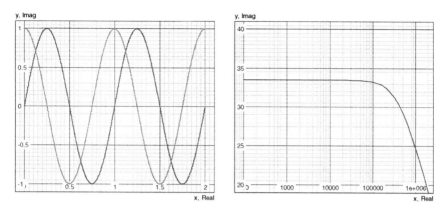

Fig. 2.25 *Left*: two manually constructed vectors. *Right*: AC response of the circuit in Fig. 2.2

We get 125.125, which is not an integer value. The `cursor` command uses linear interpolation to determine the new cursor position. We can check the value of s and c at the cursor position (with linear interpolation, of course)

```
print s[%cur] c[%cur]
```

and we get

```
s[%cur] = 9.999914e-001
c[%cur] = 2.704910e-008
```

Note that the value of the first function at the cursor position should be 1, whereas the second function should be 0. We came pretty close. The error is due to rounding and linear interpolation.

Suppose you give the `cursor` command an invalid destination (e.g., 5 in the example above).

```
cursor cur right x 5
print cur
```

The cursor value becomes negative, indicating an error. Say you want to know the value of s at the first point starting from index 0 where c is 0.5.

```
let cur=0
cursor cur right c 0.5
print cur s[%cur]
```

We get

```
cur = 8.341580e+001
s[%cur] = 8.660034e-001
```

which is quite close to the correct value 0.8660254. Again, the cause of the discrepancy is numerical error caused by linear interpolation.

Now let us look at a real-world example. We will measure the gain and the 3 dB cutoff point for the amplifier we have been dealing with since the beginning of this chapter. First perform an AC analysis.

```
ac dec 10 100 1g
```

Next create and plot (Fig. 2.25, right) a vector containing the magnitude of gain in dB.

```
let h=db(v(2))
plot h yl 20 40
```

Evaluate the maximum gain.

```
let hmax=max(h)
```

Move the cursor to the 3 dB cutoff point.

```
let cur=0
cursor cur right h hmax-3
```

Finally, evaluate the frequency at the cursor position.

```
let f3db=frequency[%cur]
```

Print the result by typing

```
print hmax f3db
```

which displays the following output:

```
hmax = 3.348050e+001
f3db = 3.884863e+005,0.000000e+000
```

Note that the frequency vector is complex, and thus the resulting f3db vector is also complex. The real part (first number) informs us about the frequency.

This procedure may look more tedious than measuring things by hand from the plot window. But keep in mind that you can put all these commands into the control block. This way they will be executed as soon as you load the circuit. The measurements you are interested in are evaluated immediately after the circuit is loaded.

Finally, we destroy the plot we created manually and all vectors in it, type

```
destroy myplot
```

If you want to destroy all plots (except for the const plot), type

```
destroy all
```

The const plot is special. It cannot be destroyed. Let us take a look at it.

```
setplot const
display
```

We get

```
Title: Constant values
Name: const (constants)
Date: Sat Aug 16 10:55:15 PDT 1986

    boltz          : notype, real, 1 long
    c              : notype, real, 1 long
    e              : notype, real, 1 long
    echarge        : notype, real, 1 long
    false          : notype, real, 1 long
    i              : notype, complex, 1 long
    kelvin         : notype, real, 1 long
    no             : notype, real, 1 long
    pi             : notype, real, 1 long
    planck         : notype, real, 1 long
    true           : notype, real, 1 long
    yes            : notype, real, 1 long [default scale]
```

As you can see, this plot contains various constants, among them pi. These constants are available regardless of the plot you are currently using.

2.11.2 Control Structures, Substitution, and Variables

Often you want to plot multiple curves on the same graph, where each curve represents the circuit's response at a different temperature. Let's look at another example:

```
setplot new
nameplot data
let tval=(-50;-25;0;25;50;75;100)
let ndx=0
plot create temp_plot
while data.ndx lt length(data.tval)
  let data.t=data.tval[data.ndx]
  set temp={data.t}
  dc vin 0.2 1.2 0.001
  if data.t ge 0
    let tfunc_{data.t}=v(2)
    plot append temp_plot tfunc_{data.t}
  else
    let tfunc_neg_{abs(data.t)}=v(2)
    plot append temp_plot tfunc_neg_{abs(data.t)}
  end
  let data.ndx=data.ndx+1
end
```

The first two lines create a plot named `data`. In this plot two vectors are created: `tval` and `ndx`. `tval` contains the values of the circuit temperature for which the circuit will be simulated. `ndx` represents the index into vector `beta` and is initialized to 0.

The line

```
plot create temp_plot
```

creates an empty plot window named `temp_plot`.

Next comes a control structure. It is a `while` loop. Note that the vectors are referenced with their full name (`data.ndx`). If we say simply `ndx` we get the `ndx` vector from the current plot. Because the `ndx` vector resides in the `data` plot, this works only if `data` is the current plot. But in the body of the `while` loop there is an analysis that creates a new plot and makes it the current plot (DC). Therefore, we must reference all vectors that do not reside in the plot created by the last analysis with their full names.

The `while` loop executes while the condition

```
data.ndx lt length(data.tval)
```

is true. `gt` is the relational "greater than" operator. This means that the loop's body is executed for `data.ndx` values from 0 to 6 (because `length(data.tval)` is 7). Note how the `data.ndx` vector is increased at the end of the body of the loop by the statement

```
let data.ndx=data.ndx+1
```

In the body of the loop vector `data.t` is set to the value of the temperature for which the circuit is to be simulated. This makes the notation much simpler as we can use `data.t` instead of `data.tval[data.ndx]`. Next the circuit temperature is set by the statement

```
set temp={data.t}
```

The expression in curly braces (`data.t`) is evaluated and inserted in place of
`{data.t}` before the `set` command is interpreted. This process is called arithmetic
substitution. Arithmetic substitution is needed because the `set` command expects a
value on the right-hand side of =. `data.t` is not a value by itself. It is an expression.

A DC analysis is performed after the circuit temperature is set. We omit the
explanation as we have dealt with DC analysis in earlier sections of this chapter.

An `if-else` statement follows the analysis. The condition is `data.t ge 0` (`ge`
is the "greater than" operator). Suppose the temperature is positive or zero. Then the
`if` branch would be chosen by the interpreter.

The line

```
let tfunc_{data.t}=v(2)
```

assigns the output voltage vector to a vector named `tfunc_xxx` where xxx is the
value stored in vector `data.t`. Enclosing an expression in curly braces invokes
expression substitution. Expression substitution takes place just before a line is executed. So if `data.t` were 100, the line would effectively look like

```
let tfunc_100=v(2)
```

Next the newly created vector is appended to the `temp_plot` plot window. In
the `else` branch (which is chosen if the `if` condition is false), something similar
happens, except that the newly created vector is called `tfunc_neg_xxx` where xxx
is the absolute value of the temperature at which the circuit was simulated.

When you enter this script in the control block of the netlist and reload the circuit,
a plot window from Fig. 2.26 pops up. Note that in the left part of the titlebar we
have the plot tag (`temp_plot`).

The list of all variables can be obtained by typing

```
set
```

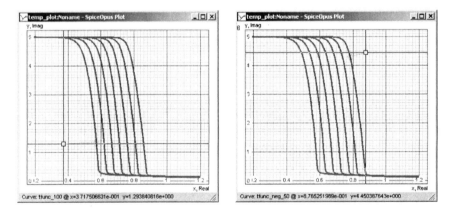

Fig. 2.26 Result of the sample script illustrating the `while` loop. *Left*: identification of the leftmost
vector. *Right*: identification of the rightmost vector

You will get a printout that looks something like this:

```
* curplot    dc7
* curplotdate    Fri Dec 24 11:29:16  2004
* curplotname    DC transfer characteristic
* curplottitle   A simple circuit
  history    100
  linewidth 2
* plots ( const data dc1 dc2 dc3 dc4 dc5 dc6 dc7 )
  plottype  line
  plotwinheight 360
  plotwinwidth  410
  pointsize 2
  pointtype o
  program   SpiceOpus (c)
  prompt    SpiceOpus (c) ! ->
  sourcepath    ( . C:/SpiceOpus/lib/scripts )
+ temp  100
  units radians
```

Lines marked with an asterisk (*) represent built-in interpreter variables. Lines marked with + represent variables that affect the simulator and the current circuit. All other lines are plain variables that affect the interpreter.

The curplot variable represents the name of the current plot. The plots variable is a list of all plots available at the moment. curplotdate, curplotname, and curplottitle are self-explanatory. The sourcepath variable lists the directories where SPICE searches for a file when a source command is issued. history specifies the depth of the command line history.

units represents the angle units used in trigonometric functions.

program is the program name and prompt is the prompt the command line interpreter presents to the user. The exclamation point is replaced by the sequential number that labels the typed command after the Enter key is pressed.

Finally, temp represents the circuit temperature as it was set by the .options directive in the netlist, or the set temp=... command on the command line or in the control block.

We can use the value of a variable in any command through variable substitution. Let us take a look at a few examples.

The following command outputs the temperature:

```
echo temperature=$(temp)
```

We used the $() character to denote that $(temp) should be replaced with the value of variable temp. Note that this outputs the temperature only if it was set by the .options directive in the netlist or by the set temp=... command. Variable substitution takes place before expression substitution.

If your variable is a list, you can select individual members by typing

```
echo $(sourcepath[0])
```

This outputs the first member of the list from variable sourcepath. If index xx refers to some member that is not in the list (the list is too short), $sourcepath[xx] expands to an empty string.

To change the value of a variable, use the set command.

```
set linewidth=3
```

Note that the string to the right of the equal sign becomes the new value of the variable. No evaluation is performed prior to assigning the string to the variable.

Variables that affect the simulator (those marked with +) can also be set using the set command

```
set temp={20+5}
```

The set command expects a value on the right-hand side of =. Because 20+5 is an expression, it must be evaluated first. This is achieved by putting it in curly braces.

To add a directory to the sourcepath type

```
set sourcepath=( $sourcepath newdirname )
```

The sourcepath variable specifies the directories where SPICE searches for circuit descriptions. By default, it is set to the current directory (.) and to the SPICE_LIB_DIR/scripts directory.

2.11.3 Libraries, Raw Files, and File Output

Often we want to divide a large circuit description into multiple smaller parts. Suppose we have the following file:

```
A simple circuit
v1 (1 0) dc=5
r1 (1 2) r=1k
q1 (2 3 0) t2n2222
r2 (3 4) r=1k
vin (4 0) dc=0.68
.model t2n2222 npn
+ is=19f bf=150 vaf=100 ikf=0.18 ise=50p
+ ne=2.5 br=7.5 var=6.4 ikr=12m isc=8.7p
+ nc=1.2 rb=50 re=0.4 rc=0.3 cje=26p tf=0.5n
+ cjc=11p tr=7n xtb=1.5 kf=0.032f af=1
.end
```

Now let's put the t2n2222 model description in a separate file named t2n2222.mod. The original netlist (netlist.cir) would now look like

```
A simple circuit
v1 (1 0) dc=5
r1 (1 2) r=1k
q1 (2 3 0) t2n2222
r2 (3 4) r=1k
vin (4 0) dc=0.68
.include t2n2222.mod
.end
```

and the t2n2222.mod file would contain

```
.model t2n2222 npn
+ is=19f bf=150 vaf=100 ikf=0.18 ise=50p
+ ne=2.5 br=7.5 var=6.4 ikr=12m isc=8.7p
+ nc=1.2 rb=50 re=0.4 rc=0.3 cje=26p tf=0.5n
+ cjc=11p tr=7n xtb=1.5 kf=0.032f af=1
```

As you can see, .include includes the contents of an external file. If the file is not found in the directory where the caller file resides (we refer to it as CALLER_DIR), SPICE tries with the directories specified in the `sourcepath` variable. Suppose you provided the following file name: `subdir/t2n2222.mod`. SPICE first checks the file `CALLER_DIR/subdir/t2n2222.mod` and then all `SOURCEPATH_DIR/subdir/t2n2222.mod`, where `SOURCEPATH_DIR` represents the directories from the `sourcepath` variable.

A more sophisticated mechanism is available through the `.lib` command. Suppose you have a file named `bjts.mod` with the following contents:

```
.lib bjt2n2222
.model t2n2222 npn
+ is=19f bf=150 vaf=100 ikf=0.18 ise=50p
+ ne=2.5 br=7.5 var=6.4 ikr=12m isc=8.7p
+ nc=1.2 rb=50 re=0.4 rc=0.3 cje=26p tf=0.5n
+ cjc=11p tr=7n xtb=1.5 kf=0.032f af=1
.endl

.lib bjt2n2907
.model t2n2907 pnp
+ is=1.1p bf=200 nf=1.2 vaf=50 ikf=0.1
+ ise=13p ne=1.9 br=6 rc=0.6 cje=23p vje=0.85
+ mje=1.25 tf=0.5n cjc=19p vjc=0.5 mjc=0.2
+ tr=34n xtb=1.5
.endl
```

To include the contents of the `bjt2n2222` section of the `bjts.mod` file, you would type

```
.lib 'bjts.mod' bjt2n2222
```

and similarly to include the `bjt2n2907` section, you would type

```
.lib 'bjts.mod' bjt2n2907
```

After an analysis is finished, you can save the results of an analysis to a file. Suppose that after performing the DC analysis from Sect. 2.6 you want to save the points from vectors `v(2)` and `v(4)` to file `data.raw`. This can be achieved by using the `write` command after the analysis is finished.

```
write data.raw v(2) v(4)
```

SPICE creates the file with the following contents:

```
Title: A simple circuit
Date: Wed Feb 02 11:21:48  2005
Plotname: DC transfer characteristic
Flags: real
Sckt. naming: from bottom to top
No. Variables: 3
No. Points: 140
Variables:
        0 sweep voltage
        1 v(2)   voltage
        2 v(4)   voltage
Values:
 0       0.000000000000000e+000
         4.999999985361407e+000
         0.000000000000000e+000

 1       1.000000000000000e-002
         4.999999985362384e+000
         1.000000000000000e-002
...
```

The file contains the default vector (sweep) and the two vectors specified with the write command. After a short header the individual component values follow.

If you omit the vectors and simply say

```
write data.raw
```

all the vectors from the current plot are saved. The default raw file format is ASCII. If your .raw files start to become very big, you can save a lot of space by switching SPICE to the binary .raw file mode by typing

```
set filetype=binary
```

Now all subsequent write commands create binary .raw files. Binary .raw files still have an ASCII header, but vector data is written in binary format. The above created .raw file would look like

```
Title: A simple circuit
Date: Wed Feb 02 11:57:59  2005
Plotname: DC transfer characteristic
Flags: real
Sckt. naming: from bottom to top
No. Variables: 3
No. Points: 140
Variables:
      0 sweep voltage
      1 v(2)  voltage
      2 v(4)  voltage
Binary:
...
```

Note the difference between the ASCII and the binary file. The ASCII file has the word Values: in the beginning of the list of vector values, whereas the binary file has Binary:.

To switch back to ASCII format, type

```
set filetype=ascii
```

Such stored results can be loaded with the load command. To load the vectors from the data.raw file, type

```
load data.raw
```

The command automatically recognizes the file format (ASCII or binary). The result is a new plot containing all the vectors that are stored in the data.raw file.

In order to output the text in a file, you can use similar output redirection as in the command shell. Say, for instance, you want to create a file containing "Hello everybody." Type

```
echo Hello everybody > file.txt
```

A file named file.txt is created containing the text that was the output of the echo command. Now say you want to append the printout of vectors v(2) and v(4) to this file. Type

```
print v(2) v(4) >> file.txt
```

Note that a single > creates an empty file and >> appends to an existing file (or creates a new one if the file does not exist). The resulting file looks like this:

```
Hello everybody
      A simple circuit
      DC transfer characteristic  Wed Feb 02 12:31:10  2005
-------------------------------------------------------------
Index   sweep            v(2)            v(4)
-------------------------------------------------------------
0     0.000000e+000   5.000000e+000   0.000000e+000
1     1.000000e-002   5.000000e+000   1.000000e-002
2     2.000000e-002   5.000000e+000   2.000000e-002
3     3.000000e-002   5.000000e+000   3.000000e-002
. . .
```

As you have probably noticed, the redirected output is not displayed in the SPICE command window.

2.11.4 Saving Device Quantities During Analyses

As we mentioned before, SPICE evaluates node voltages and currents flowing through voltage sources. However this is not completely true. Individual devices (e.g., resistors, BJTs, etc.) evaluate many auxiliary quantities. Load the circuit depicted in Fig. 2.2 and perform an operating point analysis. Next we will look at all the quantities that can be set/read for resistor r1.

```
show r1 : all
```

We get the following output:

```
Resistor: Simple linear resistor
device          r1
model           R
    resistance  1e+003
          temp  27
             l  0
             w  1e-005
             i  0.00216
             p  0.00468
      sens_dc   1.49e-313
    sens_real   1.49e-313
    sens_imag   1.49e-313
     sens_mag   1.49e-313
      sens_ph   1.49e-313
    sens_cplx   5.14e-317
                2.12e-314
             m  1
           tc1  0
           tc2  0
```

The output tells us that r1 uses the model named R. resistance, temp, l, w, m, tc1, and tc2 are parameters that can be set for a resistor. We explicitly set only resistance in the netlist, and the displayed value reflects our choice. i and p represent the current flowing through the resistor and the instantaneous power dissipated by the resistor. The last two parameters can be saved to a vector for every point evaluated by the simulator.

Now look at the parameters of a BJT. Type

```
show q1 : all
```

The output is quite long, so we are not going to go through all its details. Just note that parameters ic, ib, and ie denote the current flowing into the collector, the base, and the emitter node of the transistor.

To specify which quantities you want to save, use the save command.

```
save @r1[p] @q1[ic]
```

This will save the instantaneous power dissipated by resistor r1 and the collector current of q1. To see which quantities are going to be saved, type

```
status
```

The output looks like

```
1    save @r1[p]
2    save @q1[ic]
```

Now perform a DC analysis and take a look at the list of resulting vectors.

```
dc vin 0 1.4 0.01
display
```

The list contains only three vectors.

```
Title: A simple circuit
Name: dc1 (DC transfer characteristic)
Date: Wed Feb 02 12:58:43  2005

    @q1[ic]              : voltage, real, 140 long
    @r1[p]               : voltage, real, 140 long
    sweep                : voltage, real, 140 long [default scale]
```

SPICE saved only the specified quantities and the default vector. In order to save the node voltages, you should add all nodes to the list of save directives by issuing commands of the form

```
save v(2) v(4) ...
```

or by simply saying

```
save all
```

After rerunning the analysis, you will see that now all the usual quantities are saved along with the two device quantities @r1[p] and @q1[ic]. Now look again at the list of save directives by typing status. You should get

```
1    save @r1[p]
2    save @q1[ic]
3    save all
```

In order to remove a directive from the list, you can use the delete command. In order to remove directives 1 and 2, type

```
delete 1 2
```

To remove all directives, type

```
delete all
```

The saved quantities can be accessed by their vector names. In order to plot the collector current of q1 you should type

```
plot @q1[ic]
```

Finally, let us remove all save directives.

```
delete all
```

2.11.5 Tracing and Interactive Plotting

While an analysis is in progress, SPICE can display the quantities which are evaluated for every point in the resulting vectors. To tell SPICE which quantities should be traced, use the `trace` command.

```
trace v(2) v(4) vin#branch
```

This tells SPICE to output the node voltages of nodes 2 and 4, and the current flowing through the independent voltage source `vin`, for every point as soon as it is evaluated. The list of trace directives can be displayed by typing `status`.

```
1   trace v(2)
2   trace v(4)
3   trace vin#branch
```

Individual trace directives can be removed, just like individual save directives, by using the `delete` command.

Now start a DC analysis.

```
dc vin 0 1.4 0.2
```

Usually SPICE is silent until the analysis is finished, but now you get the following output:

```
Execution trace (remove with the "delete" command).
A simple circuit
DC transfer characteristic   Wed Feb 02 13:11:03  2005
---------------------------------------------
 sweep            V(2)            V(4)
---------------------------------------------
 0.000000e+000  5.000000e+000  0.000000e+000
 2.000000e-001  5.000000e+000  2.000000e-001
 4.000000e-001  4.999903e+000  4.000000e-001
 6.000000e-001  4.802671e+000  6.000000e-001
 8.000000e-001  1.936572e-001  8.000000e-001
 1.000000e+000  1.546093e-001  1.000000e+000
 1.200000e+000  1.388604e-001  1.200000e+000
 1.400000e+000  1.291174e-001  1.400000e+000
```

Interactive plotting is similar to tracing, except that in this case the values are plotted graphically. Interactive plotting is not available if SPICE is started in the console mode. To set up interactive plotting, first remove all save and trace directives by typing `delete all`.

Make sure the input voltage source `vin` is set to a pulse waveform staring at 0.68 V, rising to 0.78 V with 1 µs delay and 0.1 ns rise time. Now set up one interactive plot displaying node voltages `v(2)` and `v(4)` and another plot displaying the current flowing through `vin`. Type

```
iplot v(2) v(4)
iplot vin#branch
```

After you start an analysis by typing

```
tran 0.1n 2u
```

two plot windows pop up displaying the requested quantities "on the fly" as they are being evaluated. You can close a plot window while the analysis is in progress. The window does not appear until you start a new analysis.

To take a look at the interactive plot directives, type `status`. You should get something like this:

```
1 iplot v(4) v(2)
2 iplot vin#branch
```

Interactive plot directives can be deleted in the same manner as save or trace directives, with the `delete` command. Finally, note that interactive plot directives, trace directives, and save directives can be mixed so that you can save some additional quantities, trace node voltages, and interactively plot them in a single analysis.

Chapter 3
Input File Syntax

3.1 File Structure

The input file is also referred to as a netlist. A netlist has the following structure:

```
Circuit name
circuit description
.control
NUTMEG commands
.endc
.end
```

The first line of a netlist is the circuit name. Be careful not to put any element descriptions on the first line of the file, as they will be interpreted as the circuit name and therefore left out of the simulated circuit.

The block between `.control` and `.endc` (also referred to as the "control block") is the NUTMEG script that is executed after the netlist is read and processed by the simulator. The whole block is optional and can be omitted. The commands in the control block are case sensitive. Lowercase is recommended everywhere within the control block, except for strings that are printed to the command window, plot titles, and axis labels.

What remains is the circuit description. The circuit description is case insensitive. To avoid confusion lowercase is recommended for the circuit description. The netlist ends with `.end`.

Leading spaces and tabs in the beginning of every line are ignored. Comments can be added to the netlist by starting a line with an asterisk (∗). All characters in a line that follow a $ character are also regarded as comments. The latter type of comment is available only in the circuit description part of the netlist.

```
* This is a comment.
r1 (1 2) r=1k $ This is another comment.
```

Long lines can be split by using the + character. A line starting with a + character is joined with the previous line.

```
This is
+ a very long
+ line.
```

T. Tuma and Á. Bűrmen, *Circuit Simulation with SPICE OPUS*, Modeling and Simulation in Science, Engineering and Technology, DOI 10.1007/978-0-8176-4867-1_3, © Birkhäuser Boston, a part of Springer Science+Business Media, LLC 2009

Such a split line is treated by SPICE as a single logical line. It is recommended to put a space or a tab character after the + character, unless you want to join two lines without putting a space character between them. Another common mistake is when a user splits a formula in two lines and treats the + character as an operator in the formula. After the lines are joined, the + character is removed by SPICE and the formula produces a syntax error at evaluation.

If a comment is placed between two lines joined by a + character, the two lines are still joined as the comments are removed before physical lines are processed and possibly joined.

3.2 Numbers

SPICE accepts floating point numbers written in the usual C style. SPICE also supports SI prefixes. See Table 3.1 for the list of available prefixes.

A common mistake is the use of a capital M for mega. The circuit description is case insensitive, so the prefix is interpreted as milli. The correct way would be to use meg.

SI prefixes can be followed by any string. This string is ignored by SPICE as long as there is no space or tab character between the SI prefix and the string. It is a common practice to put units after the SI prefix (e.g., 5uV). This, however, can be a source of errors as the F in 5F is not interpreted as 5 farads, but rather as 5 femto. Units should be avoided when possible.

3.3 Instances and Nodes

Circuits in SPICE are collections of circuit elements (e.g., resistors) and nodes. Circuit elements are also referred to as instances. Every instance is connected to at least one node. To connect a resistor named r1 to nodes 1 and 2 the following syntax is used:

```
r1 (1 2) r=1k
```

Table 3.1 SI prefixes in SPICE

SI prefix	Notation in SPICE	Value	Example
femto-	f	10^{-15}	5f or 5e-15
pico-	p	10^{-12}	8p or 8e-12
nano-	n	10^{-9}	1.2n or 1.2e-9
micro-	u	10^{-6}	4.7u or 4.7e-6
milli-	m	10^{-3}	10m or 0.01
kilo-	k	10^{3}	100k or 100e3
mega-	meg	10^{6}	10meg or 1e7
giga-	g	10^{9}	1g or 1e9
tera-	t	10^{12}	5t or 5e12

The list of nodes to which an instance is connected is followed by a description of the instance. In the above example resistor r1 is specified to be a 1 kΩ resistor connected to nodes 1 and 2. The name of an instance parameter of a resistor is r and it is set to 1 kΩ.

Instance names may comprise letters (a–z), numbers (0–9), and underscores (_). A name must not start with a number. The first letter of the instance name is used by SPICE to recognize the instance type (e.g., r in r1 stands for resistor).

Node names follow the same conventions as instance names, except that they may start with a number. When a node name starts with a number, it may comprise only numbers.

The zero node (0) has a special meaning. It is the ground node. Its voltage is always 0 V.

3.4 Models

Every circuit element, for instance, a resistor, has parameters that describe its behavior. Some of these parameters vary widely across elements of a circuit (e.g., semiconductor resistor's width and length), but others may be the same for several resistors in a circuit (e.g., semiconductor resistor's sheet resistance). The latter share a common model in SPICE. Of course, the set of parameters that can be adjusted for a particular model is fixed and depends on the type of the device, just like the set of parameters that can be adjusted for a particular instance. A model is described using the following syntax:

.model *model_name model_type* (*parameter*[=*value*] ...)

Model names may comprise letters (a–z), numbers (0–9), and underscores (_). The name must not start with a number.

Some device types may have different model types available in SPICE, for instance, the diode. The basic (default) diode model used by SPICE is the level 1 diode model. There is also a geometric model available (level 3). The model level is chosen by supplying the level parameter in the model definition. If the level parameter is not supplied, a level 1 model is assumed.

Models of different levels differ in available model parameters. Often there are differences in available instance parameters, quantities evaluated by the simulator, internal nodes, and noise contributions. Metal oxide semiconductor (MOS) instances of different metal oxide semiconductor (MOS) model levels differ even in the number of connection nodes that are available to the user. The following example illustrates the use of different model levels for the same device type (diode):

```
* level 1 model (default)
.model d1n4148 d (level=1 ...)
* level 3 model
.model geomd d (level=3 ...)
```

Consider the following circuit description:

```
.model semires r (rsh=1e4)
r1 (1 2) semires w=1u l=100u
r1 (2 3) semires w=1u l=200u
```

Resistors r1 and r2 share the same model (semires) although they differ in length. The model named semires is a model of a resistor with sheet resistance 10 kΩ. rsh is the name of a model parameter of a resistor model.

Some devices do not require a model. A typical example is a resistor, for instance, the following example:

```
r5 (1 2) r=1k
```

Resistor r5 is a 1 kΩ resistor. SPICE uses what is called a default resistor model for such resistors. All model parameters of a default model are at their default values.

Some devices have a default model only (e.g., inductance). For such devices there are no model parameters available to the user, and thus the user cannot specify an inductance model. On the other hand, some device instances (e.g., all semiconductor devices) cannot be specified without specifying a model. For such devices there must be a corresponding .model statement in the netlist.

3.5 Instance and Model Parameters

Every instance/model (and, as will be explained later in Sect. 3.12, simulator para-meter) has a type. The parameter type can be one of

- Flag
- Integer
- Real
- String
- Expression

Flag parameters are either specified or not. Take, for instance, the off parame-ter of a diode. If you want to turn on the off parameter you simply add it to the parameter list of the diode instance.

```
d1 (1 2) d1n4001 off
```

Integer and real parameters are specified in the form *param_name* [=*param_ value*].

Some devices (e.g., independent voltage and current sources) have vector integer and real parameters. In the following example parameter pulse is a vector real parameter.

```
v1 (1 2) dc=0 pulse=(0 5.0 1m 1u 1u 10u)
```

String parameters are rare. Typically the version model parameter of the level 53 MOS model (BSIM3v3) is of type string.

```
.model nm0u18 nmos (level=53 version=3.2.4 ...)
```

The only device that has a parameter of the type expression is the nonlinear controlled source. The expression must be specified as the last parameter on the logical line.

```
b1 (1 2) i=v(10)*i(vsrc)
```

When specifying parameters, the = sign can be replaced by a space. In vector parameters the spaces can be replaced by commas or semicolons. The parentheses around vector parameters can also be omitted.

3.6 Device Syntax

For every device type the following things are listed:

- Model type (if a model can be specified by the user)
- Instance syntax
- List of instance parameters
- List of model parameters
- List of instance properties calculated by the simulator (use the save command to make them accessible in analyses other than the operating point analysis)
- List of internal nodes
- List of noise contributions

There are some common parameters and features for all instances. Instance parameter m specifies the number of instances connected in parallel. The default value of the m parameter is 1. For instance,

```
r5 (1 2) r=1k m=2
```

specifies two 1 kΩ resistors in parallel. The value of the m parameter must be integer. The m parameter can be specified for all instances except for transmission lines (T, O, and U) and inductor coupling (K). In

```
v1 (1 2) dc=10 m=2
```

the current v1#branch (or i(v1)) will be the current of a single parallel instance of v1. Instance properties, which are calculated by the simulator, also refer to a single parallel instance. In the following example:

```
r1 (1 2) r=1k m=2
```

the instance property @r1[i] which represents the current flowing through a resistor from node 1 to node 2 represents the current of a single parallel instance. To obtain the actual current caused by r1 the value @r1[i] must be multiplied by the value of m.

The SCALE simulator parameter is a scaling factor for all instance widths, lengths, and areas. Its default value is 1, so all lengths are taken to be specified in meters and all areas in square meters. To make micrometers and square micrometers the unit for instance parameters, SCALE must be set to 10^{-6}. Example:

```
.options scale=1e-6
```

All lengths that are specified as instance parameters are multiplied by SCALE. All areas that are specified as instance parameters are multiplied by SCALE^2. The SCALE parameter does not affect model parameter treatment.

The SCALE parameter affects resistors and capacitors (when specified through w and l). It also affects diodes that are using the level 3 diode model and all MOS transistors. In this chapter simulator parameters are denoted by capital letters (e.g., SCALE, ABSTOL, etc.).

Some devices (like diode, for instance) add internal nodes to the circuit. Internal nodes are named *instance#specifier* where *instance* is the name of the instance that has an internal node and *specifier* is the specifier of the internal node. In diode d1 the only internal node has specifier internal so the complete internal node name is d1#internal.

In noise analysis an element can produce several different noise contributions. A noise contribution can be identified through its postfix. For instance a diode produces three noise contributions with postfixes _rs (series resistance thermal noise), _id (shot noise), _1overf (flicker noise), and one with an empty postfix (total diode noise). The contributions of diode d1 to the output noise spectrum are named onoise_spectrum_d1_rs, onoise_spectrum_d1_id, onoise_spectrum_d1_1overf, and onoise_spectrum_d1. The last contribution is the sum of the first three contributions.

3.6.1 Resistor

Model type: r
  Syntax:
Rname (node1 node2) [model_name] param[=value] ...

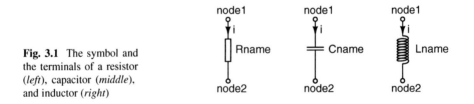

Fig. 3.1 The symbol and the terminals of a resistor (*left*), capacitor (*middle*), and inductor (*right*)

See Fig. 3.1 (left) for the symbol and the terminals of a resistor. Tables 3.2–3.4 list the resistor parameters and properties calculated by the simulator.

Resistors have no internal nodes and the noise contribution is caused by the thermal noise. The noise contribution of resistor rname is stored in vector onoise_spectrum_rname.

Table 3.2 Resistor instance parameters

Name	Unit	Default	Description
r	Ω	10^3	Resistance, overrides resistance obtained from w and l
w	SCALE m		Instance width
l	SCALE m		Instance length
tc1	1/K		First order temperature coefficient, overrides model parameter tc1
tc2	$1/K^2$		Second order temperature coefficient, overrides model parameter tc2
m	–	1	Number of parallel instances
temp	°C	TEMP	Instance temperature

Table 3.3 Resistor model parameters

Name	Unit	Default	Description
rsh	Ω/□		Sheet resistance, used in conjunction with instance parameters w and l
narrow	m	0	Narrowing due to side etching
tc1	1/K	0	First order temperature coefficient
tc2	$1/K^2$	0	Second order temperature coefficient
res	Ω		Default resistance
defw	m	10^{-6}	Default width
tnom	°C	TNOM	Parameter measurement temperature

Table 3.4 Resistor properties calculated by the simulator

Name	Unit	Description
i	A	Resistor current (for DC and transient), positive if the current flows into the resistor at its first terminal (*node1*)
p	W	Resistor power (for DC and transient)

Examples:

```
* Ordinary resistor, no temperature dependence
r1 (1 2) r=10k

* Ordinary resistor with temperature dependence
.model rmtc r (tc1=0.001)
r2 (3 4) rmtc r=5.6k

* Semiconductor resistor
.model rsemi r (rsh=10k tc1=0.001)
r3 (8 9) rsemi w=0.5u l=90u
```

3.6.2 Capacitor

Model type: c
  Syntax:
Cname (node1 node2) [model_name] param[=value] ...

Table 3.5 Capacitor instance parameters

Name	Unit	Default	Description
c	F		Capacitance, overrides capacitance obtained from w and l
w	SCALE m		Instance width
l	SCALE m		Instance length
ic	V	0	Initial voltage across capacitor
m	–	1	Number of parallel instances

Table 3.6 Capacitor model parameters

Name	Unit	Default	Description
cj	F/m^2		Bottom capacitance per area, used in conjunction with instance parameters w and l
cjsw	F/m		Sidewall capacitance per side length, used in conjunction with instance parameters w and l
defw	m	10^{-6}	Default width
narrow	m	0	Narrowing due to side etching

Table 3.7 Capacitor properties calculated by the simulator

Name	Unit	Description
i	A	Capacitor current (for transient), positive if the current flows into the capacitor at its first terminal (*node1*)
p	W	Capacitor power (for transient), positive, when energy flows into the capacitor

See Fig. 3.1 (middle) for the symbol and the terminals of a capacitor. Tables 3.5–3.7 list the capacitor parameters and properties calculated by the simulator.

Capacitors have no internal nodes and no noise contributions.

Examples:

```
* Ordinary capacitor
c1 (1 2) c=100n

* Semiconductor capacitor
*    C=50u*100u*100u+2*(100u+100u)*20p=0.508pF
.model csemi c (cj=50u cjsw=20p)
c2 (8 9) csemi w=100u l=100u
```

3.6.3 Inductor

Model type: no model parameters available
  Syntax:
Lname (node1 node2) param[=value] . . .

Table 3.8 Inductor instance parameters

Name	Unit	Default	Description
l	H		Inductance
ic	A	0	Initial current through inductor
m	–	1	Number of parallel instances

Table 3.9 Inductor properties calculated by the simulator

Name	Unit	Description
flux	Vs	Inductor flux (for transient), positive if the current flows from the first node into the inductor
i	A	Inductor current (for transient), positive if the current flows from the first node into the inductor
p	W	Inductor power (for transient), positive when energy flows into the inductor

See Fig. 3.1 (right) for the symbol and the terminals of an inductor. Tables 3.8 and 3.9 list the inductor parameters and properties calculated by the simulator.

Inductors have no noise contributions. Every inductor has a single internal current node. It is named $Lname$#branch. Its voltage represents the current flowing from the first node into one of the m parallel instances of the inductor.

3.6.4 Inductive Coupling

Model type: no model parameters available
  Syntax:
$Kname$ ($Lname1$ $Lname2$) $param$[=$value$] ...

Fig. 3.2 The symbol of an inductive coupling (*middle*) of two inductances (*left* and *right*)

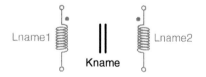

See Fig. 3.2 for the symbol of an inductive coupling. Table 3.10 lists the inductive coupling parameters.

Inductive couplings have no internal nodes and no noise contributions.

For a group of n coupled inductors a total of $n(n-1)/2$ inductive couplings must be specified.

Table 3.10 Inductive coupling instance parameters

Name	Unit	Default	Description
k	–		Coupling factor, should be between 0 and 1
m	–	1	Number of parallel instances (ignored)

Example:

```
l1 (1 2) l=1m
l2 (2 3) l=2m
l3 (4 5) l=0.5m
k12 (l1 l2) k=0.95
k12 (l2 l3) k=0.98
k12 (l1 l3) k=0.98
```

3.6.5 Independent Voltage Source

Model type: no model parameters available
   Syntax:
Vname (node1 node2) param[=value] ...

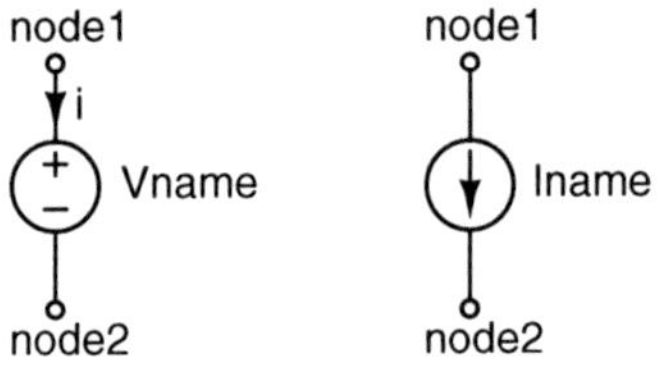

Fig. 3.3 The symbol and the terminals of an independent voltage source (*left*) and an independent current source (*right*)

   The first terminal is the positive terminal. See Fig. 3.3 (left) for the symbol and the terminals of an independent voltage source. Table 3.11 lists the independent voltage source instance parameters.

   By setting the value of one of the transient waveform parameters (pulse, sin, exp, pwl, sffm) the values of the parameters function, order, and coeffs change. The function parameter is an integer that specifies the type of transient waveform. See Table 3.12 for further explanation. The order and coeffs parameters represent the number of components in the vector describing the transient waveform and the vector itself.

   In DC analyses (operating point, DC sweep, transfer function, and all initial operating point evaluations) the voltage source is a DC voltage source with its value set to the dc parameter. If a transient waveform is specified, the DC value is the value of the transient waveform taken at $t = 0$.

Table 3.11 Independent voltage source instance parameters

Name	Unit	Default	Description
dc	V	0	DC voltage
acmag	V	0	AC magnitude
acphase	°	0	AC phase
pulse	–		Coefficient vector, sets up a pulse transient waveform
sin	–		Coefficient vector, sets up a sinusoidal transient waveform
exp	–		Coefficient vector, sets up an exponential transient waveform
pwl	–		Coefficient vector, sets up a piecewise linear transient waveform
sffm	–		Coefficient vector, sets up an FM (frequency modulated) transient waveform
function	–	0	Integer code for the transient waveform
order	–	0	Number of coefficients describing the transient waveform
coeffs	–		Coefficient vector for the transient waveform
m	–	1	Number of parallel instances

Table 3.12 Transient function codes for the function parameter

Code	Function
0	No function
1	Pulse
2	Sinusoidal
3	Exponential
4	FM
5	Piecewise linear

Table 3.13 Independent voltage source properties calculated by the simulator

Name	Unit	Description
acreal	V	Real part of the complex AC value
acimag	V	Imaginary part of the complex AC value
i	A	Independent voltage source current (for DC and transient), positive if the current flows from the first node into the source
p	W	Independent voltage source power (for DC and transient), positive when energy flows into the source

In transient analysis the transient waveform is used. If no transient waveform is specified (function parameter is set to 0), the source behaves as a DC source with its value set to the dc parameter.

In AC analysis the source behaves as a complex voltage source with its magnitude and phase specified by the acmag and acphase parameters. If the acmag parameter is not specified, the source behaves as a short for complex signals.

In noise analysis the source is a 0 V complex source (i.e., behaves as a short) for complex signals.

In pole-zero analysis the source represents a short if no acmag parameter is specified. However, if acmag is specified, the source is removed from the circuit (open circuit for complex signals).

Table 3.13 lists the independent voltage source properties calculated by the simulator.

The independent voltage source has no noise contributions and one internal current node. Its value represents the current flowing from the first node into one of the m parallel instances of the independent voltage source. Its name is V*name*#branch.

Examples:

```
v1 (1 2) dc=5
vac (5 9) dc=0 acmag=1.0 acphase=0
```

3.6.5.1 Pulse Waveform

Let V_1, V_2, t_d, t_r, t_f, t_{pw}, and T denote the components of the vector describing a pulse transient waveform. The waveform starts at V_1 and remains there until $t = t_d$ is reached. From $t = t_d$ to $t = t_d + t_r$ the waveform rises (falls) linearly to V_2. it remains at V_2 until $t = t_d + t_r + t_{pw}$, upon which it returns linearly to V_1 at $t = t_d + t_r + t_{pw}$. Finally, at $t = t_d + T$ the waveform starts to rise (fall) again, and the cycle that started at $t = t_d$ is repeated. Mathematically, a single pulse of the waveform can be described as

$$
v(t) = \begin{cases}
V_1, & 0 \le t < t_d \\
V_1 + (V_2 - V_1)\frac{t - t_d}{t_r}, & t_d \le t < t_d + t_r \\
V_2, & t_d + t_r \le t < t_d + t_r + t_{pw} \\
V_2 + (V_1 - V_2)\frac{t - (t_d + t_r + t_{pw})}{t_f}, & t_d + t_r + t_{pw} \le t < t_d + t_r + t_{pw} + t_f \\
V_1, & t_d + t_r + t_{pw} + t_f \le t < t_d + t_{per}
\end{cases}
\tag{3.1}
$$

The part between t_d and $t_d + t_{per}$ is repeated and represents one period of the signal. An example of a pulse waveform is depicted in Fig. 3.4.

The default values are $t_d = 0$, $t_r = t_f = T_{step}$, and $t_{pw} = T = T_{stop}$. T_{step} and T_{stop} are the initial step and the duration of the transient analysis. If only V_1, V_2, t_d, and t_r are specified, the waveform is a step from V_1 to V_2 that starts at $t = t_d$ and is completed at $t = t_d + t_r$. If t_f and t_{pw} are also specified, the step is extended into a single pulse. With all six values specified, the waveform is a series of pulses with period T.

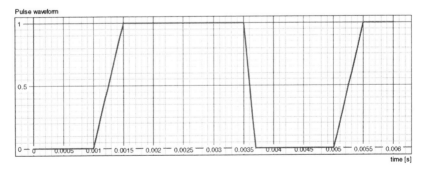

Fig. 3.4 Pulse waveform for $V_1 = 0\,\mathrm{V}$, $V_2 = 1\,\mathrm{V}$, $t_d = 1\,\mathrm{ms}$, $t_r = 0.5\,\mathrm{ms}$, $t_f = 0.2\,\mathrm{ms}$, $t_{pw} = 2\,\mathrm{ms}$, and $t_{per} = 4\,\mathrm{ms}$ (specified by pulse=(0 1 1m 0.5m 0.2m 2m 4m))

Examples:

```
* Step 0V..5V at 1ms, 0.1ms rise time
v1 (5 9) pulse=(0 5.0 1m 0.1m)
* Single pulse 0V..5V at 1ms, 0.1ms rise and fall time, 9ms pulse width
v2 (9 0) pulse=(0 5.0 1m 0.1m 0.1m 9m)
* Pulse series 0V..5V starting at 5ms with 0.1ns rise and fall time,
*    1us pulse width, and 2us period
v5 (5 0) pulse=(0 5.0 5m 0.1n 0.1n 1u 2u)
```

3.6.5.2 Sinusoidal Waveform

Let V_o, V_a, f, t_d, and θ denote the components of the vector describing a sinusoidal transient waveform. The waveform can be described mathematically as

$$v(t) = \begin{cases} V_o, & 0 \le t < t_d \\ V_o + V_a \exp(-\theta(t - t_d))\sin(2\pi f(t - t_d)), & t \ge t_d \end{cases} \quad (3.2)$$

The default values of the parameters are $f = 1/T_{stop}$, $t_d = 0$, and $\theta = 0$, where T_{stop} is the duration of the transient analysis. By specifying only the first two parameters, a single sine wave across the duration of the transient analysis is obtained. With three parameters the frequency of the sine wave can be adjusted. Note that four parameters specify a damped sine wave. An example of a sinusoidal waveform is depicted in Fig. 3.5.
Examples:

```
* Sinusoidal waveform, 0V offset, 1V amplitude, 1kHz
v1 (5 9) sin=(0 1.0 1k)
* Damped sinusoidal waveform, 0V offset, 1V amplitude,
*    1kHz, 0.001/s damping
v2 (9 0) sin=(0 1.0 1k 0.001)
```

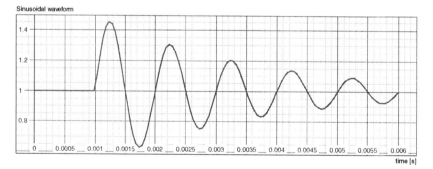

Fig. 3.5 Sinusoidal waveform for $V_o = 1\,\text{V}$, $V_a = 0.5\,\text{V}$, $f = 1\,\text{kHz}$, $t_d = 1\,\text{ms}$, and $\theta = 400\,\text{s}^{-1}$ (specified by sin=(1 0.5 1k 1m 400))

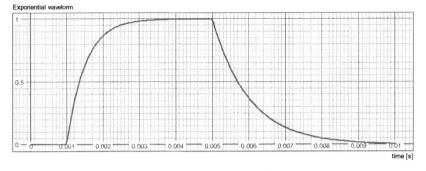

Fig. 3.6 Exponential waveform for $V_1 = 0\,\text{V}$, $V_2 = 1\,\text{V}$, $t_1 = 1\,\text{ms}$, $\tau_1 = 0.5\,\text{ms}$, $t_2 = 5\,\text{ms}$, and $\tau_2 = 1\,\text{ms}$ (specified by exp=(0 1 1m 0.5m 5m 1m))

3.6.5.3 Exponential Waveform

Let V_1, V_2, t_1, τ_1, t_2, and τ_2 denote the components of the vector describing an exponential transient waveform. Mathematically, the waveform is

$$v(t) = \begin{cases} V_1, & 0 \le t < t_1 \\ V_1 + (V_2 - V_1)(1 - \exp(-(t - t_1)/\tau_1)), & t_1 \le t < t_2 \\ V_1 + (V_2 - V_1)(1 - \exp(-(t - t_1)/\tau_1)) + \\ \quad (V_1 - V_2)(1 - \exp(-(t - t_2)/\tau_2)), & t \ge t_2 \end{cases} \tag{3.3}$$

The default values are $t_1 = 0$, $\tau_1 = T_{step}$, $t_2 = t_1 + T_{step}$, $\tau_2 = T_{step}$. An example of an exponential waveform is shown in Fig. 3.6.

Example:

```
* Single exp pulse reaching 5V, starting at 1ms, with rise tau=0.1ms,
*    starts falling at 2ms with tau=0.2ms
v1 (5 9) exp=(0 5.0 1m 0.1m 2m 0.2m)
```

3.6.5.4 Piecewise Linear Waveform

Let t_0, V_0, t_1, V_1, ... denote the consecutive components of the vector describing the piecewise linear transient waveform. The waveform starts at V_0 and remains at V_0 until $t = t_0$. At $t = t_0$ it starts to change linearly and reaches V_1 at t_1. This is followed by a linear change to V_2 at t_2. When the waveform reaches the last specified level, it remains there for the whole of the remaining time. An example of a piecewise linear waveform is depicted in Fig. 3.7.

Examples:

```
* A single triangular pulse, starts at 1ms, reaches 5V at 2ms,
*    and falls back to 0V at 5ms
v1 (5 9) pwl=(1m 0 2m 5.0 5m 0)
```

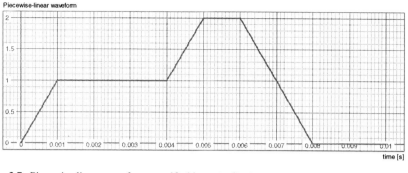

Fig. 3.7 Piecewise linear waveform specified by `pwl=(0 0 1m 1 4m 1 5m 2 6m 2 8m 0)`

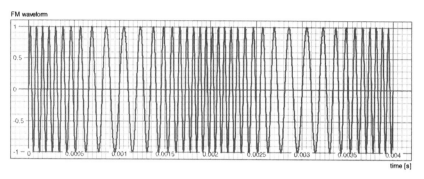

Fig. 3.8 FM waveform with $V_o = 0$ V, $V_a = 1$ V, $f_c = 10$ kHz, $m = 10$, and $f_s = 500$ Hz (specified by `sffm=(0 1 10k 10 500)`)

3.6.5.5 FM Waveform

The parameters of the FM waveform (in the same order as they appear in the vector) are V_o, V_a, f_c, m, and f_s. The mathematical description of the waveform is

$$v(t) = V_o + V_a \sin(2\pi f_c t + m \sin(2\pi f_s t)). \tag{3.4}$$

The default values are $f_c = f_s = 1/T_{stop}$, where T_{stop} is the duration of the transient analysis. An example of an FM waveform is depicted in Fig. 3.8.
   Examples:

```
* FM signal (0V offset, 10mV amplitude),
*    10MHz carrier modulated by a 1kHz sine, m=0.001
v1 (5 9) sffm=(0 10m 10meg 0.001 1k)
```

3.6.6 Independent Current Source

Model type: no model parameters available
   Syntax:
Iname (node1 node2) param[=value] ...
   The first node is the one from which the current flows into the source if the current
is positive. See the independent voltage source (Sect. 3.6.5) for the explanation of
available parameters and transient waveforms. See Fig. 3.3 (right) for the symbol
and the terminals of an independent current source.
   In AC and noise analysis the independent current source behaves as an open
circuit for complex signals if the acmag parameter is not specified. In pole-zero
analysis the source behaves as an open circuit for complex signals.
   The independent current source has no internal nodes and no noise contributions.
   Examples:

```
* Pulls 1uA from the ground into node 1
i1 (0 1) dc=1u
* 10uA current pulse, 1ms wide, starting at 0.1ms,
*   rise and fall time are 1us
i2 (5 9) pulse=(0 10uA 0.1m 1u 1u 1m)
```

3.6.7 Linear Voltage-Controlled Current Source

Model type: no model parameters available
   Syntax:
Gname (node1 node2 node_c1 node_c2) param[=value] ...

Fig. 3.9 The symbol and the
terminals of a linear voltage-
controlled current source

   The first two nodes are the positive and the negative node of the source. The
current is positive when it flows into the source at its first terminal (*node1*). The
third and the fourth node are the positive and the negative node of the controlling
voltage. See Fig. 3.9 for the symbol and the terminals of a linear voltage-controlled
current source. Tables 3.14 and 3.15 list the parameters and the properties calculated
by the simulator.
   Linear voltage-controlled current sources have no internal nodes and no noise
contributions.

Table 3.14 Linear voltage-controlled current source instance parameters

Name	Unit	Default	Description
gain	A/V		Transconductance
m	–	1	Number of parallel instances

Table 3.15 Linear voltage-controlled current source properties calculated by the simulator

Name	Unit	Description
i	A	Controlled source current (for DC and transient), positive if the current flows from the first node into the source
p	W	Controlled source power (for DC and transient), positive when energy flows into the source

Example:

```
* Controlled by v(3)-v(4)
* Pulls current from ground and pushes it into node 10
g1 (0 10 3 4) gain=10m
```

3.6.8 Linear Voltage-Controlled Voltage Source

Model type: no model parameters available
    Syntax:
Ename (node1 node2 node_c1 node_c2) param[=value] ...

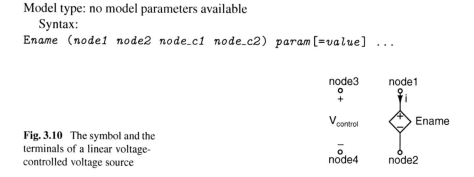

Fig. 3.10 The symbol and the terminals of a linear voltage-controlled voltage source

    The first two nodes are the positive and the negative node of the controlled source. The second two nodes are the positive and the negative node of the controlling voltage. See Fig. 3.10 for the symbol and the terminals of a linear voltage-controlled voltage source. Tables 3.16 and 3.17 list the parameters and the properties calculated by the simulator.

    Linear voltage-controlled voltage sources have no noise contributions. Every linear voltage-controlled voltage source has one internal current node named *Ename*#branch. Its voltage represents the currents flowing from the first node into one of the m parallel instances of the controlled source.

Table 3.16 Linear voltage-controlled voltage source instance parameters

Name	Unit	Default	Description
gain	–		Voltage gain
m	–	1	Number of parallel instances

Table 3.17 Linear voltage-controlled voltage source properties calculated by the simulator

Name	Unit	Description
i	A	Controlled source current (for DC and transient), positive if the current flows from the first node into the source
p	W	Controlled source power (for DC and transient), positive when energy flows into the source

Example:

```
* Controlled by v(3)-v(4)
* Connected between 1 (+) and 2 (-)
e1 (1 2 3 4) gain=100
```

3.6.9 Linear Current-Controlled Current Source

Model type: no model parameters available
  Syntax:
Fname (node1 node2 name) param[=value] ...

Fig. 3.11 The symbol and the terminals of a linear current-controlled current source

The first two nodes are the positive and the negative node of the controlled source. The current is positive when it flows from the first node into the source. The third connection (*name*) is the name of the controlling instance. The controlling instance can be any instance that has an internal current node named *name*#branch. The controlling current is the value represented by the internal current node. In practice, this means that specifying v1 as the controlling instance makes the current flowing into the positive terminal of v1 the controlling current. See Fig. 3.11 for the symbol and the terminals of a linear current-controlled current source. Tables 3.18 and 3.19 list the parameters and the properties calculated by the simulator.

Linear current-controlled current sources have no internal nodes and no noise contributions.

Table 3.18 Linear current-controlled current source instance parameters	Name	Unit	Default	Description
	gain	–		Current gain
	m	–	1	Number of parallel instances

Table 3.19 Linear current-controlled current source properties calculated by the simulator

Name	Unit	Description
i	A	Controlled source current (for DC and transient), positive if the current flows from the first node into the source
p	W	Controlled source power (for DC and transient), positive when energy flows into the source

Example:

```
* Pulls current from ground and pushes it into node 10
f1 (0 10 v10) gain=80
* The control current flows through v10. The control current
* is positive when it flows from node 50 into the positive terminal
* of v10.
v10 (50 55) dc=0
* This current-controlled current source is controlled by a current flowing
* through an inductor.
f2 (0 20 11) gain=100
* The control current flows from node 80 into the positive terminal
* of inductor 11.
11 (80 90) l=1m
```

3.6.10 Linear Current-Controlled Voltage Source

Model type: no model parameters available
 Syntax:
Hname (node1 node2 name) param[=value] ...

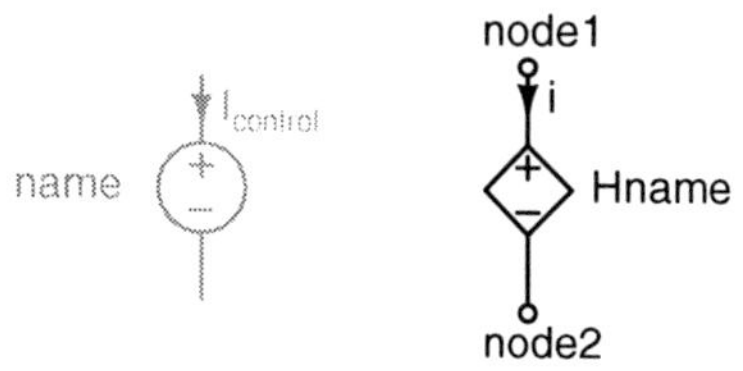

Fig. 3.12 The symbol and the terminals of a linear current-controlled voltage source

The first two nodes are the positive and the negative node of the controlled source. The third connection (*name*) is the name of the controlling instance. The controlling instance can be any instance that has an internal current node named *name* #branch. The controlling current is the value represented by the internal current node. This

Table 3.20 Linear current-controlled voltage source instance parameters

Name	Unit	Default	Description
gain	V/A		Transimpedance
m	–	1	Number of parallel instances

Table 3.21 Linear current-controlled voltage source properties calculated by the simulator

Name	Unit	Description
i	A	Controlled source current (for DC and transient), positive if the current flows from the first node into the source
p	W	Controlled source power (for DC and transient), positive when energy flows into the source

means that specifying v1 as the controlling instance makes the current flowing into the positive terminal of v1 the controlling current. See Fig. 3.12 for the symbol and the terminals of a linear current-controlled voltage source. Tables 3.20 and 3.21 list the parameters and the properties calculated by the simulator.

Linear current-controlled voltage sources have no noise contributions. Every linear current-controlled voltage source has one internal current node named H*name*#branch. Its voltage represents the currents flowing from the first node into one of the m parallel instances of the controlled source.

Example:

```
* Controlled by the current flowing from
*    the first node of v10, through v10, and into the second node of v10
* Connected between 1 (+) and 2 (-)
h1 (1 2 v10) gain=1k
```

3.6.11 Nonlinear Controlled Sources

Model type: no model parameters available
Syntax:
B*name* (*node1 node2*) *param*[=*value*] ... v=*expression*
B*name* (*node1 node2*) *param*[=*value*] ... i=*expression*

Fig. 3.13 The symbol and the terminals of a nonlinear voltage source (*left*) and current source (*right*)

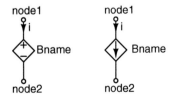

Table 3.22 Nonlinear controlled source instance parameters

Name	Unit	Default	Description
m	–	1	Number of parallel instances
i	–		Expression (controlled current source)
v	–		Expression (controlled voltage source)

Table 3.23 Nonlinear controlled source properties calculated by the simulator

Name	Unit	Description
i	A	Controlled source current (for DC and transient), positive if the current flows from the first node into the source
v	V	Controlled source voltage (for DC and transient), first node is considered to be the positive node

Table 3.24 Nonlinear controlled source operators

Name	Example	Description
unary −	-v(10)	Unary minus
+	v(2) + v(3)	Addition
−	v(2) − v(3)	Subtraction
*	v(2) * v(3)	Multiplication
/	v(2) / v(3)	Division
^	v(2)^2	Power

The first line represents a nonlinear controlled voltage source and the second line a nonlinear controlled current source. The first node is the positive node for the voltage source. For the current source it is the node from which the current flows into the source when the current is positive. See Fig. 3.13 for the symbol and the terminals of a nonlinear controlled voltage source (left) and current source (right). Tables 3.22 and 3.23 list the parameters and the properties calculated by the simulator.

A nonlinear controlled voltage source has one internal current node named B*name*#branch. Its value represents the current flowing from the first node into one of the m parallel instances of the controlled source. A nonlinear controlled current source has no internal nodes. Nonlinear controlled sources have no noise contributions.

3.6.11.1 Expression Syntax

The available operators and functions are listed in Tables 3.24 and 3.25. The syntax for numbers is the same as for parameter values. Besides numbers, two constants are available: e (the basis of the natural logarithm) and pi (π). Identifier time represents the current time in the transient analysis. In analyses other than transient, its value is 0.

Table 3.25 Nonlinear controlled source functions

Name	Example	Description
abs	abs(v(10))	Absolute value
acos	acos(v(10))	Inverse cosine (returns result in radians)
acosh	acosh(v(10))	Inverse hyperbolic cosine
asin	asin(v(10))	Inverse sine (returns result in radians)
asinh	asinh(v(10))	Inverse hyperbolic sine
atan	atan(v(10))	Inverse tangent (returns result in radians)
atanh	atanh(v(10))	Inverse hyperbolic tangent
cos	cos(v(10))	Cosine (argument in radians)
cosh	cosh(v(10))	Hyperbolic cosine
exp	exp(v(10))	Exponential e^x
ln	ln(v(10))	Base e logarithm of the absolute argument
log	ln(v(10))	Base 10 logarithm of the absolute argument
sgn	sgn(v(10))	Sign (0, 1 or -1)
sin	sin(v(10))	Sine (argument in radians)
sinh	sinh(v(10))	Hyperbolic sine
sqrt	sqrt(v(10))	Square root of the absolute argument
tan	tan(v(10))	Tangent (argument in radians)
tanh	tanh(v(10))	Hyperbolic tangent
u	u(v(10))	Unit step function, 0 for arguments less or equal zero, 1 otherwise
uramp	uramp(v(10))	Unit ramp function, integral of (u(x))

The voltage at a particular node is obtained using v(*name*). A differential voltage is obtained by typing v(*name1*,*name2*). A device current can be obtained through i(*name*). This works only if a device has an internal node named *name*#branch representing its current (voltage sources and inductors).

Examples:

```
b1 (1 8) i=v(10)*i(v1)
b2 (9 5) v=v(1)+sin(v(2))^2
```

3.6.12 Voltage-Controlled Switch

Model type: sw
  Syntax:
S*name* (*node1 node2 node_c1 node_c2*) *model_name*
+ *param*[=*value*] ...

Fig. 3.14 The symbol and the terminals of a voltage-controlled switch

The switch is connected between the first two nodes. The second two nodes represent the controlling voltage. In this sense the voltage-controlled switch is similar to the voltage-controlled voltage source. The switch has hysteresis. Both the on and off states must have finite nonzero conductance (see the list of model parameters). See Fig. 3.14 for the symbol and the terminals of a voltage-controlled switch. Tables 3.26–3.28 list the parameters and the properties calculated by the simulator.

The two flags (on and off) impact the initial state of the switch in transient analysis. In DC analysis the hysteresis is accounted for, so the result depends on the direction of the DC sweep.

The switch goes from the open to the closed state when the controlling voltage rises above vt + vh. The transition back to the open state happens at vt − vh.

The noise contribution of a voltage-controlled switch is caused by the thermal noise of the switch resistance. For switch sname it is stored in the onoise_spectrum_sname vector.

Example:

```
* Connected between nodes 1 and 0
*   controlled by voltage between nodes 2 and 0
*   initially off
s1 (1 0 2 0) swm1 off
.model swm1 sw (vt=1.5 vh=0.1 ron=0.1 roff=1e6)
```

Table 3.26 Voltage-controlled switch instance parameters

Name	Unit	Default	Description
m	–	1	Number of parallel instances
on	–	Not set	Flag that sets switch initially closed
off	–	Set	Flag that sets switch initially open

Table 3.27 Voltage-controlled switch properties calculated by the simulator

Name	Unit	Description
i	A	Switch current (for DC and transient), positive if the current flows from the first node into the switch
p	V	Switch power (for DC and transient)

Table 3.28 Voltage-controlled switch model parameters

Name	Unit	Default	Description
vt	V		Threshold voltage
vh	V	0	Hysteresis
ron	Ω	1	Resistance when closed
roff	Ω	1/GMIN	Resistance when open

3.6.13   Current-Controlled Switch

Model type: `csw`
   Syntax:
`Wname (node1 node2 name) model_name`
`+ param[=value] ...`

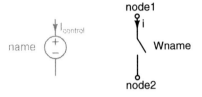

Fig. 3.15 The symbol and the terminals of a current-controlled switch

The first two nodes are the nodes to which the switch is connected. `node1` is the positive node and `node2` is the negative node. The controlling current is obtained from the instance *name*. This can be any instance with an internal current node (e.g., a voltage source or an inductor). The set of instance parameters and values calculated by the simulator is the same as with the voltage-controlled switch. See Fig. 3.15 for the symbol and the terminals of a current-controlled switch. Table 3.29 lists the model parameters.

The switch goes from the open to the closed state when the controlling current rises above it + ih. The transition back to the open state happens at it − ih.

The noise contribution of a current-controlled switch is caused by the thermal noise of the switch resistance. For switch *wname* it is stored in the `onoise_spectrum_wname` vector.

   Example:

```
* Connected between nodes 1 and 0
*    controlled by the current flowing into the + node of v10
*    initially off
w1 (1 0 v10) swm2 off
.model swm2 csw (it=2m ih=0.1m ron=0.1 roff=1e6)

* This switch is controlled by the current flowing into the
* positive node (node1) of inductance l1.
w2 (2 0 l1) swm2 off

* This is the controlling voltage source.
```

Table 3.29 Current-controlled switch model parameters

Name	Unit	Default	Description
it	A		Threshold current
ih	A	0	Hysteresis
ron	Ω	1	Resistance when closed
roff	Ω	1/GMIN	Resistance when open

```
* The control current flows from node 50 into the positive
* terminal of v10.
v10 (50 55) dc=0

* This is the controlling inductance.
* The control current flows from node 10 into the positive
* terminal of inductance l1.
l1 (10 20) l=1m
```

3.6.14 Lossless Transmission Line

Model type: no model parameters available
   Syntax:
Tname (node1 node2 node3 node4) param[=value] ...

Fig. 3.16 The symbol and
the terminals of a lossless
transmission line

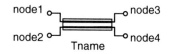

The first two nodes represent one end of the line and the second two nodes the
other end. Nodes *node1* and *node3* are the positive nodes. The current is considered to be positive if it flows from these nodes into the transmission line. See
Fig. 3.16 for the symbol of a lossless transmission line. Table 3.30 lists the instance
parameters of a lossless transmission line.
   The delay may be specified directly (through td) or through frequency (f) and
normalized length (nl). In the latter case the delay is calculated as td = nl/f. If td is
specified it takes precedence over the delay calculated from f and nl. rel and abs

Table 3.30 Lossless transmission line instance parameters

Name	Unit	Default	Description
zo	Ω		Characteristic impedance
td	s		Transmission delay
f	Hz	10^9	Frequency
nl	–	0.25	Normalized length at given frequency
v1	V	0	Initial voltage at end 1
v2	V	0	Initial voltage at end 2
i1	A	0	Initial current at end 1
i2	A	0	Initial current at end 2
m	–	1	Number of parallel instances
rel	–	1	Relative voltage derivative tolerance for timestep control
abs	V/s	1	Absolute voltage derivative tolerance for timestep control

are the tolerance criteria for time-step control. If the response of a transmission line is not smooth enough, the problem can be fixed by decreasing these two parameters.

The termination of a transmission line consists of a resistor and a voltage source. The voltage source represents the delayed signal from the remote end of the line. The resistance equals the characteristic impedance. The internal nodes between the resistor and the voltage source are named Tname#int1 and Tname#int2. The branch current flowing into *node1* (*node3*) at the end of the transmission line is named Tname#branch1 (Tname#branch2). Lossless transmission lines have no noise contributions.

Example:

```
* Input at nodes 1, 0
*   output at nodes 2, 0
*   50Ohm, 10us delay
t1 (1 0 2 0) td=10u zo=50
```

3.6.15 Lossy Transmission Line

Model type: ltra
   Syntax:
Oname (*node1 node2 node3 node4*) *model_name*
+ *param* [=*value*] ...

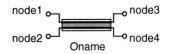

Fig. 3.17 The symbol and the terminals of a lossy transmission line

The first two nodes represent one end of the transmission line and the second two nodes the other end. The first and the third node are the positive nodes at both ends of the line. See Fig. 3.17 for the symbol of a lossy transmission line. Tables 3.31 and 3.32 list the parameters of a lossy transmission line.

A lossy transmission line has two internal current nodes named Oname#branch1 and Oname#branch2 representing the currents flowing from the positive end nodes into the transmission line. Lossy transmission lines have no noise contributions. Methods for lossy transmission line simulation are described in [43].

Table 3.31 Lossy transmission line instance parameters

Name	Unit	Default	Description
v1	V	0	Initial voltage at end 1
v2	V	0	Initial voltage at end 2
i1	A	0	Initial current at end 1
i2	A	0	Initial current at end 2
m	–	1	Number of parallel instances

Table 3.32 Lossy transmission line model parameters

Name	Unit	Default	Description
r	Ω/m	0	Resistance per length
l	H/m	0	Inductance per length
g	S/m	0	Conductance per length
c	F/m	0	Capacitance per length
len	m		Length of line
rel	–	1	Relative voltage derivative tolerance for time-step control
abs	V/s	1	Absolute voltage derivative tolerance for time-step control
nocontrol	–	Not set	Flag that disables complex time-step control
steplimit	–	Set	Limit time step to the delay of the line
nosteplimit	–	Not set	Don't limit time step to the delay of the line
lininterp	–	Not set	Use linear interpolation (set if TRYTOCOMPACT is set)
quadinterp	–	Set	Use quadratic interpolation (default)
mixedinterp	–	Not set	Use linear interpolation when quadratic interpolation works badly
truncnr	–	Not set	Use Newton–Raphson iterations for step calculation
truncdontcut	–	Not set	Don't limit time step to keep impulse response calculation errors low
compactrel	–	RELTOL	Reltol for straight line checking
compactabs	–	ABSTOL	Abstol for straight line checking

Example:

```
* Lossy transmission line
*    r=0.2Ohm/m l=9nH/m g=0H/m c=4pF/m len=10m
o1 (1 0 2 0) lossy1
.model lossy1 ltra (r=0.2 l=9n g=0 c=4p len=10)
```

3.6.16 Uniformly Distributed RC Line

Model type: urc
   Syntax:
Uname (node1 node2 node3) model_name param[=value] ...

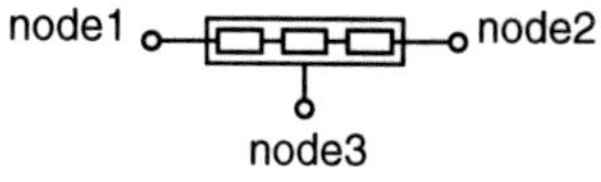

Fig. 3.18 The symbol and the terminals of a uniformly distributed RC line

The first two nodes represent the two inputs. The third node is the common node for both inputs. See Fig. 3.18 for the symbol. Tables 3.33 and 3.34 list the parameters of a uniformly distributed RC transmission line.

The uniformly distributed RC line (URC line) consists of discrete sections (see Fig. 3.19). The number of sections (when not given on the model definition) is obtained from

$$N = \max(3, \ln(\text{fmax} \cdot \text{rperl} \cdot \text{cperl} \cdot 2\pi l^2 ((k-1)/k)^2) / \ln(k)). \qquad (3.5)$$

If isperl is given, the capacitors are replaced by diodes (level 1 model). The names of the diodes follow the same key as the names of the capacitors (e.g., $Uname\#cxx$ changes to $Uname\#dxx$).

The capacitor, diode, and resistor models used by the elements of a URC line are named $Uname\#$capmod, $Uname\#$diodemod, and $Uname\#$resmod, respectively.

Table 3.33 Uniformly distributed RC line instance parameters

Name	Unit	Default	Description
l	m		Length
n	–		Number of sections
m	–	1	Number of parallel instances

Table 3.34 Uniformly distributed RC line model parameters

Name	Unit	Default	Description
k	–	1.5	Resistance per length
fmax	Hz	10^9	Inductance per length
rperl	Ω/m	1,000	Conductance per length
cperl	F/m	10^{-12}	Capacitance per length
isperl	A/m		Diode saturation current per length
rsperl	Ω/m	0	Diode series resistance per length

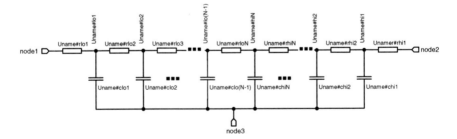

Fig. 3.19 The internal structure of a URC line

The diode model's `cjo`, `is`, and `rs` parameters are set to the section's capacitance, diode saturation current, and diode series resistance. The size of the sections is not uniform. It increases geometrically (k) from the end nodes toward the center of the line.

Note that every diode introduces an additional internal node if the value of `rperl` is greater than zero. See Sect. 3.6.17 for details. Every resistor and diode introduces its own noise contributions. See the sections on diodes and resistors for details.

Example:

```
* Inputs at nodes 2 and 3, common node is the ground
*    automatically calculate the number of sections
*    8kOhm/m, 4pF/m, use diodes with 1uA/m
u1 (2 3 0) rcline1
.model rcline urc (rperl=8k cperl=4p isperl=1p)
```

3.6.17 Semiconductor Diode

Model type: d
    Syntax:
Dname (node1 node2) model_name param[=value] ...

Fig. 3.20 The symbol and the terminals of a semiconductor diode

node1

Dname

node2

The direction of the forward current is from first node into the diode. See Fig. 3.20 for the symbol.

3.6.17.1 Level 1 Diode Model

Tables 3.35–3.38 list the level 1 diode parameters, properties calculated by the simulator, and noise contributions.

The diode is scaled with its `area` and `pj` parameters. The effective values of `is`, `expli`, and `cjo` are obtained by multiplying the corresponding model parameter with the value of `area` diode parameter. `rs` is divided by `area`. `jsw` and `cjsw` are multiplied by diode parameter `pj`.

A diode with model parameter `rs` set to a nonzero value has a single internal node between the actual diode and the series resistance. The node is named Dname#internal (see Fig. 3.21).

More details on the SPICE diode model can be found in [31].

Table 3.35 Level 1 diode instance parameters

Name	Unit	Default	Description
off	–	Not set	Flag that turns off the diode in the first Newton–Raphson (NR) iteration of DC analysis
temp	°C	TEMP	Device temperature
ic	V	0	Initial voltage across diode capacitance
area	–	1	Area factor
pj	–	0	Junction periphery
m	–	1	Number of parallel instances

Table 3.36 Level 1 diode model parameters

Name	Unit	Default	Description
is	A	10^{-14}	Saturation current
jsw	A	0	Sidewall saturation current
tnom	°C	TNOM	Parameter measurement temperature
rs	Ω	0	Series resistance
n	–	1	Emission coefficient
expli	A	10^{15}	Current explosion parameter (for diode current in excess of expli the $i(v)$ dependency is linearized)
tt	s	0	Transit time
cjo	F	0	Zero-bias bottom junction capacitance
cjsw	F	0	Zero-bias sidewall junction capacitance
vj	V	0.8	Junction potential
php	V	vj	Sidewall junction contact potential
m	–	0.5	Area junction grading coefficient
mjsw	–	0.33	Sidewall junction grading coefficient
eg	eV	1.11	Activation energy
xti	–	3	Saturation current temperature exponent
kf	–	0	Flicker noise coefficient
af	–	1	Flicker noise exponent
fc	–	0.5	Forward bias depletion bottom capacitance coefficient
fcs	–	0.5	Forward bias depletion sidewall capacitance coefficient
bv	V	0	0 indicates no breakdown
ibv	A	10^{-3}	Current at reverse breakdown voltage

3.6.17.2 Level 3 Diode Model

The level 3 diode model uses the same equations as the level 1 model. The only difference is that the diode parameters area and junction periphery pj are in m^2 and m, respectively. Their effective values are affected by the SCALE parameter (effective area is obtained by multiplying area with SCALE^2 and effective junction periphery is obtained by multiplying pj with SCALE).

Table 3.37 Level 1 diode properties calculated by the simulator

Name	Unit	Description
vd	V	Voltage across diode (excluding the voltage across series resistance), DC and transient
id	A	Internal diode current (excluding the current flowing into the capacitance), DC and transient
gd	A/V	Internal diode conductance (DC and transient)
cd	F	Diode capacitance (transient)
charge	As	Diode charge stored in the capacitance (transient)
capcur	A	The current flowing into the capacitance (transient)
p	W	Diode power (excluding the power dissipated at the capacitance and series resistance), DC and transient

Table 3.38 Level 1 diode noise contributions

Postfix	Description
_rs	Series resistance thermal noise
_id	Diode shot noise
_1overf	Diode flicker noise
no postfix	Total diode noise contribution

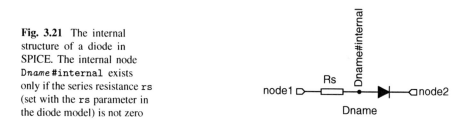

Fig. 3.21 The internal structure of a diode in SPICE. The internal node Dname#internal exists only if the series resistance rs (set with the rs parameter in the diode model) is not zero

Diode model parameters is, expli, and cjo must be given in their respective units per square meter (m²). rs is given in Ωm^2. jsw and cjsw must be given in their respective units per meter (m).

All values calculated by the simulator, noise contributions, and the internal node behave the same as those in the level 1 model.

3.6.18 Bipolar Junction Transistor (BJT)

Model type: npn or pnp
Syntax:
```
Qname (nodeC nodeB nodeE [nodeS]) model_name
+ param[=value] ...
```

Fig. 3.22 The symbol and
the terminals of an NPN (*left*)
and a PNP (*right*) BJT

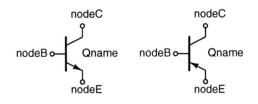

The first three nodes are required (collector, base, and emitter). The fourth node (substrate) is optional. If it is not specified, the substrate is assumed to be connected to the ground. The type (NPN or PNP) is determined in the model definition. See Fig. 3.22 for the NPN (left) and PNP (right) BJT symbol. Tables 3.39–3.42 list the parameters, properties calculated by the simulator, and noise contributions.

The BJT model in SPICE is based on the Gummel–Poon model [21]. More details on the SPICE implementation of the Gummel–Poon model can be found in [31].

Effective parameter values for is, ikf, ise, ikr, isc, irb, cje, cjc, cjs, and itf are obtained by multiplying the respective model parameter values with the area instance parameter value. Effective values for rb, rbm, re, and rc are obtained by dividing the respective model parameter values by the value of the area instance parameter.

BJTs have up to three internal nodes. Nodes Qname#base, Qname#collector, and Qname#emitter are created if model parameters rb, rc, and re are greater than zero, respectively. These nodes are needed for connecting the series resistances of the base, the collector, and the emitter.

Examples:

```
* Vertical npn
q1 (1 5 0) bjtnpn1
.model bjtnpn1 npn (bf=200)
* Vertical pnp (if subs=1 is omitted, it becomes lateral)
q2 (5 9 0) bjtpnpvert
.model bjtpnpvert pnp (bf=50 subs=1)
```

Table 3.39 BJT instance parameters

Name	Unit	Default	Description
off	–	Not set	Flag that turns off the BJT in the first NR iteration of DC analysis
icvbe	V	0	Initial B–E voltage (internal BJT instance)
icvce	V	0	Initial C–E voltage (internal BJT instance)
temp	°C	TEMP	Device temperature
area	–	1	Area factor
m	–	1	Number of parallel instances

Table 3.40 BJT model parameters

Name	Unit	Default	Description
is	A	10^{-16}	Saturation current
bf	–	100	Ideal forward beta
nf	–	1	Forward emission coefficient
vaf	V	∞	Forward Early voltage
ikf	A	∞	Forward beta roll-off corner current
ise	A	0	B–E leakage saturation current
ne	–	1.5	B–E leakage emission coefficient
br	–	1	Ideal reverse beta
nr	–	1	Reverse emission coefficient
var	V	∞	Reverse Early voltage
ikr	A	∞	Reverse beta roll-off corner current
isc	A	0	B–C leakage saturation current
nc	–	2	B–C leakage emission coefficient
rb	Ω	0	Zero bias base resistance
irb	A	∞	Current where base resistance becomes (rb + rbm)/2
rbm	Ω	rb	Minimum base resistance
re	Ω	0	Emitter resistance
rc	Ω	0	Collector resistance
cje	F	0	Zero bias B–E depletion capacitance
vje	V	0.75	B–E built-in potential
mje	–	0.33	B–E junction grading coefficient
tf	s	0	Ideal forward transit time
xtf	–	0	Coefficient for bias dependence of forward transit time
vtf	V	0	Voltage giving V_{bc} dependence of forward transit time
itf	A	0	High current dependence of forward transit time
ptf	$^\circ$	0	Excess phase
cjc	F	0	Zero bias B–C depletion capacitance
vjc	V	0.75	B–C built-in potential
mjc	–	0.33	B–C junction grading coefficient
xcjc	–	1	Fraction of B–C capacitance to internal base
tr	s	0	Ideal reverse transit time
cjs	F	0	Zero bias C–S (vertical geometry) or B–S (lateral geometry) capacitance
vjs	V	0.75	Substrate built-in junction potential
mjs	–	0	Substrate junction grading coefficient
xtb	–	0	Forward and reverse beta temperature exponent
eg	eV	1.11	Energy gap for is temperature dependency
xti	–	3	Exponent for is temperature dependency
fc	–	0.5	Forward bias junction fit parameter
tnom	$^\circ$C	TNOM	Parameter measurement temperature
kf	–	0	Flicker noise coefficient
af	–	1	Flicker noise exponent
subs	–	1 or −1	Substrate connection selector (1 means vertical BJT, −1 means lateral BJT), default is 1 for NPN and −1 for PNP

Table 3.41 BJT properties calculated by the simulator

Name	Unit	Description
ic	A	Collector current
ib	A	Base current
ie	A	Emitter current
is	A	Substrate current
vbe	V	B–E voltage
vbc	V	B–C voltage
gm	A/V	Small signal transconductance
gpi	A/V	Small signal input conductance (pi)
gmu	A/V	Small signal conductance (mu)
gx	A/V	Conductance from base to internal base
go	A/V	Small signal output conductance
geqcb	A/V	$d\,I_{be}/d\,V_{bc}$
cpi	F	Internal base-to-emitter capacitance
cmu	F	Internal base-to-collector capacitance
cbx	F	Base-to-collector capacitance
ccbs	F	Collector (vertical geometry) or base (lateral geometry)-to-substrate capacitance
ccs	F	Collector-to-substrate capacitance
cbs	F	Base-to-substrate capacitance
cqbe	F	Capacitance due to charge storage in B–E junction
cqbc	F	Capacitance due to charge storage in B–C junction
cqcbs	F	Capacitance due to charge storage in C–S (vertical geometry) or B–S (lateral geometry) junction
cqcs	F	Capacitance due to charge storage in C–S junction
cqbs	F	Capacitance due to charge storage in B–S junction
cqbx	F	Capacitance between the base node and the internal collector node
cexbc	F	Total capacitance due to forward transit time when B–E junction is forward-polarized
qbe	As	Charge storage in B–E junction
qbc	As	Charge storage in B–C junction
qcbs	As	Charge storage in C–S (vertical geometry) or B–S (lateral geometry) junction
qcs	As	Charge storage in C–S junction
qbs	As	Charge storage in B–S junction
qbx	As	Charge storage in the capacitance between the base node and the internal collector node
p	W	Power dissipation

Table 3.42 BJT noise contributions

Postfix	Description
_rc	Collector series resistance thermal noise
_rb	Base series resistance thermal noise
_re	Emitter series resistance thermal noise
_ic	Collector current shot noise
_ib	Base current shot noise
_1overf	Flicker noise (1/f noise)
No postfix	Total BJT noise contribution

3.6.19 *Junction Field Effect Transistor (JFET)*

Model type: njf or pjf
   Syntax:
Jname (nodeD nodeG nodeS) model_name param[=value] ...

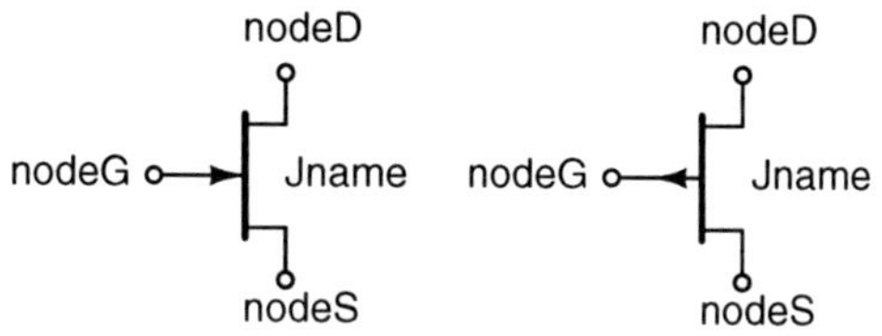

Fig. 3.23 The symbol and
the terminals of an n-channel
(*left*) and a p-channel (*right*)
JFET

njf sets the model to be an n-channel JFET model and pjf makes it a p-channel
JFET model. When specifying a JFET instance, all three nodes must be supplied in
the following order: drain, gate, and source. See Fig. 3.23 for the n-channel (left)
and p-channel (right) JFET symbols.

3.6.19.1 Level 1 JFET Model

The level 1 JFET model is derived from the FET model of Shichman and Hodges
[45]. Tables 3.43–3.46 list the parameters, properties calculated by the simulator,
and noise contributions of a level 1 JFET.

   Effective values of parameters beta, cgs, cgd, and is are obtained by multiplying their respective model parameter values by the area instance parameter
value. Effective values of parameters rd and rs are obtained by dividing the model
parameter values by the area instance parameter value.

   JFETs have up to two internal nodes. Nodes Jname#drain and Jname#source
are created if model parameters rd and rs are greater than zero, respectively. They
are needed for connecting the drain and the source series resistance to the internal
JFET instance.

Table 3.43 JFET instance parameters

Name	Unit	Default	Description
off	–	Not set	Flag that turns off the JFET in the first NR iteration of DC analysis
icvds	V	0	Initial D–S voltage (internal JFET instance)
icvgs	V	0	Initial G–S voltage (internal JFET instance)
temp	°C	TEMP	Device temperature
area	–	1	Area factor
m	–	1	Number of parallel instances

Table 3.44 JFET level 1 model parameters

Name	Unit	Default	Description
vto	V	-2	Threshold voltage
beta	A/V^2	10^{-4}	Transconductance parameter
lambda	$1/V$	0	Channel length modulation parameter
rd	Ω	0	Drain resistance
rs	Ω	0	Source resistance
cgs	F	0	G–S junction capacitance
cgd	F	0	G–D junction capacitance
pb	V	1	Gate junction potential
is	A	10^{-14}	Gate junction saturation current
fc	–	0.5	Forward bias junction fit parameter
b	–	1	Doping tail parameter
tnom	$^\circ$C	TNOM	Parameter measurement temperature
kf	–	0	Flicker noise coefficient
af	–	1	Flicker noise exponent
betatce	$1/^\circ$C	0	Beta temperature coefficient
bex	–	0	Mobility temperature exponent

Table 3.45 JFET level 1 properties calculated by the simulator

Name	Unit	Description
vgs	V	G–S voltage
vgd	V	G–D voltage
ig	A	Gate current
id	A	Drain current
is	A	Source current
igd	A	Gate-drain current
gm	A/V	Transconductance
gds	A/V	D–S conductance
ggs	A/V	D–S conductance
ggd	A/V	G–D conductance
qgs	As	Charge storage in G–S junction
qgd	As	Charge storage in G–D junction
cqgs	F	Capacitance due to charge storage in G–S junction
cqgd	F	Capacitance due to charge storage in G–D junction
p	W	Power dissipation

Table 3.46 JFET noise contributions

Postfix	Description
_rd	Drain series resistance thermal noise
_rs	Source series resistance thermal noise
_id	Drain current shot noise
_1overf	Flicker noise
No postfix	Total JFET noise contribution

3.6.19.2 Level 2 JFET Model

The JFET level 2 model is the Parker–Skellern model [38].

The level 2 JFET instances have the same set of parameters as the level 1 instances. Tables 3.47 and 3.48 list the level 2 JFET parameters and additional properties calculated by the simulator.

Level 2 JFETs have two additional instance properties calculated by the simulator.

Table 3.47 JFET level 2 model parameters

Name	Unit	Default	Description
acgam	–	0	Capacitance modulation
af	–	1	Flicker noise exponent
beta	–	10^{-4}	Transconductance parameter
cds	F	0	D–S capacitance
cgd	F	0	Zero-bias G–D junction capacitance
cgs	F	0	Zero-bias G–S junction capacitance
delta	1/W	0	Coefficient of thermal current reduction
hfeta	–	0	Drain feedback modulation
hfe1	1/V	0	hfgam modulation by V_{gd}
hfe2	1/V	0	hfgam modulation by V_{gs}
hfg1	1/V	0	hfgam modulation by V_{sg}
hfg2	1/V	0	hfgam modulation by V_{dg}
mvst	1/V	0	Modulation index for subthreshold current
mxi	–	0	Saturation potential modulation parameter
fc	–	0.5	Forward bias junction fit parameter
ibd	A	0	Breakdown current of diode junction
is	A	10^{-14}	Gate junction saturation current
kf	–	0	Flicker noise coefficient
lambda	1/V	0	Channel length modulation parameter
lfgam	–	0	Drain feedback parameter
lfg1	1/V	0	lfgam modulation by V_{sg}
lfg2	1/V	0	lfgam modulation by V_{dg}
n	–	1	Gate junction ideality factor
p	–	2	Power law (triode region)
pb	V	1	Gate junction potential
q	–	2	Power law (saturated region)
rd	Ω	0	Drain ohmic resistance
rs	Ω	0	Source ohmic resistance
taud	s	0	Thermal relaxation time
taug	s	0	Drain feedback relaxation time
vbd	V	1	Breakdown potential of diode junction
ver	–	0	Version number of PS model
vst	V	0	Subthreshold potential
vto	V	−2	Threshold voltage
xc	–	0	Capacitance pinch-off reduction factor
xi	–	1,000	Saturation-knee potential factor
z	–	1	Knee transition parameter
hfgam	–	lfgam	High-frequency V_{gd} feedback parameter
tnom	°C	−2	Parameter measurement temperature

Table 3.48 Additional JFET level 2 properties calculated by the simulator

Name	Unit	Description
vtrap	V	Quiescent drain feedback potential
vpave	W	Quiescent power dissipation

The set of internal nodes and the set of noise contributions are the same as in the level 1 model.

Examples:

```
* n-channel JFET (level 1)
j1 (2 5 0) nch1
.model nch1 njf (beta=9e-4)
* n-channel JFET with twice the area of j1
j2 (4 9 0) nch1 area=2
* n-channel JFET (level 2)
j9 (11 99 0) nch2
.model nch2 njf (level=2 beta=9e-4)
```

3.6.20 Metal Semiconductor Field Effect Transistor (MESFET)

Model type: nmf or pmf

Syntax:
Zname (nodeD nodeG nodeS) model_name param[=value] ...

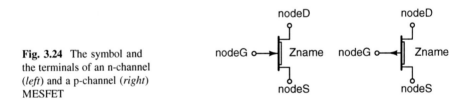

Fig. 3.24 The symbol and the terminals of an n-channel (*left*) and a p-channel (*right*) MESFET

nmf sets the model to be an n-channel MESFET model and pmf makes it a p-channel MESFET model. When specifying a MESFET instance, all three nodes must be supplied in the following order: drain, gate, and source. See Fig. 3.24 for the symbol of a MESFET. Tables 3.49–3.51 list the MESFET parameters and properties calculated by the simulator.

The MESFET model used in SPICE is derived from the GaAs FET model of Statz et al. [52].

Effective values of beta, is, cgs, and cgd are obtained by scaling the values from the netlist by the value of instance parameter area. Effective values of rs and rd are obtained by dividing the values from the netlist by area.

MESFETs have up to two internal nodes. Nodes Zname#drain and Zname#source are created if model parameters rd and rs are greater than zero, respectively. They are needed for connecting the drain and the source series resistance to

Table 3.49 MESFET instance parameters

Name	Unit	Default	Description
off	–	Not set	Flag that turns off the MESFET in the first NR iteration of DC analysis
icvds	V	0	Initial D–S voltage (internal MESFET instance)
icvgs	V	0	Initial G–S voltage (internal MESFET instance)
area	–	1	Area factor
m	–	1	Number of parallel instances

Table 3.50 MESFET model parameters

Name	Unit	Default	Description
vto	V	-2	Threshold voltage
alpha	1/V	2	Saturation voltage parameter
beta	A/V^2	$2.5 \cdot 10^{-3}$	Transconductance parameter
lambda	1/V	0	Channel length modulation parameter
b	1/V	0.3	Doping tail parameter
rd	Ω	0	Drain resistance
rs	Ω	0	Source resistance
cgs	F	0	G–S junction capacitance
cgd	F	0	G–D junction capacitance
pb	V	1	Gate junction potential
is	A	10^{-14}	Gate junction saturation current
fc	–	0.5	Forward bias junction fit parameter
kf	–	0	Flicker noise coefficient
af	–	1	Flicker noise exponent

Table 3.51 MESFET properties calculated by the simulator

Name	Unit	Description
vgs	V	G–S voltage
vgd	V	G–D voltage
cg	F	Gate capacitance
cd	F	Drain capacitance
cgd	F	G–D capacitance
gm	A/V	Transconductance
gds	A/V	D–S conductance
ggs	A/V	D–S conductance
ggd	A/V	G–D conductance
cqgs	F	Capacitance due to charge storage in G–S junction
cqgd	F	Capacitance due to charge storage in G–D junction
qgs	As	Charge storage in G–S junction
qgd	As	Charge storage in G–D junction
is	A	Source current
p	W	Power dissipation

the internal MESFET instance. MESFETs have the same set of noise contributions
as JFETs.

Example:

```
* n-channel MES
z1 (4 5 0) mes1
.model mes1 nmf (beta=5m lambda=1m)
```

3.6.21 Metal Oxide Semiconductor Field Effect Transistor (MOSFET)

Model type: nmos or pmos

Syntax:

```
Mname (nodeD nodeG nodeS nodeB) model_name
+ param[=value] ...
```

Fig. 3.25 The symbol and
the terminals of an n-channel
(*left*) and a p-channel (*right*)
MOSFET

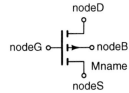

nmos sets the model to be an n-channel MOSFET model and pmos makes it a
p-channel MOSFET model. When specifying a MOSFET instance, all four nodes
must be supplied in the following order: drain, gate, source, and bulk. See Fig. 3.25
for the symbol of a MOSFET.

3.6.22 Legacy MOSFET Models (Levels 1–6)

3.6.22.1 Level 1 MOSFET Model

The level 1 model is also known as the Shichman–Hodges model [45]. Tables 3.52–
3.55 list the level 1 MOSFET parameters, properties calculated by the simulator,
and noise contributions.

Effective values of instance parameters l, w, pd, and ps are obtained by scaling
the values from the netlist with simulator parameter SCALE. ad and as are scaled
with $SCALE^2$.

All these quantities are evaluated in the transient analysis. In DC analysis capacitances and charges are not evaluated.

Table 3.52 MOSFET level 1 instance parameters

Name	Unit	Default	Description
l	SCALE m	DEFL	Channel length
w	SCALE m	DEFW	Channel width
ad	SCALE^2 m^2	DEFAD	Drain area
as	SCALE^2 m^2	DEFAS	Source area
pd	SCALE m	DEFPD	Drain perimeter
ps	SCALE m	DEFPS	Source perimeter
nrd	–	DEFNRD	Drain squares (quotient of drain length and width)
nrs	–	DEFNRS	Source squares (quotient of source length and width)
m	–	1	Number of parallel instances
temp	°C	TEMP	Device temperature
off	–	Not set	Flag that turns off the device in the first NR iteration of DC analysis
icvds	V	0	Initial D–S voltage (internal MOS instance)
icvgs	V	0	Initial G–S voltage (internal MOS instance)
icvbs	V	0	Initial B–S voltage (internal MOS instance)

Level 1 MOSFETs have up to two internal nodes. Nodes `Mname#drain` and `Mname#source` are created if the drain and source resistances are greater than zero, respectively.

Example:

```
* A CMOS inverter circuit
m1 (out in vdd vdd) pm w=10u l=2u
m2 (out in vss vss) nm w=5u l=2u
.model nm nmos (level=1 kp=4e-5)
.model pm pmos (level=1 kp=2e-5)
```

3.6.22.2 Level 2 MOSFET Model

The level 2 MOSFET model is also known as the Grove–Frohman model [18].

The set of instance parameters is the same as in the level 1 model. All level 1 model parameters are supported along with some additional parameters. See Tables 3.56 and 3.57.

The set of values calculated by the simulator is the same as in the level 1 model. Six additional quantities are calculated.

The set of noise contributions and internal nodes is the same as in the level 1 model.

Table 3.53 MOSFET level 1 model parameters

Name	Unit	Default	Description
vto	V	0	Threshold voltage
kp	A/V^2	$2 \cdot 10^{-5}$	Transconductance parameter
gamma	$V^{1/2}$	0	Bulk threshold parameter
phi	V	0.6	Surface potential
lambda	1/V	0	Channel length modulation
rd	Ω	0	Drain resistance
rs	Ω	0	Source resistance
cbd	F	0	B–D junction capacitance
cbs	F	0	B–S junction capacitance
is	A	10^{-14}	Bulk junction saturation current
n	–	1	Emission coefficient
pb	V	0.8	Bulk junction potential
cgso	F/m		G–S overlap capacitance
cgdo	F/m		G–D overlap capacitance
cgbo	F/m		G–B overlap capacitance
rsh	$\Omega/\square$	0	Sheet resistance
cj	F/m^2	0	Bottom junction capacitance per area
mj	–	0.5	Bottom grading coefficient
cjsw	F/m	0	Side junction capacitance per area
mjsw	–	0.5	Side grading coefficient
js	A/m^2	0	Bulk junction saturation current density
tox	m		Oxide thickness
ld	m	0	Lateral diffusion
wd	m	0	Lateral diffusion into channel width from bulk
xl	m	0	Length bias accounts for masking and etching effects
xw	m	0	Width bias accounts for masking and etching effects
delvto	V	0	Zero-bias threshold voltage shift
uo	cm^2/Vs	600	Surface mobility
bex	–	−1.5	Low field mobility temperature exponent
fc	–	0.5	Forward bias junction fit parameter
nsub	$1/cm^3$		Substrate doping
tpg	–	1	Gate type (0 .. Al gate, 1 .. same as source-drain diffusion, −1 .. opposite to source-drain diffusion)
nss	$1/cm^2$	0	Surface state density
tnom	°C	TNOM	Parameter measurement temperature
kf	–	0	Flicker noise coefficient
af	–	1	Flicker noise exponent

3.6.22.3 Level 3 MOSFET Model

This model is based on [12,55]. The set of instance parameters is the same as in the level 2 MOSFET model. All level 2 model parameters except lambda, utra, uexp, ucrit, and neff are supported, and some additional parameters are available. See Table 3.58.

Table 3.54 MOSFET level 1 properties calculated by the simulator

Name	Unit	Description
id	A	Drain current
is	A	Source current
ig	A	Gate current
ib	A	Bulk current
ibd	A	B–D junction current
ibs	A	B–S junction current
vgs	V	G–S voltage
vds	V	D–S voltage
vbs	V	B–S voltage
vbd	V	B–D voltage
gm	A/V	Transconductance
gds	A/V	D–S conductance
gmb	A/V	B–S transconductance
gbd	A/V	B–D conductance
gbs	A/V	B–S conductance
cbd	F	B–D capacitance
cbs	F	B–S capacitance
cgs	F	G–S capacitance
cgd	F	G–D capacitance
cgb	F	G–B capacitance
cqgs	F	Capacitance due to G–S charge storage
cqgd	F	Capacitance due to G–D charge storage
cqgb	F	Capacitance due to G–B charge storage
cqbd	F	Capacitance due to B–D charge storage
cqbs	F	Capacitance due to B–S charge storage
cbd0	F	Zero-bias B–D junction capacitance
cbdsw0	F	Zero-bias B–D sidewall junction capacitance
cbs0	F	Zero-bias B–S junction capacitance
cbssw0	F	Zero-bias B–S sidewall junction capacitance
qgs	As	G–S charge storage
qgd	As	G–D charge storage
qgb	As	G–B charge storage
qbd	As	B–D charge storage
qbs	As	B–S charge storage
p	W	Power dissipation

Table 3.55 MOSFET level 1 noise contributions

Postfix	Description
_rd	Drain series resistance thermal noise
_rs	Source series resistance thermal noise
_id	Drain current shot noise
_1overf	Flicker noise
No postfix	Total MOSFET noise contribution

Table 3.56 Additional MOSFET level 2 model parameters

Name	Unit	Default	Description
utra	–	0	Transverse field coefficient
delta	–	0	Width effect on threshold
uexp	–	0	Critical field exponent for surface mobility degradation
ucrit	V/cm	10^4	Critical field for mobility degradation
vmax	m/s	0	Maximum drift velocity of carriers
xj	m	0	Junction depth
neff	–	1	Total channel charge coefficient
nfs	$1/\mathrm{cm}^2$	0	Fast surface state density

Table 3.57 Additional MOSFET level 2 properties calculated by the simulator

Name	Unit	Description
vth	V	Threshold voltage
vdsat	V	Saturation drain voltage
sourcevcrit	V	Critical source voltage
drainvcrit	V	Critical drain voltage
rs	Ω	Source resistance
rd	Ω	Drain resistance

Table 3.58 Additional MOSFET level 3 model parameters

Name	Unit	Default	Description
eta	–	0	Static feedback factor for adjusting threshold
theta	$1/V$	0	Mobility degradation factor
kappa	$1/V$	0.2	Saturation field factor

The set of values calculated by the simulator, noise contributions, and internal nodes are the same as in the level 2 model.

3.6.22.4 Level 4 MOSFET Model (BSIM1)

BSIM is short for Berkeley short-channel IGFET (insulated gate FET) model. This is the beginning of a long line of BSIM models used mostly in IC design.

The set of instance parameters is the same as in the level 1 model.

BSIM models are designed for use with a process characterization system. Therefore, model parameters have no default values. Omitting a parameter is considered to be an error. A more detailed description of the model parameters can be found in [44].

BSIM models have many parameters. See Table 3.59. These parameters are usually unimportant to an IC designer, as they are extracted at the foundry which

Table 3.59 BSIM1 MOSFET (level 4) model parameters

Name	Unit	Description
vfb	V	Flatband voltage
lvfb	$V \cdot \mu m$	Length dependence of vfb
wvfb	$V \cdot \mu m$	Width dependence of vfb
phi	V	Strong inversion surface potential
lphi	$V \cdot \mu m$	Length dependence of phi
wphi	$V \cdot \mu m$	Width dependence of phi
k1	$V^{1/2}$	Bulk effect coefficient 1
lk1	$V^{1/2} \cdot \mu m$	Length dependence of k1
wk1	$V^{1/2} \cdot \mu m$	Width dependence of k1
k2	–	Bulk effect coefficient 2
lk2	μm	Length dependence of k2
wk2	μm	Width dependence of k2
eta	–	V_{ds} dependence of threshold voltage
leta	μm	Length dependence of eta
weta	μm	Width dependence of eta
x2e	$1/V$	V_{bs} dependence of eta
lx2e	$\mu m/V$	Length dependence of x2e
wx2e	$\mu m/V$	Width dependence of x2e
x3e	$1/V$	V_{ds} dependence of eta
lx3e	$\mu m/V$	Length dependence of x3e
wx3e	$\mu m/V$	Width dependence of x3e
dl	μm	Channel length reduction
dw	μm	Channel width reduction
muz	cm^2/Vs	Zero field mobility at $V_{ds} = 0$, $V_{gs} = V_{th}$
x2mz	$cm^2/V^2 s$	V_{bs} dependence of muz
lx2mz	$\mu m \cdot cm^2/V^2 s$	Length dependence of x2mz
wx2mz	$\mu m \cdot cm^2/V^2 s$	Width dependence of x2mz
mus	cm^2/Vs	Mobility at $V_{ds} = V_{dd}$, $V_{gs} = V_{th}$, channel length modulation
lmus	$\mu m \cdot cm^2/Vs$	Length dependence of mus
wmus	$\mu m \cdot cm^2/Vs$	Width dependence of mus
x2ms	$cm^2/V^2 s$	V_{bs} dependence of mus
lx2ms	$\mu m \cdot cm^2/V^2 s$	Length dependence of x2ms
wx2ms	$\mu m \cdot cm^2/V^2 s$	Width dependence of x2ms
x3ms	$cm^2/V^2 s$	V_{ds} dependence of mus
lx3ms	$\mu m \cdot cm^2/V^2 s$	Length dependence of x3ms
wx3ms	$\mu m \cdot cm^2/V^2 s$	Width dependence of x3ms
u0	$1/V$	V_{gs} dependence of mobility
lu0	$\mu m/V$	Length dependence of u0
wu0	$\mu m/V$	Width dependence of u0
x2u0	$1/V^2$	V_{bs} dependence of u0
lx2u0	$\mu m/V^2$	Length dependence of x2u0
wx2u0	$\mu m/V^2$	Width dependence of x2u0
u1	$\mu m/V$	V_{ds} dependence of mobility, velocity saturation
lu1	$\mu m^2/V$	Length dependence of u1
wu1	$\mu m^2/V$	Width dependence of u1
x2u1	$\mu m/V^2$	V_{bs} dependence of u1
lx2u1	$\mu m^2/V^2$	Length dependence of x2u1
wx2u1	$\mu m^2/V^2$	Width dependence of x2u1
x3u1	$\mu m/V^2$	V_{ds} dependence of u1

(continued)

Table 3.59 (continued)

Name	Unit	Description
lx3u1	$\mu m^2/V^2$	Length dependence of x3u1
wx3u1	$\mu m^2/V^2$	Width dependence of x3u1
n0	–	Subthreshold slope
ln0	–	Length dependence of n0
wn0	–	Width dependence of n0
nb	–	V_{bs} dependence of subthreshold slope
lnb	–	Length dependence of nb
wnb	–	Width dependence of nb
nd	–	V_{ds} dependence of subthreshold slope
lnd	–	Length dependence of nd
wnd	–	Width dependence of nd
tox	μm	Gate oxide thickness
temp	$^\circ C$	Temperature at which parameters were measured
vdd	V	Supply voltage to specify mus
cgso	F/m	G–S overlap capacitance per unit channel width
cgdo	F/m	G–D overlap capacitance per unit channel width
cgbo	F/m	G–B overlap capacitance per unit channel length
xpart	–	Flag for channel charge partitioning
rsh	$\Omega/\square$	Source drain diffusion sheet resistance
js	A/m^2	Source drain junction saturation current per unit area
pb	V	Source drain junction built-in potential
mj	–	Source drain bottom junction capacitance grading coefficient
pbsw	V	Source drain side junction capacitance built-in potential
mjsw	–	Source drain side junction capacitance grading coefficient
cj	F/m^2	Source drain bottom junction capacitance per unit area
cjsw	F/m	Source drain side junction capacitance per unit area
wdf	m	Default width of source drain diffusion
dell	m	Length reduction of source drain diffusion

supplies the models for the devices that are created by its manufacturing processes. IC designers focus mostly on choosing the device geometry and circuit layout.

BSIM1 instances have no instance properties calculated by the simulator. The set of internal nodes is the same as in level 1 MOSFETs. BSIM1 instances have no noise contributions.

3.6.22.5   Level 5 MOSFET Model (BSIM2)

BSIM2 is an improved variant of the BSIM1 model [13]. It has the same set of instance parameters as the level 1 MOSFET model.

All BSIM1 model parameters are supported except eta, leta, weta, x2e, lx2e, wx2e, x3e, lx3e, wx3e, muz, x2mz, lx2mz, wx2mz, mus, lmus, wmus, x2ms, lx2ms, wx2ms, x3ms, lx3ms, wx3ms, u0, lu0, wu0, x2u0, lx2u0, wx2u0, u1, lu1, wu1, x2u1, lx2u1, wx2u1, x3u1, lx3u1, and wx3u1. BSIM2 models also support some additional parameters. See Table 3.60.

Table 3.60 Additional BSIM2 MOSFET (level 5) model parameters

Name	Unit	Description
eta0	–	V_{ds} dependence of threshold voltage at $V_{dd} = 0$
leta0	μm	Length dependence of eta0
weta0	μm	Width dependence of eta0
etab	1/V	V_{bs} dependence of eta
letab	μm/V	Length dependence of etab
wetab	μm/V	Width dependence of etab
mu0	cm^2/Vs	Low-field mobility, at $V_{ds} = 0$, $V_{gs} = V_{th}$
mu0b	cm^2/V^{2}s	V_{bs} dependence of low-field mobility
lmu0b	μm $\cdot$ cm^2/V^{2}s	Length dependence of mu0b
wmu0b	μm $\cdot$ cm^2/V^{2}s	Width dependence of mu0b
mus0	cm^2/Vs	Mobility at $V_{ds} = V_{dd}$, $V_{gs} = V_{th}$
lmus0	μm $\cdot$ cm^2/Vs	Length dependence of mus0
wmus0	μm $\cdot$ cm^2/Vs	Width dependence of mus0
musb	cm^2/V^{2}s	V_{bs} dependence of mus0
lmusb	μm $\cdot$ cm^2/V^{2}s	Length dependence of musb
wmusb	μm $\cdot$ cm^2/V^{2}s	Width dependence of musb
mu20	–	V_{ds} dependence of mu in tanh term
lmu20	μm	Length dependence of mu20
wmu20	μm	Width dependence of mu20
mu2b	1/V	V_{bs} dependence of mu20
lmu2b	μm/V	Length dependence of mu2b
wmu2b	μm/V	Width dependence of mu2b
mu2g	1/V	V_{gs} dependence of mu20
lmu2g	μm/V	Length dependence of mu2g
wmu2g	μm/V	Width dependence of mu2g
mu30	cm^2/V^{2}s	V_{ds} dependence of mu in linear term
lmu30	μm $\cdot$ cm^2/V^{2}s	Length dependence of mu30
wmu30	μm $\cdot$ cm^2/V^{2}s	Width dependence of mu30
mu3b	cm^2/V^{3}s	V_{bs} dependence of mu3
lmu3b	μm $\cdot$ cm^2/V^{3}s	Length dependence of mu3b
wmu3b	μm $\cdot$ cm^2/V^{3}s	Width dependence of mu3b
mu3g	cm^2/V^{3}s	V_{gs} dependence of mu3
lmu3g	μm $\cdot$ cm^2/V^{3}s	Length dependence of mu3g
wmu3g	μm $\cdot$ cm^2/V^{3}s	Width dependence of mu3g
mu40	cm^2/V^{3}s	V_{ds} dependence of mu in linear term
lmu40	μm $\cdot$ cm^2/V^{3}s	Length dependence of mu40
wmu40	μm $\cdot$ cm^2/V^{3}s	Width dependence of mu40
mu4b	cm^2/V^{4}s	V_{bs} dependence of mu40
lmu4b	μm $\cdot$ cm^2/V^{4}s	Length dependence of mu4b
wmu4b	μm $\cdot$ cm^2/V^{4}s	Width dependence of mu4b
mu4g	cm^2/V^{4}s	V_{gs} dependence of mu40
lmu4g	μm $\cdot$ cm^2/V^{4}s	Length dependence of mu4g
wmu4g	μm $\cdot$ cm^2/V^{4}s	Width dependence of mu4g
ua0	1/V	Linear V_{gs} dependence of mobility
lua0	μm/V	Length dependence of ua0
wua0	μm/V	Width dependence of ua0
uab	1/V^2	V_{bs} dependence of ua0
luab	μm/V^2	Length dependence of uab
wuab	μm/V^2	Width dependence of uab
ub0	1/V^2	Quadratic V_{gs} dependence of mobility
lub0	μm/V^2	Length dependence of ub0

(continued)

Table 3.60 (continued)

Name	Unit	Description
wub0	$\mu m/V^2$	Width dependence of ub0
ubb	$1/V^3$	V_{bs} dependence of ub0
lubb	$\mu m/V^3$	Length dependence of ubb
wubb	$\mu m/V^3$	Width dependence of ubb
u10	$1/V$	V_{ds} dependence of mobility
lu10	$\mu m/V$	Length dependence of u10
wu10	$\mu m/V$	Width dependence of u10
u1b	$1/V^2$	V_{bs} dependence of u10
lu1b	$\mu m/V^2$	Length dependence of u1b
wu1b	$\mu m/V^2$	Width dependence of u1b
u1d	$1/V^2$	V_{ds} dependence of u10
lu1d	$\mu m/V^2$	Length dependence of u1d
wu1d	$\mu m/V^2$	Width dependence of u1d
vof0	–	Threshold voltage offset at $V_{ds} = 0$, $V_{bs} = 0$
lvof0	μm	Length dependence of vof0
wvof0	μm	Width dependence of vof0
vofb	$1/V$	V_{bs} dependence of vof0
lvofb	$\mu m/V$	Length dependence of vofb
wvofb	$\mu m/V$	Width dependence of vofb
vofd	$1/V$	V_{ds} dependence of vof0
lvofd	$\mu m/V$	Length dependence of vofd
wvofd	$\mu m/V$	Width dependence of vofd
ai0	–	Impact ionization coefficient
lai0	μm	Length dependence of ai0
wai0	μm	Width dependence of ai0
aib	$1/V$	V_{bs} dependence of ai0
laib	$\mu m/V$	Length dependence of aib
waib	$\mu m/V$	Width dependence of aib
bi0	V	Impact ionization exponent
lbi0	$\mu m \cdot V$	Length dependence of bi0
wbi0	$\mu m \cdot V$	Width dependence of bi0
bib	–	V_{bs} dependence of bi0
lbib	μm	Length dependence of bib
wbib	μm	Width dependence of bib
vghigh	V	Upper bound of the weak-strong inversion transition region
lvghigh	$\mu m \cdot V$	Length dependence of vghigh
wvghigh	$\mu m \cdot V$	Width dependence of vghigh
vglow	V	Lower bound of the weak-strong inversion transition region
lvglow	$\mu m \cdot V$	Length dependence of vglow
wvglow	$\mu m \cdot V$	Width dependence of vglow
vgg	V	Maximum V_{gs}
vbb	V	Maximum V_{bs}

BSIM2 instances have no instance properties calculated by the simulator. The set of internal nodes is the same as in level 1 MOSFETs. BSIM2 instances have no noise contributions.

Table 3.61 Additional MOSFET level 6 model parameters

Name	Unit	Default	Description
kv	V	2	Saturation voltage factor
nv	–	0.5	Saturation voltage coefficient
kc	A	$5 \cdot 10^{-5}$	Saturation current factor
nc	–	1	Saturation current coefficient
nvth	–	0.5	Threshold voltage coefficient
ps	–	0	Saturation current modification parameter
gamma1	–	0	Bulk threshold parameter 1
sigma	–	0	Static feedback effect parameter
lambda0	1/V	lambda	Channel length modulation parameter 0
lambda1	$1/V^2$	0	Channel length modulation parameter 1

3.6.22.6 Level 6 MOSFET Model

The level 6 model is a simplified MOSFET model developed at UC Berkeley. All level 1 model parameters are supported except kp, n, wd, xl, xw, delvto, bex, kf, and af. Some additional model parameters are also supported. See Table 3.61.

The set of values calculated by the simulator and the set of internal nodes are the same as those in the level 2 model. Level 6 MOSFETs have no noise model and therefore no noise contributions.

3.6.23 BSIM3 and BSIM4 MOSFET Models (Levels 47, 53, and 60)

Model type: nmos or pmos

These models are used in IC design. The set of instance parameters is the same as in level 1 MOSFETs.

3.6.23.1 BSIM3v2 (Level 47) MOSFET Model

Syntax:
Mname (nodeD nodeG nodeS nodeB) model_name
+ param [=value] ...

BSIM3v2 models have over 100 parameters. Most of these parameters are interesting only to engineers involved in the process of device characterization (i.e., extraction of BSIM3v2 model parameter values from experimental data). For details on this model consult [3, 9].

BSIM3v2 instances have the following properties that can be saved during simulation. See Table 3.62.

Table 3.62 BSIM3v2 MOS-
FET (level 47) properties
calculated by the simulator

Name	Unit	Description
vbs	V	B–S voltage
vgs	V	G–S voltage
vds	V	D–S voltage
id	A	Drain current
gm	A/V	Transconductance
gmbs	A/V	B–S transconductance
gds	A/V	D–S conductance
vth	V	Threshold voltage
vdsat	V	Saturation D–S voltage

Table 3.63 BSIM3v3 MOSFET (level 53) additional instance parameters

Name	Unit	Default	Description
nqsmod	–	0	Non-quasi-static (NQS) model selector (0..disabled, 1..enabled)
check	–	DEFINSTCHECK	Check instance parameters (1..yes, 0..no)
geo	–	0	Source/drain sharing selector (0..no sharing, 1..drain shared, 2..source shared, 3..drain and source shared)

BSIM3v2 instances have the same set of internal nodes and noise contributions as level 1 MOSFETs.

3.6.23.2 BSIM3v3 (Level 53) MOSFET Model

BSIM3v3 is a very popular MOSFET model for submicrometer IC design. Many different versions are available. The model version is chosen with the `version` model parameter. The following versions are available: 3.0, 3.1, 3.2, 3.2.2, 3.2.3, and 3.2.4. The latest version (3.2.4) is the default.

Example:

```
.model nm nmos (level=53 version=3.2.3 ...)
```

All level 1 instance parameters are supported, and the BSIM3v3 instances have some additional instance parameters available to the user. See Table 3.63.

The list of BSIM3v3 model parameters is very long. Interested readers can refer to [3, 9].

For BSIM3v3 models the simulator evaluates all properties that are evaluated for BSIM3v2 instances. Some additional properties are evaluated, too, as shown in Table 3.64.

BSIM3v3 instances have the same internal nodes as level 1 MOSFETs. Also, an internal charge node named *Mname*#charge is created when the instance parameter nqsmod is set to a nonzero value. The set of noise contributions is the same as for level 1 MOSFETs.

Table 3.64 Additional BSIM3v3 MOSFET (level 53) properties calculated by the simulator

Name	Unit	Description
idrain	A	Drain current (absolute value in AC analysis)
idraincplx	A	Drain ac current (complex)
igate	A	Gate current (absolute value in AC analysis)
igatecplx	A	Gate ac current (complex)
isource	A	Source current (absolute value in AC analysis)
isourcecplx	A	Source ac current (complex)
ibulk	A	Bulk current (absolute value in AC analysis)
ibulkcplx	A	Bulk ac current (complex)

Table 3.65 BSIM4 MOSFET (level 60) additional instance parameters

Name	Unit	Default	Description
nf	–	1	Number of fingers
min	–	0	Minimize number of D or S diffusions for even-number fingered devices (1..yes)
rbdb	Ω	From model	Resistance between B and dbody internal node
rbsb	Ω	From model	Resistance between B and sbody internal node
rbpb	Ω	From model	Resistance between B and body internal node
rbps	Ω	From model	Resistance between sbody and body internal nodes
rbpd	Ω	From model	Resistance between dbody and body internal nodes
trnqsmod	–	From model	Transient NQS model selector (1..enabled)
acnqsmod	–	From model	AC NQS model selector (1..enabled)
rbodymod	–	From model	Distributed body R model selector (1..enabled)
rgatemod	–	From model	Gate resistance model selector (1..enabled)
geomod	–	From model	Geometry-dependent parasitics model selector (integer between 0 and 10)
rgeomod	–	0	S/D resistance and contact model selector (integer between 0 and 8)

3.6.23.3 BSIM4 (Level 60) MOSFET Model

There are multiple versions of the BSIM4 model available. The choice of version is made by specifying the version model parameter. Possible values are 4.0, 4.1, 4.2, and 4.2.1. The default version is the latest one (4.2.1).

BSIM4 instances support all level 1 instance parameters, and additional parameters are supported, as shown in Table 3.65.

BSIM4 models have many parameters. As these parameters as usually interesting only to process engineers, the list is omitted. Interested readers may refer to the BSIM4 manual available at [4] for details.

All quantities that are calculated for BSIM3v2 models are also calculated for BSIM4 models. Some additional quantities are also available (only in version 4.2.1). See Table 3.66.

Table 3.66 Additional BSIM4 MOSFET (level 60) instance
properties calculated by the simulator

Name	Unit	Description
ibd	A	B–D current
ibs	A	B–S current
isub	A	Impact ionization substrate current
igidl	A	Gate-induced drain leakage current
igisl	A	Gate-induced source leakage current
igs	A	G–S diffusion region current
igd	A	G–D diffusion region current
igb	A	G–B tunneling current
igcs	A	Gate-channel source current
igcd	A	Gate-channel drain current
cgg	F	Gate capacitance
cgs	F	G–S overlap capacitance
cgd	F	G–D overlap capacitance
cbg	F	B–G capacitance
cbd	F	B–D capacitance
cbs	F	B–S capacitance
cdg	F	D–G overlap capacitance
cdd	F	Drain capacitance
cds	F	D–S capacitance
csg	F	S–G capacitance
csd	F	S–D capacitance
css	F	Source capacitance
cgb	F	G–B capacitance
cdb	F	D–B capacitance
csb	F	S–B capacitance
cbb	F	Bulk capacitance
capbd	F	Body-drain capacitance
capbs	F	Body-source capacitance
qg	As	Gate charge
qb	As	Bulk charge
qd	As	Drain charge
qs	As	Source charge
qinv	As	Inversion charge

BSIM4 instances have several internal nodes. Nodes M*name*#drain and M*name*#
source are used for connecting the drain and source series resistances. Nodes
M*name*#gate and M*name*#midgate are created if the rgatemod instance parameter
is set to a nonzero value. Nodes M*name*#dbody, M*name*#body, and M*name*#sbody
are created if the rbodymod instance parameter is set to a nonzero value. Finally,
internal node M*name*#charge is created if the instance parameter trnqsmod is set
to a nonzero value. The noise contributions are shown in Table 3.67.

Table 3.67 BSIM4
(level 60) MOSFET noise
contributions

Postfix	Description
_rd	Drain series resistance thermal noise
_rs	Source series resistance thermal noise
_rg	Gate series resistance thermal noise
_rbps	sbody-body resistance thermal noise
_rbpd	dbody-body resistance thermal noise
_rbpb	B-body resistance thermal noise
_rbsb	B-sbody resistance thermal noise
_rbdb	B-dbody resistance thermal noise
_id	Drain current shot noise
_igs	G–S current shot noise
_igd	G–D current shot noise
_igb	G–B current shot noise
_1overf	Flicker noise
no postfix	Total MOSFET noise contribution

3.6.24 Silicon on Insulator (SOI) MOSFET Models (Levels 55–58)

3.6.24.1 BSIM3SOIv1 (Level 55) MOSFET Model

Syntax:
```
Mname (nodeD nodeG nodeS nodeE [nodeB]) model_name
+ param[=value] ...
```
The first and third nodes have the same meaning as in ordinary MOS devices (drain and source). Nodes $nodeG$ and $nodeE$ are the front and the back gates. The optional fifth node ($nodeB$) is the external body contact node.

There are multiple versions of the BSIM3SOIv1 model. A version is chosen with the version model parameter. The following versions are available: 1.0, 1.1, 1.2, and 1.3. The default version is 1.3.

BSIM3SOIv1 supports all level 1 instance parameters and some additional ones, as shown in Table 3.68.

The list of model parameters is available in the BSIM3SOIv1 manual [19]. Table 3.69 lists the properties calculated by the simulator.

Internal nodes $Mname$#drain and $Mname$#source are created for connecting the drain and source series resistance to the internal MOSFET instance. Node $Mname$#temp represents the device temperature and is created if the shmod model parameter is set to 1 (enable self-heating model) and the value of instance parameter rth0 is nonzero. An internal charge node (only in version 1.0) is created if the instance parameter nqsmod is set to a nonzero value. Internal node $Mname$#vbs represents the true B–S voltage (only in version 1.0).

If the fifth node is not specified (floating body), an internal body node $Mname$#body is created. If the body is not floating, the behavior differs for different

Table 3.68 BSIM3SOIv1 MOSFET (level 55) additional instance parameters

Name	Unit	Default	Description
nrb	–	1	Number of body squares
icves	V	0	Initial back gate vs. source voltage
icvps	V	0	Initial external voltage vs. source
nqsmod	–	0	NQS model selector (1..enabled)
bjtoff	–	0	Turn off BJT if value is nonzero
rth0	m°C/W	From model	Thermal resistance per unit width
cth0	m°C/Ws	From model	Thermal capacitance per unit width

Table 3.69 BSIM3SOIv1 MOSFET (level 55) instance properties calculated by the simulator

Name	Unit	Description
gmbs	A/V	B–S transconductance
gm	A/V	Transconductance
gmids	1/V	g_m/I_{ds}
gds	A/V	D–S conductance
vdsat	V	Saturation D–S voltage
vth	V	Threshold voltage
ids	A	Drain current
vbs	V	B–S voltage
vgs	V	G–S voltage
vds	V	D–S voltage
ves	V	E–S voltage

versions. In version 1.0 an internal body node Mname#body is always created. In other versions it is created only if either the rbody or the rbsh model parameter is nonzero.

The set of noise contributions is the same as for the level 1 model.

3.6.24.2 BSIMSOIv2 (Level 56) MOSFET Model

Syntax:
Mname (nodeD nodeG nodeS nodeE [nodeP [nodeB [nodeT]]])
+ model_name param[=value] ...

Nodes nodeD and nodeS are the drain and the source nodes. nodeG and nodeE represent the front and the back gates. nodeP and nodeB are the optional external and internal body contacts. nodeT is the optional temperature node.

The following versions of the BSIMSOIv2 model are available: 2.0, 2.0.1, 2.1, 2.2, 2.2.1, 2.2.2, and 2.2.3. The user can choose the version by specifying the version model parameter. The default version is 2.2.3.

The depletion type can be chosen with the model parameter soimod. The model can be partially depleted (soimod=1), dynamically depleted (soimod=2), or fully

Table 3.70 Additional BSIMSOIv2 MOSFET (level 56) instance parameters

Name	Unit	Default	Description
nrb	–	1	Number of body squares
frbody	–	1.0	Coefficient of distributed body resistance effects
nbc	–	0.0	Number of body contact isolation edge
nseg	–	1	Number of segments for channel width partitioning
pdbpc	m	0	Parasitic perimeter length for the body contact at drain side
psbpc	m	0	Parasitic perimeter length for the body contact at source side
agbpc	m²	0	Parasitic gate-to-body overlap area for body contact
aebpc	m²	0	Parasitic body-to-substrate overlap area for body contact
vbsusr	V	0	Optional initial value of V_{bs} for transient analysis
tnodeout	–	Not set	Flag indicating the use of temperature node
icves	V	0	Initial back gate vs. source voltage
icvps	V	0	Initial external voltage vs. source
bjtoff	–	0	Turn off BJT if value is nonzero
rth0	m°C/W	From model	Normalized thermal resistance
cth0	m°C/Ws	From model	Normalized thermal capacitance

Table 3.71 BSIMSOIv2 MOSFET (level 56) instance properties calculated by the simulator

Name	Unit	Description
ids	A	Drain current
gm	A/V	Transconductance
gmbs	A/V	B–S transconductance
gmids	A/V	g_m/I_{ds}
gds	A/V	D–S conductance
vdsat	V	Saturation D–S voltage
vth	V	Threshold voltage
vgs	V	G–S voltage
vds	V	D–S voltage
vbs	V	B–S voltage
ves	V	E–S voltage

depleted (other soimod values). The default value is 2 (dynamically depleted). Only versions 2.0 and 2.1 of the dynamically and fully depleted models are available. The default version for fully or dynamically depleted models is 2.1.

BSIMSOIv2 supports all level 1 instance parameters and some additional parameters, as shown in Table 3.70.

The details on model parameters are available in the documentation from the archive at [5]. Table 3.71 lists the calculated instance properties.

Internal nodes Mname#drain and Mname#source are created for connecting the series resistances. Internal nodes Mname#body and Mname#temp are created when necessary. Table 3.72 lists the noise contributions.

Table 3.72 BSIMSOIv2 (level 56) MOSFET noise contributions

Postfix	Description
_rd	Drain series resistance thermal noise
_rs	Source series resistance thermal noise
_id	Drain current shot noise
_1overf	Flicker noise
_fb	Noise due to floating body
No postfix	Total MOSFET noise contribution

Table 3.73 STAGSOI3 MOSFET (level 57) instance parameters

Name	Unit	Default	Description
w	m	DEFW	Device width
l	m	DEFL	Device length
nrd	–	DEFNRD	Number of squares in drain
nrs	–	DEFNRS	Number of squares in source
off	–	Not set	Flag that sets the device initially off
m	–	1	Number of parallel instances
icvbs	V	0	Initial B–S voltage
icvds	V	0	Initial D–S voltage
icvgfs	V	0	Initial front gate-to-source voltage
icvgbs	V	0	Initial back gate-to-source voltage
temp	°C	TEMP	Instance temperature
rt	W/K		Instance lumped thermal resistance
ct	J/K		Instance lumped thermal capacitance
rt1	W/K		Second thermal resistance
ct1	J/K		Second thermal capacitance
rt2	W/K		Third thermal resistance
ct2	J/K		Third thermal capacitance
rt3	W/K		Fourth thermal resistance
ct3	J/K		Fourth thermal capacitance
rt4	W/K		Fifth thermal resistance
ct4	J/K		Fifth thermal capacitance

3.6.24.3 STAGSOI3 (Level 57) MOSFET Model

Syntax:
Mname (nodeD nodeGF nodeS nodeGB [nodeB [nodeT]])
+ model_name param[=value] ...

STAG in STAGSOI3 [30] stands for Southampton Thermal AnaloGue. This is a partially depleted SOI model. The meaning of the nodes is the same as in the BSIM3SOIv1 model. Nodes GF and GB are the front and the back gates. nodeT is the optional temperature node.

Two versions are available (use the version model parameter to choose between them): 2.1 and 2.6. The default version is 2.6. Tables 3.73 and 3.74 list the instance parameters and calculated properties.

Internal nodes Mname#drain and Mname#source are created for connecting the series resistances to the internal MOSFET's drain and source. Version 2.6 also

Table 3.74 STAGSOI3 MOSFET (level 57) instance properties calculated by the simulator

Name	Unit	Description
id	A	Drain current
ibs	A	B–S junction current
ibd	A	B–D junction current
gmbs	A/V	B–S transconductance
gmf	A/V	Front transconductance
gmb	A/V	Back transconductance
gds	A/V	D–S conductance
gbd	A/V	B–D conductance
gbs	A/V	B–S conductance
capbd	F	B–D capacitance
capbs	F	B–S capacitance
cbd0	F	Zero-bias B–D junction capacitance
cbs0	F	Zero-bias B–S junction capacitance
qgf	As	Front gate charge storage
iqgf	A	Current due to front gate charge storage
qd	As	Drain charge storage
iqd	A	Current due to drain charge storage
qs	As	Source charge storage
iqs	A	Current due to source charge storage
qbd	As	Bulk-drain charge storage
iqbd	A	Current due to bulk-drain charge storage
qbs	As	Bulk-source charge storage
iqbs	A	Current due to bulk-source charge storage
vdsat	V	Saturation D–S voltage
vth	V	Threshold voltage
vfbf	V	Temperature adjusted flatband voltage
vfbb	V	Temperature adjusted back flatband voltage
sourcevcrit	V	Critical source voltage
drainvcrit	V	Critical drain voltage
vbd	V	B–D voltage
vbs	V	B–S voltage
vgfs	V	GF–S voltage
vgbs	V	GB–S voltage
vds	V	D–S voltage

creates internal thermal nodes M*name*#tout1, M*name*#tout2, M*name*#tout3, and M*name*#tout4. Internal current node M*name*#branch is created for connecting the thermal node to the ground when rt=0.

STAGSOI3 instances have the same set of noise contributions as level 1 MOSFETs.

3.6.24.4   UFSOI (Level 58) MOSFET Model

Syntax:
M*name* (*nodeD nodeG nodeS nodeE* [*nodeB*])
+ *model_name param* [=*value*] ...

UFSOI stands for University of Florida SOI model [16]. The meaning of the nodes is the same as in the BSIM3SOIv1 model. Tables 3.75–3.77 list the parameters, properties, and noise contributions.

Besides the basic quantities, the following current derivatives with respect to voltages are calculated: diddvgf, diddvd, diddvgb, diddvb, diddvs, dirdvb, dirdvs, digtdvgf, digtdvd, digtdvgb, digtdvb, digtdvs, digidvgf, digidvd, digidvgb, digidvb, digidvs, digbdvgf, digbdvd, digbdvgb, digbdvb, and digbdvs.

The following charge derivatives with respect to voltages are calculated: dqgfdvgf, dqgfdvd, dqgfdvgb, dqgfdvb, dqgfdvs, dqddvgf, dqddvd, dqddvgb, dqddvb, dqddvs, dqgbvgf, dqgbdvd, dqgbdvgb, dqgbdvb, dqgbdvs, dqbdvgf, dqbdvd, dqbdvgb, dqbdvb, dqbdvs, dqsdvgf, dqsdvd, dqsdvgb, dqsdvb, and dqsdvs.

The following derivatives with respect to temperature are calculated: diddt, dirdt, digtdt, digidt, digbdt, dqgfdt, dqddt, dqgbdt, dqbdt, and dqsdt.

The following power derivatives with respect to voltages are calculated: dpdt, dpdvgf, dpdvd, dpdvgb, dpdvb, and dpdvs.

Internal nodes Mname#drain and Mname#source are created for connecting the series resistances. Mname#body is the internal body node, and Mname#temp is the internal temperature node.

Table 3.75 UFSOI MOSFET (level 58) instance parameters

Name	Unit	Default	Description
w	m	10^{-6}	Device width
l	m	10^{-6}	Device length
m	–	1	Number of parallel instances
ad	m^2	0	Drain area
as	m^2	0	Source area
ab	m^2	0	Body area
nrd	–	0	Number of squares in drain
nrs	–	0	Number of squares in source
nrb	–	0	Number of squares in body
rth	°C/V	1,000	Thermal resistance
cth	Ws/°C	0	Thermal capacitance
pdj	m	0	Drain junction perimeter
psj	m	0	Source junction perimeter
off	–	1	Flag that sets the device initially off
icvbs	V	0	Initial B–S voltage
icvds	V	0	Initial D–S voltage
icvgfs	V	0	Initial GF–S voltage
icvgbs	V	0	Initial GB–S voltage

Table 3.76 UFSOI MOSFET (level 58) basic instance properties calculated by the simulator

Name	Unit	Description
ich	A	Channel current
ibjt	A	Parasitic bipolar current
ir	A	Recombination current (B–S)
igt	A	Thermal generation current (B–D)
igi	A	Impact ionization current
igb	A	Gate-to-body tunneling current
vbs	V	Internal B–S voltage
vbd	V	Internal B–D voltage
vgfs	V	Internal GF–S voltage
vgfd	V	Internal GF–D voltage
vgbs	V	Internal GB–S voltage
vds	V	Internal D–S voltage
qgf	As	Front gate charge
qd	As	Drain charge
qgb	As	Back gate charge
qb	As	Body charge
qs	As	Source charge
qn	As	Inversion layer charge
ueff	cm^2/Vs	Field effect mobility
vtw	V	Weak inversion threshold voltage
vts	V	Strong inversion threshold voltage
vdsat	V	Drain saturation voltage
le	A	Modulated channel length
power	W	Power dissipation
temp	$^\circ$C	Device temperature

Table 3.77 UFSOI (level 58) MOSFET noise contributions

Postfix	Description
_rd	Drain series resistance thermal noise
_rs	Source series resistance thermal noise
_rb	Body series resistance thermal noise
_id	Drain current shot noise
_irgts	Shot noise at source junction
_irgtd	Shot noise at drain junction
_ibjt	Shot noise of transport current I_{bjt}
_1overf	Flicker noise
no postfix	Total MOSFET noise contribution

3.6.25 UFET and EKV MOSFET Models (Levels 7 and 44)

3.6.25.1 UFET (Level 7) MOSFET Model

UFET is a MOSFET model from the University of Florida [17]. The instance syntax is the same as in level 1 MOSFETs. Tables 3.78 and 3.79 list the parameters and calculated properties.

Table 3.78 UFET (level 7) MOSFET instance parameters

Name	Unit	Default	Description
w	m	10^{-6}	Device width
l	m	10^{-5}	Device length
m	–	1	Number of parallel instances
ad	m^2	0	Drain area
as	m^2	0	Source area
pd	m	0	Drain perimeter
ps	m	0	Source perimeter
off	–	Not set	Device initially off
temp	°C	TEMP	Device temperature
icvgs	V	0	Initial G–S voltage
icvbs	V	0	Initial B–S voltage
icvds	V	0	Initial D–S voltage

The list of internal nodes is the same as for the level 1 MOSFET. UFET instances have no noise contributions.

3.6.25.2 EKV (Level 44) MOSFET Model

The EKV (Enz–Krummenacher–Vittoz) model [14] is a MOSFET model. SPICE OPUS offers version 2.6 of the model. The syntax of EKV instances is the same as in level 1 MOSFETs. EKV instances support all level 1 instance parameters. A wealth of additional information is available at the EKV model homepage [15]. Table 3.80 lists the instance properties.

EKV instances have the same set of internal nodes and noise contributions as level 1 MOSFETs. More details on EKV model parameters can be found in [8].

3.7 Binning of MOS Models

It is not difficult to find the set of model parameters that will accurately model the behavior of a MOSFET with a fixed channel length and width (e.g., w = 10u l = 0.5u). But if one wants to use the same model for MOSFETs with different channel dimensions, this task becomes much more complicated. Some models are not capable of providing a good fit across a wide range of channel dimensions (e.g., level 1 MOSFET).

Binning is the solution to such problems. Instead of using a single model for all combinations of channel width and length, we use multiple models where every model is responsible for a particular range of widths and lengths (also referred to as a bin). Such partial models are referred to as binned models. The choice of the model now depends not only on the technology but also on the channel dimensions. A transistor with $W/L = 100\,\mu m/5\,\mu m$ uses a different model than a transistor with $W/L = 5\,\mu m/10\,\mu m$.

Table 3.79 UFET (level 7) MOSFET instance properties calculated by the simulator

Name	Unit	Description
vgs	V	Internal G–S voltage
vds	V	Internal D–S voltage
vbs	V	Internal B–S voltage
vbd	V	Internal B–D voltage
ich	A	Channel current
igi	A	Impact ionization current
ibd	A	Body-drain leakage current
ibs	A	Body-source leakage current
gm	A/V	Transconductance
gmb	A/V	Body transconductance
gms	A/V	Source transconductance
gds	A/V	Conductance
gim	A/V	Impact ionization transconductance
gimb	A/V	Impact ionization body transconductance
gims	A/V	Impact ionization source transconductance
gids	A/V	Impact ionization conductance
gbd	A/V	Body-drain conductance
gbs	A/V	Body-drain conductance
qg	As	Gate charge
qb	As	Body charge
qd	As	Drain charge
qs	As	Source charge
iqg	A	Dynamic gate current
iqb	A	Dynamic body current
iqd	A	Dynamic drain current
iqs	A	Dynamic source current
cgg	F	Gate capacitance
cgb	F	G–B capacitance
cgd	F	G–D capacitance
cgs	F	G–S capacitance
cbg	F	B–F capacitance
cbb	F	Body capacitance
cbd	F	B–D capacitance
cbs	F	B–S capacitance
cdg	F	D–G capacitance
cdb	F	D–B capacitance
cdd	F	Drain capacitance
cds	F	D–S capacitance
csg	F	S–G capacitance
csb	F	S–B capacitance
csd	F	S–D capacitance
css	F	Source capacitance
capbd	F	Body-drain capacitance
capbs	F	Body-source capacitance

Table 3.80 EKV (level 44) MOSFET instance properties calculated by the simulator

Name	Unit	Description
id	A	Drain current
isub	A	Substrate current
is	A	Source current
ig	A	Gate current
ib	A	Bulk current
ibd	A	B–D junction current
ibs	A	B–S junction current
vgs	V	Internal G–S voltage
vds	V	Internal D–S voltage
vbs	V	Internal B–S voltage
vbd	V	Internal B–D voltage
vth	V	Threshold voltage
vgeff	V	Effective gate voltage
vp	V	Pinch-off voltage
vdsat	V	Saturation drain voltage
sourcevcrit	V	Critical source voltage
drainvcrit	V	Critical drain voltage
gm	A/V	Transconductance
gmb	A/V	B–S transconductance
gms	A/V	Source transconductance
gds	A/V	D–S conductance
gbd	A/V	B–D conductance
gbs	A/V	B–S conductance
if	A	Forward current
ir	A	Reverse current
irprime	A	Reverse current (prime)
n_slope	–	Slope factor
tau	s	NQS time constant
cbd	F	B–D capacitance
cbs	F	B–S capacitance
cgs	F	G–S capacitance
cgd	F	G–D capacitance
cgb	F	G–B capacitance
cqgs	F	Capacitance due to G–S charge storage
cqgd	F	Capacitance due to G–D charge storage
cqgb	F	Capacitance due to G–B charge storage
cqbd	F	Capacitance due to B–D charge storage
cqbs	F	Capacitance due to B–S charge storage
cbd0	F	Zero-bias B–D junction capacitance
cbdsw0	F	Zero bias B–D junction sidewall capacitance
cbs0	F	Zero-bias B–S junction capacitance
cbssw0	F	Zero bias B–S junction sidewall capacitance
qgs	As	G–S charge storage
qgd	As	G–D charge storage
qgb	As	G–B charge storage
qbd	As	B–D charge storage
qbs	As	B–S charge storage
power	W	Instantaneous power

When binned models are used, the model parameters are not constant, but depend on the channel width and length. BSIM1 and BSIM2 models use three values to define most model parameters. Take, for instance, model parameter vfb. The following values must be given in the model description: vfb0, lvfb, and wvfb. They describe the dependence of parameter vfb on the channel width and length. The actual value of vfb used for a particular transistor is obtained from the following formula:

$$vfb = vfb0 + lvfb/Leff + wvfb/Weff, \qquad (3.6)$$

where Leff and Weff are the effective channel length and width.

In BSIM3v3 model a different kind of interpolation is used. Suppose you have a model parameter X that depends on effective channel length (Leff) and width (Weff). The dependency is described by four values: X0, X1, Xw, and Xp. The interpolated value of X is obtained using the following formula (supported by BSIM3v3, BSIM3SOIv1, BSIMSOIv2, and BSIM4 models):

$$X = X0 + X1/Leff + Xw/Weff + Xp/(Leff \cdot Weff). \qquad (3.7)$$

Every BSIM3v3 model has four additional parameters: wmin, wmax, lmin, and lmax. These parameters specify the range of channel widths and lengths for which the model is valid. SPICE OPUS uses them to automatically assign a model to a transistor based on its channel dimensions.

To describe a MOSFET for a wide range of channel dimensions, integrated circuit manufacturers provide multiple binned BSIM3v3 models. Say, for instance, that there are four different width intervals and three different length intervals. The whole range of applicable channel dimensions is now divided in 12 bins. Every bin is described by a separate model.

```
.model nm_1 nmos (level=53 wmin=0.3u wmax=1u    lmin=0.18u lmax=0.5u ...)
.model nm_2 nmos (level=53 wmin=0.3u wmax=1u    lmin=0.5u  lmax=1u    ...)
.model nm_3 nmos (level=53 wmin=0.3u wmax=1u    lmin=1u    lmax=10u   ...)
.model nm_4 nmos (level=53 wmin=1u   wmax=10u   lmin=0.18u lmax=0.5u ...)
.model nm_5 nmos (level=53 wmin=1u   wmax=10u   lmin=0.5u  lmax=1u    ...)
...
.model nm_10 nmos (level=53 wmin=50u  wmax=100u lmin=0.18u lmax=0.5u ...)
.model nm_11 nmos (level=53 wmin=50u  wmax=100u lmin=0.5u  lmax=1u    ...)
.model nm_12 nmos (level=53 wmin=50u  wmax=100u lmin=1u    lmax=10u   ...)
```

When creating an instance, the instance must be put in the correct bin. This can be done manually.

```
m1 (4 5 9 9) nm_1 w=2u l=0.18u
m2 (5 5 9 9) nm_5 w=8u l=0.9u
```

SPICE OPUS can automatically choose the correct model for a MOS transistor. You only have to specify the common part of the model name (for the example above this is nm).

```
* This one goes into bin 1 (model nm_1 applies)
m1 (4 5 9 9) nm w=2u l=0.18u
* This one goes into bin 5 (model nm_5 applies)
m2 (5 5 9 9) nm w=8u l=0.9u
```

To take advantage of the automatic binning feature the model names for a set of models describing the behavior across a range of channel dimensions must be named with a common prefix followed by an underscore (_) and a bin name (*modelName_binName*). Automatic bin selection is currently supported only by the BSIM3v3 model.

3.8 Including Files in the Netlist

Netlists can become extremely large (for instance, binned models can span thousands of lines of text). In such cases it is of great advantage if a netlist can be split into multiple files. There are two mechanisms for including files in a netlist. The first is the .include statement and the second is the .lib statement. The syntax of the former is

.include *fileName*

or also

.include '*fileName*'

When the netlist is processed by SPICE, the .include statement is replaced by the contents of the specified file. SPICE searches for the file in the current directory and in the directories specified in the sourcepath SPICE variable.

If the file is still not found, the directory of the calling file (the file with the .include statement) is used as the prefix for the *fileName* specified in the .include statement. This makes it possible to create trees of model libraries that can be relocated on the disk without changing the paths in the .include and .lib statements within the library. For such easy relocation all paths in the library must be relative.

In the Linux version a tilde character ˜ in the filename expands to the home directory of the user that is running SPICE and ˜*usrname* expands to the home directory of user *usrname*.

The syntax of the .lib statement is

.lib '*fileName*' *sectionName*

SPICE searches for the file in the same way as it does in the case of an .include statement. A .lib statement includes a part of the specified file (called a section). A section starts with

.lib *sectionName*

and ends with

.endl

Say, for instance, that we have file mos.mod with the following contents:

```
.lib nmosmodel
.model nm nmos (level=53 ...)
.endl

.lib pmosmodel
.model pm pmos (level=53 ...)
.endl
```

Consider the following examples:

```
* This includes the section nmosmodel from file mos.mod
* .lib and .endl are omitted.
.lib 'mos.mod' nmosmodel

* This would include the whole file mos.mod literally and produces an
* error, as the beginning of every section in mos.mod is denoted by
* '.lib sectionName' which gets interpreted as a .lib statement that
* has no section name specified (syntax error).
.include mos.mod
```

3.9 Describing Circuits Hierarchically

SPICE OPUS provides means for describing a circuit hierarchically in the form of nested subcircuit blocks. The hierarchy can be parametrized, meaning that the same block can be reused with different parameter values. All this can be achieved by using subcircuits.

3.9.1 Defining a Subcircuit, Subcircuit Instantiation

Subcircuits make it possible to define a module which can then be inserted anywhere in the circuit. The definition of a subcircuit has the following syntax:
```
.subckt defName node1 node2 ...
subcircuit element and model definitions
.ends
```
The definition name (*defName*) must obey the same rules as model names. The nodes that follow the *defName* are the external nodes of a subcircuit. A subcircuit definition does not affect the remaining circuit.

Example:

```
.subckt inverter in out vdd vss
m1 (out in vdd vdd) pm w=8u l=0.5u
m2 (out in vss vss) nm w=4u l=0.5u
.ends
```

To use (instantiate) a subcircuit the following syntax is used:
```
Xname (con1 con2 ...) defName
```
A subcircuit instantiation connects the elements specified in the subcircuit definition between nodes *con1*, *con2*, ... Connection nodes appear in the same order as the external nodes in the subcircuit definition.

Example:

```
vpow (v5 0) dc=5

* Connect an inverter named x1 between nodes
*    10  ... input
*    20  ... output
*    v5  ... vdd
*    0   ... vss
x1 (10 20 v5 0) inverter
```

3.9.2 Global Nodes

The ground node (0) is called a global node. Take, for instance, the following netlist:

```
.subckt rnet in1 in2 com
r1 (in1 5) r=1k
r2 (in2 5) r=1k
r3 (com 5) r=10k
rleak (5 0) r=100meg
.ends
```

Node 5 is an internal node of the subcircuit. Any two instances of subcircuit rnet will have their own internal node 5. With node 0 it is different. Because it is a global node, all instances of subcircuit rnet have a resistor rleak connected between their respective internal node 5 and the ground node which is common for all subcircuits.

Normally only node 0 is global. However, the user can make any node global by adding a .global statement to the netlist. Its syntax is

```
.global gnode1 gnode2 ...
```

Take the following example:

```
.global vdd vss

.subckt inverter1 in out
m1 (out in vdd vdd) pm w=8u l=0.5u
m2 (out in vss vss) nm w=4u l=0.5u
.ends

vpow (vdd 0) dc=5
vsscon (vss 0) dc=0
x1 (10 20) inverter1
```

Nodes vdd and vss are now global. This means that all instances of inverter1 are connected to nodes vdd and vss. Note how we had to connect the power supply to node vdd in order to supply power to the inverter. A zero voltage source vsscon is used for connecting the global node vss to the ground. Essentially this is still the same circuit as it was before, except that now the user has much less flexibility when connecting a subcircuit.

All nodes in a subcircuit definition that are neither global nodes nor external nodes are local nodes. If a model definition appears in a subcircuit definition, the model definition is considered local and is available only in the subcircuit definition. Models defined outside a subcircuit definition are global. Global models are also available in all subcircuit definitions.

3.9.3 Nesting Subcircuits and Understanding the Flat Circuit

Subcircuits can be nested. Take, for instance, the following definition of a subcircuit:

```
.subckt ring out vdd vss
x1 (10 20 vdd vss) inverter
c1 (20 vss) c=10p
```

```
x2 (20 30 vdd vss) inverter
c2 (30 vss) c=10p
x3 (30 10 vdd vss) inverter
c3 (10 vss) c=10p
.ends
```

In this definition subcircuit `ring` comprises subcircuits `x1`, `x2`, and `x3` which are instances of `inverter`. The circuit is unaffected until you create an instance of `ring`. Then the `inverter` instances are created along with their complementary MOS transistors.

Subcircuit definitions cannot be nested. Thus there can be only one level of subcircuit definitions, but infinitely many levels of subcircuit instantiation.

SPICE flattens the circuit hierarchy before any simulations take place. The resulting circuit is referred to as a flat circuit. In the process of flattening, instance, model, and internal node names are expanded with a postfix. The postfix contains the path from the top level circuit through all subcircuit instances to the particular instance, model, or internal node. The path separator is the colon character (:). Take, for instance, the last example and instantiate it as

```
xosc (sig vdd 0) ring
```

The following primitives are created:

- MOS instances: `m1:x1:xosc`, `m2:x1:xosc`, `m1:x2:xosc`, `m2:x2:xosc`, `m1:x3:xosc`, and `m2:x3:xosc`
- Capacitor instances: `c1:xosc`, `c2:xosc`, and `c3:xosc`
- Nodes: `10:xosc`, `20:xosc`, and `30:xosc`

If the nm model was defined in the `inverter` definition, three flat models would be created: `nm:x1:xosc`, `nm:x2:xosc`, and `nm:x3:xosc`. Model `nm:x1:xosc` would be used only by MOS instance `m2:x1:xosc`. If a model (name it `capm`) was defined in the `ring` definition, it would result in a single flat model (`capm:xosc`).

Note that only local nodes are subject to name expansion when the circuit is flattened. Sometimes an instance refers to another instance or node (i.e., a nonlinear controlled source may be controlled by some node voltage or instance current). If such an instance lies in a subcircuit definition, the instance or node to which it refers is sought among the local instances and nodes. Take, for instance,

```
.subckt somesub 1 2 3
v1 (3 0) dc=0
b1 (1 2) i=i(v1)*v(1,2)
.ends
```

The expression represents the product of the current flowing through independent voltage source `v1` with the voltage between nodes 1 and 2. In all three cases the instances and nodes are considered to be referring to local instances and nodes (within the subcircuit instance).

All instances, models, and nodes that are not internal (not within a subcircuit definition) are considered to be within the top level subcircuit. Top level models are all global (accessible in all subcircuit definitions). Top level nodes are global (and accessible in all subcircuit definitions) only if they are made global with a `.global` statement. Top level instances are not accessible in subcircuit definitions.

There is a single top level subcircuit definition named `topdef_` and a single top level subcircuit instance named `xtopinst_`. The top level subcircuit definition has no external nodes. Note that `xtopinst_` is never added to the path when the circuit is flattened.

3.9.4 Parametrization of Subcircuits

Often it is the case that one wants a subcircuit with the same topology and different instance/model parameter values. This is where parametrized subcircuits can be helpful. A parametrized subcircuit is defined with the following syntax:

`.subckt` *defName node1 node2* ...
`+ param:` *pname1* `[=`*default1*`]` *pname2* `[=`*default2*`]` ...
subcircuit element and model definitions
`.ends`

If the default value is not specified for some subcircuit parameter, that parameter has no default value and a value must be specified at instantiation. In order to use the subcircuit parameters in instance and model definitions, the expression containing parameters must be enclosed in curly braces (`{`*expression*`}`). *expression* can be any NUTMEG expression and can refer to any subcircuit parameter.

Example:

```
.subckt div top out bot param: r=10k fac
r1 (top out) r={r*(1-fac)}
r2 (out bot) r={r*fac}
.ends
```

Subcircuit `div` is a voltage divider. It has two parameters (`r` and `fac`). The default value for `r` is 10k. `fac` has no default value.

A parametrized subcircuit is instantiated using the following syntax:

`X`*name* `(`*con1 con2* ...`)` *defName*
`+ param:` *pname1*`=`*value1* *pname2*`=`*value2* ...

Example:

```
v1 (10 0) dc=5
x1 (10 5 0) div k=0.2
```

The above example instantiates a voltage divider with input at node 10 and output at node 5. The divider ratio is 1 : 0.2 meaning that the output voltage is 1 V. The divider pulls 5 mA current as its resistance is 10 kΩ (the default value). If parameter `fac` is not specified, SPICE will signal an error as `fac` is a parameter with no default value.

Within a subcircuit definition auxiliary quantities can be defined using the following syntax:

`.param` *auxName*`=`*expression*

The *expression* can be any NUTMEG expression. It can refer to any subcircuit parameter or previously defined auxiliary value. Auxiliary values can be used in expressions for instance and model parameter values.

Example:

```
.subckt div top out bot param: r=10k fac
.param r1val=r*(1-fac)
.param r2val=r*fac
r1 (top out) r={r1val}
r2 (out bot) r={r2val}
.ends
```

Auxiliary quantities can also be defined in the top level subcircuit. Such auxiliary quantities are considered global and are available in all subcircuit definitions.

Example:

```
* Supply power to an amplifier and put a load at its output
.param vpow=5
.param rl=1meg

vsrc (vdd 0) dc={vpow}
x1 (in out vdd 0) amp
rload (out 0) r={rl}
```

Parametrizing nonlinear controlled sources can be a source of errors. Let us look at an example:

```
b1 (10 0) i={coeff}*v(10)*log(1+v(10)^2)
```

The nonlinear controlled source's controlling expression can be parametrized. The expressions in curly braces are evaluated before any simulation is started, and the resulting values are used in the controlling expression.

The following circuit will not work:

```
b1 (10 0) i={coeff*v(10)}*log(1+v(10)^2)
```

Note that now the controlling voltage is a part of the expression in curly braces. As the simulator tries to evaluate it, it cannot find v(10) as the circuit has not been simulated yet. After evaluation, all expressions in curly braces result in a real number without resorting to quantities that are calculated during simulation.

3.10 Initial Operating Point Solution and Initial Conditions

An initial operating point for the Newton–Raphson algorithm can be specified in the netlist.

```
.nodeset v(node)=value ...
```

This initial operating point solution is used to speed up the evaluation of the circuit's operating point. Considerable speedups can be obtained if the nodesets are close to the circuit's operating point. Nodesets can also be set using the nodeset NUTMEG command.

Example:

```
.nodeset v(1)=10.0 v(2)=0.7 v(9)=-0.8
```

When a transient analysis is performed without an initial operating point (when the uic switch is specified in the transient analysis), the circuit's initial condition is assumed to be 0 (capacitors have no charge). This can be changed by adding initial conditions to the netlist. The syntax is as follows:

`.ic v(node)=value ...`

Initial conditions can also be set using the `ic` NUTMEG command. The initial voltage (charge) across a capacitor is calculated from the corresponding initial conditions of the nodes to which the capacitor is connected. If a node has no initial condition specified, its initial condition is assumed to be 0. The capacitor's initial condition specified through the `ic` parameter overrides the initial condition that is obtained from the circuit's `.ic` values.

```
.ic v(5)=10 v(9)=8
* This capacitor has initial voltage 2V (.ic values v(5)-v(9))
c1 (5 9) c=1n
* This capacitor overrides the initial conditions specified by .ic
c2 (5 9) c=1n ic=4
```

3.11 Joining Multiple Topologies in a Netlist

Netclasses are a SPICE OPUS feature for joining multiple different circuit topologies in one netlist. The topology can be switched without reloading the circuit. A topology is described by means of a `.netclass` block.

`.netclass groupName blockName`
`netlist statements`
`.endn`

In a group of `.netclass` blocks with the same `groupName` only one block at a time can be active. Other blocks from the same group are disabled and treated as if they were not in the netlist. By default (after the circuit is loaded), the first block in every group is active. All netlist statements that are not in a `.netclass` block are always active.

Example:

```
v1 (5 0) dc=5
r1 (5 9) r=1k

.netclass r2status normal
r2 (9 0) r=1k
.endn
.netclass r2status open
r2 (9 0) r=100meg
c2 (9 0) c=10p
.endn
.netclass r2status shorted
r2 (9 90) r=1m
l2 (90 0) l=10n
.endn

.netclass coutstatus normal
cout (9 0) c=100n
.endn
.netclass coutstatus shorted
rout (9 0) r=1m
.endn
```

In the above netlist `r1` is outside any `netclass` block. This means that `r1` is always active (i.e., always in the circuit). Group `r2status` has three blocks (`normal`, `open`, and `shorted`). By default, block `normal` is active, while the other two are disabled. In group `coutstatus` there are two blocks: `normal` and `shorted`. By default, the first one (`normal`) is active.

The active block for any given group can be changed. To select the `open` block from the `r2status` group, type in the command window

```
netclass select r2status::open
```

Selecting a block does not apply changes to the circuit in the memory. To apply the changes, type

```
netclass rebuild
```

Whether or not the circuit is rebuilt when the `netclass rebuild` command is issued depends on the `forcerebuild` variable (see Sect. 5.5.5). If it is not set, the circuit is rebuilt only when the active block of some group is changed using the `netclass select` command. If it is set, a rebuild occurs even if it is unnecessary. The values of circuit parameters after the rebuild are set to the values specified in the netlist. All changes to the parameter values are invalidated by a rebuild.

3.12 Simulator Parameters

Simulator parameters are set with the `.options` netlist statement.
`.options` *optionName1* `[=`*value1*`]` *optionName2* `[=`*value2*`]` `...`
These parameters are listed in Table 3.81.

The following simulator parameters are aliases: `imin` (for `itl3`), `imax` (for `itl4`), `mu` (for `xmu`), and `mumult` (for `xmumult`).

Simulator parameters `defw`, `defl`, `defpd`, `defps`, `defad`, and `defas` are affected by the `scale` simulator parameter. If `scale` is set to 1, the units for these parameters are meters (square meters for `ad` and `as`). If `scale` is set to 10^{-6}, the units are micrometers (square micrometers for `ad` and `as`).

Setting the `method` simulator parameter to `gear` automatically disables the `bypass` parameter. If the `lvltim` parameter is not given, it is set to 2 (step control through local truncation error). The `rmax` and `slopetol` simulator parameters are set to 2 and 0.5, respectively, if they are not specified by the user.

Setting the `lvltim` simulator parameter to values other than 1 (dVdt time-step control) sets `rmax` and `slopetol` to 2 and 0.5, respectively, if they are not specified by the user.

Setting the `newtrunc` simulator parameter disables the `nopredictor` parameter. Setting the `nopredictor` simulator parameter disables the `newtrunc` parameter.

A more detailed explanation of simulator parameters can be found in Chaps. 4 and 6.

Table 3.81 Simulator parameters

Name	Default	Description
noopiter	Not set	Skip initial NR iterations and go directly to stepping/lifting (flag)
nosrclift	Not set	Skip source value lifting (flag)
noinitsrcl	Not set	Skip initial source value lifting without *cmin* capacitors (flag)
opdebug	Not set	Print debug information during an operating point analysis (flag)
sollimdebug	Not set	Print debug information during damped NR iterations (flag)
matrixcheck	Not set	Check matrix for NaN[a] and Inf[b] values after every load (flag)
nofloatnodescheck	Not set	Skip floating nodes check (flag)
noautoconv	Not set	Disable automated setting of simulator parameters that control the convergence (flag)
dcap	1	Depletion capacitance equation selector (1..3), currently only dcap=1 is supported
gmin	10^{-12} A/V	Minimum conductance
gmindc	10^{-12} A/V	Minimum conductance in DC domain
cmin	10^{-12} F	Initial capacitance for *cmin* value stepping
cshunt	10^{-12} F	Shunt capacitance to ground at every node
reltol	10^{-3}	Relative error tolerance
abstol	10^{-12} A	Absolute current error tolerance
vntol	10^{-6} V	Absolute voltage error tolerance
slopetol	0.75	Maximum slope change for dVdt time-step control algorithm
absvar	0.5	Maximum absolute solution change in dVdt time-step control algorithm
relvar	0.3	Maximum relative solution change in dVdt time-step control algorithm
voltagelimit	10^{30} V	Voltage limit in per-iteration voltage limit functions
trtol	1.5	Allowed truncation error overestimation factor
chgtol	10^{-14} As	Charge error tolerance
pivtol	10^{-13}	Minimum acceptable pivot
pivrel	10^{-3}	Minimum acceptable ratio of pivot
sollim	10	Largest allowed step in damped NR algorithm
sollimiter	10	Number of iterations in damped NR algorithm
noconviter	Not set	Do not double-check convergence in NR algorithm (flag)
tnom	27°C	Parameter measurement temperature
temp	27°C	Circuit temperature
itl1	100	Number of NR iterations in OP analysis
itl2	50	Number of NR iterations in DC transfer function analysis
itl3	3	Lower transient iteration limit for iteration count time-step control algorithm
itl4	8	Upper transient iteration limit for iteration count time-step control algorithm
srcsteps	10	Number of source steps
srcspriority	2	Source value stepping priority
gminsteps	10	Number of GMIN steps

(continued)

Table 3.81 (continued)

Name	Default	Description
gminpriority	1	GMIN stepping priority
cminsteps	2	Number of *cmin* value steps
opts	Not set	Print a list of the simulator parameters
oldlimit	Not set	Use SPICE2 MOSFET limiting
lvltim	2	Type of time-step control (0..iteration count, 1..dVdt, 2..local truncation); setting it to values other than 1 can affect `rmax` and `slopetol`
newtrunc	Not set	Use predictor-corrector method to calculate local truncation error; turns of `nopredictor`
nopredictor	Not set	Disable predictor-corrector integration; turns off `newtrunc`
method	trap	Integration method (`trap` for trapezoidal, `gear` for gear); setting it to `gear` turns off `bypass` and can affect `lvltim`, `rmax`, and `slopetol`
maxord	2	Maximum integration order (1..2 for trapezoidal integration, 1..6 for gear integration)
xmu	0.5	Coefficient for trapezoidal integration algorithm (between 0 (force first order) and 0.5 (second order))
xmumult	0.8	Multiplier for `xmu` in case numerical oscillations are detected
trapratio	10	Ratio for detecting numerical oscillations
defl	10^{-4}	Default MOSFET channel length
defw	10^{-4}	Default MOSFET channel width
defad	0	Default MOSFET drain area
defas	0	Default MOSFET source area
defpd	0	Default MOSFET drain perimeter
defps	0	Default MOSFET source perimeter
defnrd	0	Default MOSFET drain squares
defnrs	0	Default MOSFET source squares
defmodcheck	0	Default value for BSIM3v3 check model parameter
definstcheck	0	Default value for BSIM3v3 check instance parameter
bypass	1	Bypass unchanging elements (0..no, 1..yes)
trytocompact	Not set	Try compaction for lossy transmission lines
badmos3	Not set	Use old MOSFET level 3 model (discontinuities with respect to `kappa`)
keepopinfo	Not set	Record operating point for small signal analysis
icstep	0	Fraction of initial time step for solving the circuit at $t = 0$ (0..disabled)
nsfactor	10^{10}	Initial factor used for forcing the nodesets
nssteps	1	Number of NR iterations in which the nodeset factor goes to 0
rmax	5	Maximal transient step multiplier
rmin	10^{-9}	Minimal transient step multiplier
fs	0.25	Fraction of time step for first time point
ft	0.25	Factor for multiplying the time step in case of nonconvergence
scale	1	Metric multiplier for width, length, perimeter, and area (SCALE^2 for area)
pziter	200	Number of iterations for pole-zero analysis

(continued)

Table 3.81 (continued)

Name	Default	Description
srclpriority	3	Source value lifting priority
srclmaxiter	5,000	Maximal number of transient iterations for source value lifting
srclconviter	50	Number of transient iterations to assume convergence in source value lifting
srclriseiter	50	Number of transient iterations in which sources reach their initial value
srclrisetime	0	Time in which sources reach their initial value (if greater than zero it takes precedence over srclriseiter)
srclmaxtime	10 s	Total time for source value lifting (if srclmaxiter is reached sooner, source value lifting is stopped before this time is reached)
srclminstep	10^{-15} s	Minimal time step for source value lifting
relq	0.01	Relative charge tolerance for truncation error
ltereltol	0.01	Relative current tolerance for truncation error
lteabstol	10^{-6} A or V	Absolute tolerance for truncation error (volts are used when newtrunc is set)
newtrunc is set		
resmin	10^{-5} Ω	Minimum resistance value for resistors
rshunt	∞ Ω	Shunt resistance to ground at every node

[a]IEEE floating point value for "Not a number". It is obtained as the result of 0/0.
[b]IEEE floating point value representing infinity. It is obtained when a nonzero value is divided by zero.

Examples:

```
* Make tolerances more strict
.options reltol=1e-6 abstol=1e-15 vntol=1e-9
* Try to compact LTRA lines
.options trytocompact
* Set gear integration, maximal order is 4
.options method=gear maxord=4
```

Chapter 4
Analyzing the Circuit

All analyses in SPICE are started from the SPICE command window by typing the appropriate command. The command can also be a part of the `.control` block in the input file. In the latter case, the command is executed along with the rest of the `.control` block automatically after the circuit is loaded.

All analyses result in one or more groups of results called plots. Every plot consists of vectors. A vector is a one-dimensional array of real or complex values. Every plot has a default vector. The default vector usually represents the scale (i.e., time, frequency, sweep) for the remaining vectors in the plot.

4.1 Operating Point (OP) Analysis

The operating point (OP) analysis evaluates the circuit's operating point. All capacitors (inductors) behave as open (short) circuits. When no transient waveform is specified, the value of the dc parameter is used in all independent sources. If some transient waveform is specified, then its value at $t = 0$ is used. In nonlinear controlled sources the `time` variable is assumed to be zero.

An operating point analysis creates one plot. All node voltages and currents through voltage sources and inductors are calculated. All vectors are real with length 1 because the operating point of the circuit is represented by a single real value per node or current.

The operating point analysis is started by

op

The details on how the operating point is calculated can be found in Sects. 6.1–6.3.2. Convergence detection and convergence helpers for operating point analysis are described in Sects. 6.3.3–6.3.4.

T. Tuma and Á. Bűrmen, *Circuit Simulation with SPICE OPUS*, Modeling and
Simulation in Science, Engineering and Technology, DOI 10.1007/978-0-8176-4867-1_4,
© Birkhäuser Boston, a part of Springer Science+Business Media, LLC 2009

4.1.1 Handling of Initial Solutions (Nodesets)

An initial solution can be supplied to the operating point analysis by issuing a
nodeset NUTMEG command or by adding a .nodeset specification into the
netlist. The supplied information can significantly speed up the operating point analysis. If this information is not specified, the initial point is assumed to be 0 for all
circuit quantities. The nodesets are forced onto the circuit nodes by connecting a
large conductance (g) in parallel with a current source representing the nodeset between a node and the ground. The current source is set to a value $g \cdot NS$, where NS
represents the value of the nodeset. The nsfactor simulator parameter specifies
the value for g. Larger values of g mean that the nodeset is forced more strongly
onto the node. The conductance (g) is then gradually decreased toward 0, where
the nodeset disappears after it has (one hopes) affected the Newton–Raphson algorithm. The number of Newton–Raphson iterations in which the nodeset disappears
is controlled by the nssteps simulator parameter.

All other circuit analyses are preceded by an operating point analysis. An exception is the transient analysis if the uic parameter is specified. It makes sense to
perform a single operating point analysis and then set the nodesets to the results of
that analysis by issuing the following command:

nodeset

This way all analyses that follow will use the resulting operating point as the
initial point for their corresponding operating point analysis, which will be finished
much faster with a good starting point.

4.1.2 Solving Convergence Problems

The OP analysis precedes almost every other analysis. It is the most common source
of convergence problems. SPICE OPUS offers a wide variety of algorithms for en-
suring convergence (convergence helpers).

Convergence helpers are based on the algorithms described in Chap. 6.

Detecting floating nodes
Nodes with no DC path to the ground node are a common cause of convergence
problems. If a circuit contains such a node, it always results in convergence problems. SPICE can detect such nodes and notify the user. The check is on by default.
To disable it, you can use the nofloatnodescheck simulator parameter.

Putting a lower limit on resistor values
Small values of resistors correspond to large conductivities and, consequently, large
entries in the matrix solved by SPICE. The simulator parameter resmin puts a lower
limit on resistor values.

Limiting the junction voltage
A relatively small voltage across p–n junctions can cause extremely large currents
due to the exponential characteristic of the junction. Furthermore, large values of

the voltage across junctions cause current values exceeding the range of floating point numbers and result in convergence problems. The `voltagelimit` simulator parameter sets the limit on junction voltage.

Detecting *inf* and *nan* values in the circuit matrix

Sometimes convergence problems are caused by infinite (inf) or undefined (nan) values in the circuit matrix. SPICE can check for the values every time the matrix is updated and reports inf and nan entries. The `matrixcheck` simulator parameter enables this checking.

Setting the number of Newton–Raphson iterations

Often convergence problems are solved if the number of iterations that are available to the Newton–Raphson algorithm is increased. The simulator parameter that sets the number of Newton–Raphson iterations in the operating point analysis is `itl1`.

Damped Newton–Raphson algorithm

Sometimes SPICE fails to converge after a small step in GMIN stepping or source value stepping. In such cases the damped Newton–Raphson algorithm is enabled for the particular step that causes problems. The algorithm limits the change of circuit quantities between consecutive Newton–Raphson iterations to a fraction of the maximal step. The maximal step for every right-hand side quantity is determined by the `reltol`, `vntol` (voltages), and `abstol` (currents) simulator parameters and the largest value of the quantity from the last two iterations. If the maximal step is exceeded, SPICE limits the step to `sollim` times the maximal step. As such small steps require a larger number of iterations, the number of Newton–Raphson iterations is increased by the factor specified through simulator parameter `sollimiter`. The damped Newton–Raphson algorithm is considered only in GMIN stepping and source value stepping, after the step in the second part of the stepping becomes too small. It is also used in the final operating point analysis after source value lifting if the ordinary Newton–Raphson algorithm fails to converge.

GMIN stepping

The name originates from the fact that in this algorithm a conductance is connected between every node and the ground. The conductances are gradually increased until the circuit converges. The solution of the resulting circuit is used as the initial point for the next operating point calculation, which is conducted with all the conductances decreased. This procedure of decreasing the conductances and performing an operating point analysis with the result of the previous step as the initial point for the Newton–Raphson algorithm is repeated until the conductances become smaller than the value of simulator parameter `gmindc`. If a GMIN step fails, the rate by which the conductances are decreased is reduced and the last successful GMIN step is repeated. The rate is increased after every successful GMIN step. If it happens that the rate becomes very small, the damped Newton–Raphson algorithm is employed. The total number of steps in the GMIN stepping algorithm is limited by the `gminsteps` simulator parameter. After the GMIN stepping algorithm is finished, one more Newton–Raphson iteration is performed with all the conductances removed and the solution from the last step of the GMIN stepping used as the initial point.

Source value stepping

Source value stepping works similarly to GMIN stepping. All the independent sources in the circuit are gradually decreased toward 0 until convergence is obtained. Then the sources are increased gradually until they reach their respective values used in the operating point analysis. The stepping procedure is similar to the GMIN stepping algorithm. If the amount of increase applied in one step of the source value stepping algorithm becomes too small, the damped Newton–Raphson algorithm is employed. The upper limit on the number of source steps is determined by the `srcsteps` simulator parameter.

Source value lifting

In source value lifting all of the independent sources are set to 0. Then a time-domain analysis is conducted during which the values of the sources are gradually increased toward their respective values used in the operating point analysis. The time domain analysis is an ordinary transient analysis where the initial operating point is not calculated. Instead, all circuit states are assumed to be 0 (i.e., capacitances have no initial charge). The final result is used as the initial point for the Newton–Raphson algorithm. If convergence is not obtained, the damped Newton–Raphson algorithm is employed.

The time-domain analysis is stopped after the time point `srclmaxtime` is reached or after `srclmaxiter` time-domain analysis iterations are performed. The analysis is also stopped if `srclconviter` time-domain iterations produce results within tolerances specified by `reltol`, `vntol`, and `abstol`. `srclrisetime` specifies the time point where the independent sources reach their final value. If it is not specified, `srclriseiter` specifies the number of time-domain iterations after which the sources reach their final values. Finally, `srclminstep` sets a lower limit on the time step in the time-domain analysis.

cmin **value stepping (source value lifting continued)**

If the time-domain analysis in source value lifting fails to produce an initial point that obtains convergence in the Newton–Raphson algorithm, it is repeated. But this time capacitors are connected between every node and the ground. The value of the capacitors is specified by the `cmin` simulator parameter. If this fails, the capacitors are increased by the value of `cmin` and the analysis is repeated. The number of *cmin* steps is limited by the `cminsteps` simulator parameter. If it is set to 0, *cmin* value stepping is disabled.

Shunting

This is the only simulator parameter that permanently changes the circuit. Setting the `rshunt` simulator parameter connects resistors between every node and the ground. The resistance is specified by the `rshunt` simulator parameter. This parameter always obtains a convergent circuit, albeit at the price of the quality of results. `rshunt` (if set) should be set to a value as large as possible.

Choosing the convergence helper

Simulator parameter `noopiter` disables the initial normal Newton–Raphson algorithm and goes straight to the convergence helpers. `nosrclift` disables the source value lifting algorithm. `noinitsrcl` skips the initial source value lifting that is performed without `cmin` capacitances at circuit nodes and goes straight to *cmin* value stepping. *cmin* value stepping can be disabled by setting the `cminstep` parameter to 0. GMIN and source value stepping are disabled by setting `gminsteps` and `srcsteps` to 0.

The order in which convergence helpers are tried is specified by the `srcspriority`, `gminpriority`, and the `srclpriority` parameters. These parameters are assigned values between 1 and 3 (where 1 means the highest priority). By default, GMIN stepping has the highest priority, followed by source value stepping and source value lifting.

Automated setting of convergence parameters

SPICE automatically sets the convergence simulator parameters (for GMIN stepping, source value stepping, source value lifting, *cmin* value stepping, and damped Newton–Raphson algorithm) in order to obtain convergence. The `noautoconv` simulator parameter disables the automatic convergence parameter setting algorithm and leaves the choice of convergence parameters to the user.

Checking for convergence

The convergence is checked by comparing two consecutive Newton–Raphson iterations. If the difference is within the tolerances prescribed by `reltol`, `vntol`, and `abstol` for all of the circuit quantities, the algorithm stops iterating. This is the case when the `noconviter` parameter is set. If it is not set, convergence checking is more strict. The last three iterations (denoted by 1, 2, and 3) are used in the convergence checking. The two consecutive differences (between 1 and 2, and 2 and 3) are compared to the tolerances. In general, the three solutions form a triangle in n-dimensional space. The angle of the triangle at the point corresponding to iteration 2 must be smaller than 90°.

Monitoring the convergence

SPICE can print a lot of debugging information as the operating point is sought. Printing of these debugging messages can be turned on by choosing the `opdebug` simulator parameter. Similarly, the `sollimdebug` simulator parameter turns on the debugging message output for the solution limiting algorithm.

4.2 Operating Point Sweep Analysis (DC Analysis)

The DC analysis performs an operating point sweep. Any circuit parameter, the global circuit temperature, and the global parameter measurement temperature (the temperature for which the model parameters were extracted) can be swept. A DC analysis is significantly faster than the corresponding number of independent operating point analyses because it uses the result obtained with a particular value of the parameter as the initial point for the calculation of the operating point at the next value of the parameter. This effectively reduces the number of Newton–Raphson iterations required to reach a solution.

A DC analysis creates one plot. The set of vectors (except for the `sweep` vector) is the same as in the operating point analysis, except that now every vector contains multiple values. Every value belongs to some value of the swept parameter. The values of the parameter for which the operating point was evaluated are in the `sweep` vector, which is the default scale of the plot.

A DC analysis is started by one of the following two commands:

dc *parameter start stop step*
dc *parameter start stop sweeptype nsteps*

The first dc command sweeps the given parameter from *start* to *stop* with *step* as the step size. If *start* is greater than *stop*, *step* must be negative.

The second dc command allows for logarithmic sweeps. The meaning of the *nsteps* parameter and the available values for the *sweeptype* parameter are listed in Table 4.1.

The various ways in which a parameter can be specified are listed in Table 4.2.

When a DC analysis is started, the operating point is evaluated in the same manner as in the operating point analysis. The succeeding points, however, always use the operating point from the previous step of the DC analysis, and the number of Newton–Raphson iterations is specified by the `itl2` simulator parameter. `itl2` is used until the DC analysis is finished, or until convergence problems are encountered. In the latter case, convergence helpers are invoked, and the number of Newton–Raphson iterations is taken from the `itl1` simulator parameter. `itl1` is used only until the convergence problem is solved, upon which `itl2` is used again.

Table 4.1 Sweep specification in DC analysis (second form)

sweeptype	Meaning of *nsteps*
lin	Linear sweep, number of points between *start* and *stop*
dec	Logarithmic sweep, number of points per decade (range from x to $10x$)
oct	Logarithmic sweep, number of points per octave (range from x to $2x$)

Table 4.2 Parameter specification in DC analysis

parameter	Description
vname	dc parameter of independent voltage source *vname*
iname	dc parameter of independent voltage source *iname*
@*name* [*par*]	Parameter *par* of device *name*
@@*name* [*par*]	Parameter *par* of model *name*
@@@temp	Global circuit temperature in degrees centigrade
@@@tempk	Global circuit temperature in kelvins
@@@tnom	Global parameter measurement temperature in degrees centigrade
@@@tnomk	Global parameter measurement temperature in kelvins

Examples:

```
* sweep iload from 0mA to 10mA in 0.01mA steps
dc iload 0 10m 0.01m
* sweep vin from 0V to 10V in 0.1V steps
dc vin 0 10 0.1
* sweep parameter w of m1 from 10um to 50um, linear sweep, 100 points
dc @m1[w] 10u 50u lin 100
* logarithmic sweep of r10 resistance from 10 to 100k, 50 points per decade
dc @r10[r] 10 100k dec 50
* sweep parameter vth0 of model nm from 0.20V to 0.28V, 0.001V step
dc @@nm[vth0] 0.20 0.28 0.001
* linear sweep of global circuit temperature from -60oC to 200oC, 10oC step
dc @@@temp -60 200 10
```

4.3 DC Transfer Function Analysis (TF Analysis)

The TF analysis evaluates the small signal transfer function at 0 Hz between two ports of a circuit and the respective input impedances at those two ports. The transfer function and impedances are evaluated at the circuit's operating point. TF analysis is invoked by the following command:

tf *output input*

The output can be a voltage (v(*node*)), a differential voltage (v(*node1, node2*)), or a current flowing through a voltage source or inductance (i(*vname*), i(*ename*), i(*hname*), or i(*lname*)).

The input is specified as an independent voltage or current source (*vname* or *iname*). If the input is specified as an independent current source, the input current is considered to be positive when it flows in the direction dictated by the current source.

There are four possible transfer functions, as listed in Table 4.3.

TF analysis results in a single plot containing three vectors with one component. transfer_function contains the transfer function value. input_impedance and output_impedance contain both impedances. The output impedance is correct only if the output is a voltage (v(*node*) or v(*node1, node2*)).

Before the actual TF analysis takes place, an operating point analysis is performed. The conductance matrix generated in the last Newton–Raphson iteration of the operating point analysis is then used in the TF analysis. This matrix accurately describes relations between differential signals around the operating point of the circuit. The input and output impedances and the transfer function can be obtained by LU factorization of this matrix (which is performed in every Newton–Raphson iteration of the operating point analysis that preceded the TF analysis).

Table 4.3 Transfer functions in TF analysis

Function	Output	Input
Voltage/voltage	v(*node*) or v(*node1, node2*)	*vname*
Voltage/current	v(*node*) or v(*node1, node2*)	*iname*
Current/voltage	i(*name*)	*vname*
Current/current	i(*name*)	*iname*

Examples:

```
* Transfer function, input is between 1 and 2 where vin is connected,
* output is between 10 and 20.
*    vin (1 2) dc=...
tf v(10,20) vin

* Transfer function from the same input to the current flowing through vload.
* The output impedance here is not correct.
tf i(vload) vin
```

4.4 Small Signal Analysis (AC Analysis)

In the small signal analysis the response of the linearized circuit is evaluated. First
the operating point is solved and the circuit is linearized. The AC analysis is conducted in the frequency domain, so the results are complex voltages and currents.
The circuit is excited by the independent sources with their ac parameter set to a
nonzero value. It is not customary, but the user can also specify the phase of the AC
excitation for every source. Take, for instance, an independent voltage source

vname (np nn) dc=0 ac=mag [phase]

The phase is optional. Its value is zero by default. However, if it is specified, it
must be specified in degrees. Such a source is represented by the following complex
excitation: $\mathrm{mag} \cdot e^{j \cdot \mathrm{phase} \cdot \pi/180}$. All sources with no ac parameter specified are disabled
(i.e., voltage sources are shorted and current sources removed from the circuit).

The AC analysis is invoked with the following command:

ac sweeptype nsteps start stop

The meaning of sweeptype (either dec, oct, or lin) and nsteps is the same
as in DC analysis. Values start and stop represent the beginning and the end of
the frequency range.

The AC analysis produces one plot with the frequency vector as the default
scale. All vectors are complex (including frequency), albeit the imaginary part of
frequency is zero. The remaining set of vectors is the same as in the operating
point analysis.

AC analysis is fast and experiences almost no convergence problems. The ac
command usually spends most of its time on the initial operating point evaluation.
Details regarding the operating point analysis can be found in Sect. 6.4.1.

Examples:

```
* 10 points per decade between 1Hz and 1MHz
ac dec 10 1 1meg
* linear sweep (100 points) from 0.99MHz to 1.01MHz
ac lin 100 0.99meg 1.01meg
```

If the input of a circuit is excited by an independent source with the ac parameter
set to 1, the transfer function does not have to be a quotient of the complex output
and input signal. Take, for instance, an amplifier with input between nodes 1 and 2.
By connecting an independent voltage source between these two nodes and setting
its ac parameter to 1, the complex voltage at the input equals 1 for all frequencies.
The transfer function becomes equal to the output complex signal.

4.5   Pole-Zero (PZ) Analysis

Pole-zero (PZ) analysis evaluates the poles and/or the zeros of the transfer function in the frequency domain. First the operating point of the circuit is evaluated, and the circuit is linearized. All independent current sources are removed from the circuit, including those for which the ac parameter is set to a nonzero value. The independent voltage sources lacking an ac parameter are replaced by short circuits. Independent voltage sources with an ac parameter specified in the netlist are removed from the circuit. The reasoning is as follows. An independent voltage source with an ac parameter is almost always intended to be a signal source. During transfer function evaluation such a source causes a short at the circuit's input and thus corrupts the transfer function. On the other hand, independent voltage sources with no ac parameter represent DC power sources and appear as short circuits after the linearization.

The syntax of the PZ analysis is the following:

pz *in+ in- out+ out- type results*

The nodes *in+* and *in-* represent the input, and nodes *out+* and *out-* represent the output of the circuit.

The parameter *type* specifies the type of the transfer function. vol means that the transfer function is considered to be of the type output voltage divided by the input voltage. cur means that the transfer function is of the type output voltage divided by the input current. Although it may appear important which current is considered to be the input current (the one that flows into node *in+* or the one flowing in the opposite direction), the set of poles and zeros is not affected by it.

The *results* parameter chooses what the PZ analysis will calculate. pol means that only the poles will be determined, zer determines only the zeros, and pz determines both.

A PZ analysis produces one plot. It contains four complex vectors. All vectors have the same length. Vectors pole_valid and zero_valid determine the multiplicity of every pole and zero. A multiplicity of 0 means that the respective pole (zero) is not valid and should therefore be ignored. The entries in the pole and zero vectors are the actual complex poles and zeros.

If there are four poles and two zeros, all vectors have four components. The four poles occupy all four entries of the pole vector, and all four components of the pole_valid vector are nonzero. The two zeros occupy the first two components of the zero vector, and the first two components of the zero_valid vector are nonzero. The remaining two components in the zero_valid vector are equal to 0. The last two components of the zero vector have no meaning and should be ignored.

PZ analysis cannot handle transmission lines. The number of iterations available to the PZ analysis is set by the pziter simulator parameter. More details on the PZ analysis can be found in Sect. 6.4.2.

Examples:

```
* Poles and zeros of transfer function v(3,4)/v(1,2)
pz 1 2 3 4 vol pz
* Poles of transfer function v(3,4)/i where i represents the current
* flowing from node 1 into node 2
pz 1 2 3 4 cur pol
```

4.6   Noise (NOISE) Analysis

The NOISE analysis evaluates the small signal noise spectra at the circuit's operating point. First the operating point is evaluated. The circuit is linearized, and the effect of every noise contribution on the output signal is simulated.

Every noise contribution is simulated by a small signal current source connected to the corresponding part of the circuit where the noise is generated. The current source's magnitude depends on the circuit's operating point and the frequency. It represents the spectral density of the generated noise. This noise is propagated through the linearized circuit and appears as a noise contribution with modified spectral density at the circuit's output. Because these noise contributions are uncorrelated, the total output power spectrum density is the sum of power spectrum densities belonging to individual contributions.

Besides output noise contributions and total output noise, the equivalent input noise is also evaluated. It represents the noise that must be fed into an ideal circuit (one that has no noise sources) in order to produce the same output noise spectrum as the real-world circuit does.

The NOISE analysis produces results in the frequency domain. It is started by the following command:

`noise output input sweeptype nsteps start stop [ptssum]`

The output can be a differential voltage (v(`outp,outn`)) or a node voltage (v(`out`)). The input is either an independent voltage source (`vname`) or an independent current source (`iname`). In order to obtain correct results for equivalent input noise, the ac parameter of the source must be set to 1. The meaning of parameters `sweeptype`, `nsteps`, `start`, and `stop` is the same as in AC analysis. The *ptssum* parameter is optional and determines how the vectors representing noise spectra are constructed.

The NOISE analysis produces two plots. The first one contains the noise power spectra. All vectors are real and generally contain more than one component. Vectors named onoise_*device*_*contrib* represent the contribution named *contrib* (e.g., id, 1overf, etc.) of device *device* (e.g., qname if the device is a BJT) to the output noise spectrum. Note that for simple devices (those with a single noise contribution, e.g., resistors) the _*contrib* part is missing. The onoise_spectrum vector contains the output noise spectrum. The unit for output spectra is V^2/Hz. The equivalent input noise spectrum is in the input_spectrum vector. The unit is either V^2/Hz (input is a voltage source) or A^2/Hz (input is a current source). The default scale for the first plot is the frequency vector.

The second plot contains the integrated noise. The *ptssum* parameter specifies how many points are calculated before one point is stored in the noise spectra vectors. Usually this parameter is set to 1. If it is set to a higher value, the noise spectra contain only every *ptssum*-th calculated point. The vectors in the second plot are named onoise_total for the total output noise and onoise_total_*device*_*contrib* for individual output noise contributions. Note that for simple devices (those with a single noise contribution, e.g., resistors) the _*contrib* part is missing. The unit for these vectors is V^2. Vectors inoise_total and `inoise_total_device_contrib` represent the equivalent input noise and the contributions of individual noise sources to the equivalent input noise. The unit is either V^2 (input is a voltage source) or A^2 (input is a current source).

If the `ptssum` parameter is omitted, the NOISE analysis does not produce vectors for individual noise source contributions. Only the input and output noise are stored in the plots. No contributions from individual devices are stored. Regardless of the fact that `ptssum` is not specified, SPICE still stores every calculated point in noise spectra vectors.

A more detailed explanation of the NOISE analysis can be found in Sect. 6.4.3.

Examples:

```
* Noise, output is between nodes 5 and 9, input is at vin
* frequency range 1Hz..1MHz, 10 points per decade, store every point
* Note that the ac parameter of vin must be set to 1.
noise v(5,9) vin dec 10 1 10meg 1
* Noise, same output, but without storing individual contributions
noise v(5,9) vin dec 10 1 10meg
```

4.7 Transient (TRAN) Analysis

Transient analysis evaluates the circuit in the time domain. Independent sources that have no transient waveform specified (one of `pulse`, `sin`, `exp`, `pwl`, or `sffm`) behave as DC sources with their values determined by the `dc` parameter. The transient analysis is started with the following command:

`tran step stop [start [maxstep [uic]]]`

If the `uic` switch is not specified, an initial operating point analysis is performed. The results of this analysis are used as initial conditions for the transient analysis. If the `uic` switch is specified, the initial operating point analysis is omitted and the initial conditions are taken from the `.ic` netlist specification and from the `ic` commands that were issued for the current circuit.

Parameters `step` and `stop` specify the initial time step and the final time at which the simulation stops. The optional `start` parameter (0 by default) specifies the time at which the transient analysis results start to be stored in the resulting plot. The `maxstep` parameter specifies the upper limit on the time step. By default, it equals simulator parameter `rmax` times `step`.

The transient analysis produces one plot with the same set of vectors as the operating point analysis. There is one additional vector named `time` that contains the time scale for the results. It is the default vector of the plot. All results are real vectors.

The algorithms used in transient analysis are described in Sect. 6.5.

Choosing the integration algorithm

The integration algorithm is chosen using the `method` simulator parameter. The order of integration is chosen using the `maxord` simulator parameter. The trapezoidal integration algorithm with `maxord` set to 1 is the backward Euler algorithm [10]. SPICE automatically adapts the order of the integration algorithm to the circuit's dynamics. By default, the predictor-corrector algorithm is used. This can be disabled by setting the nopredictor simulator parameter.

The trapezoidal algorithm [10] can sometimes exhibit ringing (the sign of the error alternates on every time step) [29]. The solution for this problem is either to use the backward Euler algorithm (often not precise enough) or Gear integration [10]. Another way to bypass the problem is to temporarily modify the integration formula so that it becomes similar to backward Euler integration. Simulator parameter xmu is the modification factor for the trapezoidal algorithm. When set to 0.5 (default) the trapezoidal algorithm behaves just as it is described in the literature. When it is set to 0, the algorithm behaves like backward Euler integration. In between the algorithm is a mixture of trapezoidal and backward Euler integration. When ringing is detected, SPICE changes the xmu value used in the algorithm to xmu · xmumult. When ringing disappears, the original xmu value is restored. The trapratio simulator parameter sets the threshold for detection of ringing. Higher values mean that the detection happens more rarely. Setting xmumult to 1 disables the modifications of the trapezoidal algorithm.

Time-step control

SPICE automatically adapts the time step to the circuit's dynamics. There are three time-step control algorithms available: iteration count, dVd, and local truncation error (LTE) (which is the default) time-step control. The algorithm is chosen with the lvltim simulator parameter. All algorithms decrease the time step (using a factor specified by the ft simulator parameter) if the number of Newton–Raphson iterations reaches imax (an alias for it14).

The iteration count algorithm decides whether to increase the time step on the basis of Newton–Raphson iteration count for the current time point. If this count falls below the value specified by the imin simulator parameter (an alias for it13), the time step is increased (by a factor of 2).

The dvdt time-step control algorithm uses the changes in the node voltages and the slopes of these changes to decide whether to increase or decrease the time step. The simulator parameters absvar, relvar, and slopetol control the behavior of this algorithm. The algorithm is implemented for historical reasons and is not recommended for use with integration methods of order greater than 1.

The LTE algorithm (the default) makes its decisions based on the LTE estimate and compares it to a tolerance based on currents and charges of the capacitances in the circuit. Simulator parameters lteabstol, ltereltol, chgtol, and relq specify the tolerance. Simulator parameter lteabstol represents the absolute current tolerance. The trtol parameter is the LTE underestimation factor. Larger values of trtol make the time-step control less strict.

If the newtrunc simulator parameter is enabled, the LTE is estimated from the difference between the predicted and the corrected node voltage values. In this case the tolerance is specified by the lteabstol and the ltereltol parameter, where lteabstol specifies the absolute voltage tolerance. This algorithm estimates LTE slightly faster for circuits with many devices and few nodes, but is otherwise equivalent to the algorithm that is used when newtrunc is not enabled (albeit the time steps are not necessarily identical). Enabling `newtrunc` turns off the `nopredictor` simulator parameter (enables predictor-corrector integration).

First time point problems when `uic` is used
When the `uic` switch is specified, the first time point is usually not correct as it is a copy of the initial conditions specified on the circuit's nodes. This is due to the way in which transient analysis is performed in SPICE. In order to obtain the correct values for the first time point, the `icstep` simulator parameter must be specified. The first time step is multiplied by `icstep` and the results at this new time point are stored as the solution at $t = 0$. The value of `icstep` should be small (i.e., 10^{-9}). To disable this behavior, `icstep` must be set to 0.

First time step and limiting the time step
`fs` specifies a factor used in the calculation of the first time step. The first time step is obtained from the minimum of analysis parameters *step* and *maxstep*, the position of the first break point, and $1/2$ of the period of the sinusoidal source with the largest frequency. This minimum is multiplied by `fs` giving the first time step.

   The maximal time step is limited to the `rmax` simulator parameter times the *step* analysis parameter. If the upper bound specified by the *maxstep* analysis parameter is lower, the latter is used.

   If, however, the time step becomes smaller than the `rmin` simulator parameter times the *step* analysis parameter, the time-step control is temporarily switched to the iteration count algorithm. In the next time point SPICE switches back to the original time-step control. If the time step reaches the lower precision bound (around 10^{-14} relative precision, 10^{-300} absolute precision), SPICE aborts the simulation.

Initial conditions
When the `uic` switch is set, the transient analysis skips the initial operating point evaluation and starts the transient simulation with user-specified initial conditions. The initial conditions can be specified in two ways: either by specifying initial values for circuit nodes or by providing initial values for reactive elements (e.g., using the `ic` parameter for capacitors and inductors). Take, for instance, a capacitor.

```
c1 (5 9) c=1n
```

   The initial voltage for `c1` is calculated as the difference between initial conditions set for nodes 5 and 9. If not specified, all initial conditions are assumed to be 0. To specify initial conditions for some nodes, use the `.ic` netlist statement.

```
* This sets initial conditions for nodes 5 and 9 to 7V and 8.5V, respectively.
.ic v(5)=7 v(9)=8.5
```

   For the upper example the initial condition for `c1` is $8.5\,V - 7\,V = 1.5\,V$. If, however, the `ic` parameter is specified for `c1`, its value overrides the initial condition that is calculated from the initial conditions set by the `.ic` netlist statement.

```
* This overrides the initial condition for c1 and sets it to 2.5V.
c1 (5 9) c=1n ic=2.5
```

   Initial conditions for nodes can also be set from the NUTMEG scripting language by using the `ic` command.

Chapter 5
NUTMEG Scripting Language

SPICE OPUS has a built-in scripting language called NUTMEG.[1] NUTMEG is a SPICE specific language developed at Berkeley as part of the SPICE3 software [36]. The NUTMEG in SPICE OPUS is an extended version of the original Berkeley NUTMEG. All commands that are entered in the command window are, in fact, NUTMEG commands. NUTMEG is also the language of the `.control` block, which is an optional part of the netlist. The initialization script (`spinit`) that is run at SPICE startup is also written in NUTMEG.

A NUTMEG session can be closed by issuing the `quit` command. The command is affected by the `noaskquit` variable. If this variable is set, the session is exited immediately after the `quit` command is issued. Otherwise, the user may be asked for a confirmation.

If the `quit` command appears in the control block or in a script, the session that loaded that particular netlist or script file is closed.

5.1 Input Handling

The command input is read line by line. Every line of input handled by the command interpreter (whether it comes as interactive input from the command window or is read from an input file) is first split into words.

5.1.1 Splitting the Input Line into Words, Escaped Characters

The input is split into words:

- When a space or tab character outside square and curly braces is encountered
- When a comma or semicolon outside parentheses, square, and curly braces is encountered
- At closing double quotes outside square and curly braces

---

[1] Nutmeg is the seed of the trees from the *Myristica* genus. Among other uses, it is used as a food flavoring.

T. Tuma and Á. Bűrmen, *Circuit Simulation with SPICE OPUS*, Modeling and
Simulation in Science, Engineering and Technology, DOI 10.1007/978-0-8176-4867-1_5,
© Birkhäuser Boston, a part of Springer Science+Business Media, LLC 2009

The character that breaks a word is omitted from the word. The normal behavior of NUTMEG can be disabled by escaping a character. Escaped characters are taken literally. There are two ways for escaping a character: using the backslash and using single quotes.

$\backslash character$

or

$' <char1> <char2> \ldots '$

A backslash is used for escaping a single character, whereas single quotes are used for escaping multiple consecutive characters.

Double quotes force the text between them to behave as a single word, regardless of the characters that cause word splitting. The closing double quote causes a word break when it is outside square and curly braces.

Example:

```
* There are only 2 words in the following command. The normal function
* of the space character is disabled by escaping it.
echo first\ second\ third

* This can also be done with
echo first' 'second' 'third

* Here we also have two words.
echo "first second third"

* Now this causes an error (there exists no command named
* "echo first second third" in SPICE).
"echo first second third"

* Note how the comma and semicolon are omitted.
* Here we have 4 words: echo, a, b, and c.
* The output is
*    a b c
echo a,b;c

* But not this time. Here the commas and semicolons
* are not dropped as they are within parenthesis.
* The output is
*    (a,b,c)
* This time we have only two words: echo and (a,b,c).
echo (a,b,c)
```

To obtain literal single and double quote characters escaping must be used.

```
* Prints the text in single quotes. The result is
*    'hello world'
* There are 3 words in this command:
*   echo
*   'hello
*   world'
echo \'hello world\'

* This will print the text in double quotes. The output will be
*    "hello world"
* Note how double quotes loose their functionality (inhibiting
* word break) when they are escaped.
* Again there are 3 words in the command.
echo \"hello world\"

* This time the double quotes are not escaped. They are not printed and
* force SPICE to treat "hello world" as a single word. The output is
*    hello world
* This time we have only two words.
echo "hello world"
```

```
* Note how the closing double quote causes a word break.
* We have 3 words here: ec, ho, and hello.
* The result is an error as there is no command named ec in SPICE.
"ec"ho hello

* But this still gets interpreted as "echo hello".
* Escaping ordinary characters has no special effect.
'ec'ho hello
```

5.1.2 History and History Substitution

This section covers the `history` command. Every input entered interactively in the command window is stored in a history list immediately after history substitution is performed on it. The history list can be obtained with the `history` command. The output of the history command looks something like this.

```
1    echo hello world
2    echo "hello world"
3    source whatever.txt
4    op
5    print v(9)
```

The oldest entry has the lowest number. The number that is assigned to an entry is displayed in the prompt. This is what the prompt looked like when the entry numbered as 5 was entered.

```
SpiceOpus (c) 5 ->
```

The syntax of the history command is

`history [-r [n]]`

By specifying the `-r` argument, the history list is printed in reverse order. The optional argument n specifies how many of the newest history entries should be printed.

The length of the history is specified with the `history` variable. By default, it is set to 100.

In graphical user interface (GUI) mode entries from the history can be retrieved by pressing the `Cursor Up` and `Cursor Down` keys. Pressing `Ctrl+Cursor Up` or `Ctrl+Cursor Down` moves to the oldest or newest entry in the history, respectively.

Entries from the history can also be retrieved through history substitution. History substitution takes place immediately after the interactive input is split into words. History substitution is triggered by the ! character.

The last entry in the history is obtained with `!!`. The entry numbered n is obtained with `!n`, whereas the n-th newest entry is obtained with `!-n`. So `!-1` represents the last command and is equivalent to `!!`. To get the newest history entry that contains substring str type `!?str`. To search for the newest history entry with prefix str enter `!str`.

All of the above-mentioned history substitutions can be suffixed with one or more modifiers. Modifier `:$` chooses only the last word from the specified history entry (e.g., `!!:$` is the last word of the latest entry). Modifier `:*` retrieves all words except for the first one (e.g., `!!:*` is a b c if the last command entered was echo a b c).

Modifier : n retrieves the n-th word where the first word is denoted by 0 (e.g., !!:0 is echo if the last command was echo a b c).

: $n*$ retrieves all words starting at n-th up to and including the last word (e.g., !!:2* is b c if the last command was echo a b c).

: $-n$ retrieves all words starting at the first and up to and including the n-th. If n is omitted, all words are returned (e.g., !!:-1 is echo a if the last command entered was echo a b c).

: $n-m$ retrieves words starting at the n-th up to and including the m-th (e.g., !!:1-2 is a b if the last command was echo a b c).

Modifier :p forces the printout of the substitution. The actual command that contains the substitution is not executed.

Modifier :s/pat/sub/ forces the substitution of pattern pat with sub (e.g., if the last command was print a b c, !!:s/print/display/ results in display a b c). Instead of / any other character can be used. Only the first occurrence of pattern is substituted.

To prevent history substitution from occurring, the ! character must be escaped (e.g., \!5 results in !5 and not in the history entry marked with 5).

5.1.3 Aliases and Alias Substitution

This section covers the alias and unalias commands.

After the input is split into words and history substitution is completed, alias substitution takes place. In alias substitution the first word in a line is matched against the alias table. If a match is found, the word is replaced by its alias. History substitution is performed on the alias before it replaces the first word.

An alias is added to the alias table with the alias command. Its syntax is

alias [$name$ [$alias$]]

If an alias for some name exists, it is replaced with a new one.

Examples:

```
* From now on disp behaves as echo
alias disp echo
* print Hello world.
disp Hello world.
```

If the second argument to alias is omitted, the alias for the given name is printed. If no arguments are given, the alias table is printed.

An alias is removed with the unalias command.

unalias $name$

History substitution is performed on alias before it replaces the first word. The following example defines an alias for the last command entered:

```
* Note that we escape the ! character so that history
* substitution is not performed before the alias is
* stored in the alias table.
alias last '!!'
```

Alias substitution is repeated until there is nothing left to be replaced by an alias. Recursive aliases are detected and an error is reported.

```
alias b a
alias a b
* This results in an error (alias recursion).
a
```

5.1.4   Control Lines and Statements

Every line is either a statement or a control line. Control lines begin with one of the following words: if, elseif, else, end, while, dowhile, repeat, break, and continue. Statements wrapped in control lines are control structures. Section 5.9 is devoted to control structures.

Control structures control the execution of statements.

```
* Print Hello world. 9 times

* A 'repeat' control structure.
* Next line is a control line.
repeat 9
  * The following line is a statement.
  echo Hello world.
* Next line is a control line.
end
```

5.1.5   Evaluation of Statements

In statements the first word is the command name and the remaining words are the arguments to the command. Optionally, every statement can have an output redirection specification at the end.

command_name arg1 arg2 ... [redirection]

Before possible redirection is handled, arithmetic substitution is performed followed by variable substitution.

The following is an example of arithmetic substitution:

```
echo {2+5}
```

First the expression in the curly braces is evaluated. Next the curly braces and the expression between them are replaced by the result of the evaluation. This results in

```
echo 7
```

Variable substitution is similar. It takes place after arithmetic substitution.

```
set a=Hello
echo $(a) world.
```

The echo command changes to

```
echo Hello world.
```

Arithmetic substitution is triggered by the { character. To disable it, the { character must be escaped.

```
* This prints: { within braces }
echo \{ within braces }
```

Variable substitution is triggered by the $ character. It can be disabled by escaping the $ character.

```
* This prints: Price: $(100)
echo Price: \$(100)
```

Arithmetic and variable substitution are described in more detail in the sections that follow. After arithmetic and variable substitution are completed, the redirection is handled. Several kinds of redirection are available.

Output redirection to a new file

command_name arg1 arg2 ... > file_name

If the file does not exist, it is created. If it already exists, its contents are cleared and the output of the command is written to the file.

Output redirection appending to a file

command_name arg1 arg2 ... >> file_name

If the file does not exist, it is created. If it already exists, the output of the command is appended to the file.

Output and error redirection to a file

command_name arg1 arg2 ... >& file_name

or

command_name arg1 arg2 ... >>& file_name

These two are the same as the previous two redirections, except that the error stream is joined with the output stream.

Example:

```
* An echo command outputs its arguments to the output stream
echo First second third
* This redirects the output to a file named out.txt.
* If the file exists its contents are lost. If the file
* doesn't exist it is created.
echo First second third > out.txt
* This appends the output to file out.txt. If the file
* doesn't exist it is created.
echo third fourth fifth >> out.txt
```

At this point the command is executed. If the command is not a NUTMEG internal command, it is treated as a script. The list of directories where NUTMEG searches for scripts is specified by the `sourcepath` variable. If a script with the corresponding name is found, it is executed. The script can retrieve the passed arguments from the `argv` variable. The number of passed arguments can be found in the argc variable. Running a script by typing

script_name arg1 arg2 ...

is equivalent to invoking a file with the `source` command

source *script_name arg1 arg2 ...*

See Sect. 5.5.1 on circuit and script input for more details on the `source` command and the usage of scripts.

5.2 Text Output

In this section the echo command is covered.

Displaying text
Any text can be forced to appear in the command window. The echo command can be used for this purpose. The syntax of the echo command is the following:
echo [-n] *text*
    The -n option instructs the echo command not to jump to a new line after it outputs the text.

```
* Print a line of text.
echo This is a line of text.

* This time we use the -n option to suppress the newline after echo.
* The result should be the same.
* Note how we explicitly add a space after every echo (double quotes).
echo -n "This "
echo -n "is "
echo -n "a line "
echo "of text."
```

5.3 Plots, Vectors, and Expressions

Plots are groups of vectors. By combining vectors, operators, and functions, expressions can be built.

5.3.1 Organization of Numeric Data

Numeric data in NUTMEG is organized as vectors. A vector is a one-dimensional collection of values. All values in a vector are of the same type, either real or complex. The number of values in a vector is the vector length. A vector also has a name by which the user can refer to it. Every value in a vector has an index associated with it. The first value has index 0, the second value index 1, etc.

    Vectors are grouped into plots. A plot with no vectors is an empty plot. Every plot has a name. All plots reside in memory and can be deleted, except for the const plot, which is always present. See Table 5.1 for the default contents of the const plot. The contents of the const plot can be changed, but the plot itself cannot be deleted. In NUTMEG there is always exactly one plot active (referred to as active plot or also current plot).

    Names of vectors and plots created by the user must follow the same rules as names of variables in C. A name cannot start with a number and it may contain English letters, numbers, and underscores. NUTMEG is case sensitive. All names in NUTMEG must be written in lowercase.

Table 5.1 NUTMEG con-
stants in the const plot

Constant	Explanation	Value
boltz	Boltzmann constant	$1.380620 \cdot 10^{-23}$
c	Speed of light	$2.997925 \cdot 10^{8}$
e	Base of natural logarithm	2.718282
echarge	Elementary charge	$1.602190 \cdot 10^{-19}$
kelvin	Absolute zero in °C	-273.15
planck	Planck constant	$6.626200 \cdot 10^{-34}$
pi	π	3.141593
i	Imaginary unit	i
false	Logical false	0
no	Same as false	
true	Logical true	1
yes	Same as true	

5.3.2 Plot Management

This section covers the setplot, destroy, destroyto, nameplot, and copyplot commands.

Listing all plots
To print the list of all plots, type
setplot
    An output that looks something like this is printed:

```
     new New plot
Current unknown1    Anonymous (unknown)
    const   Constant values (constants)
```

    From the above output one can see that there are two plots in the memory: const which is always present and unknown1. The active plot is the one named unknown1.

Creating a new plot
To create a new plot, type
setplot new
    A new plot is created, named unknown*number*, and made active.

Selecting the active plot
The plot that is at the lowermost position on the printed list is the oldest plot. The topmost plot (below new) is the newest plot. To change the active plot from the current active plot to its predecessor (previous plot), type
setplot previous
    To switch to the newer plot (next plot), type
setplot next
    If there is no previous or next plot, a warning is printed.

Deleting plots
To delete a plot and all its vectors, the destroy command is used.
destroy [*plot_name*]

If no additional arguments are specified, the active plot is destroyed. The next plot becomes active. If, however, there is no next plot, the previous plot becomes active. This is always possible as the oldest plot is the `const` plot, which cannot be destroyed.

To delete all plots (except the `const` plot) use

`destroy all`

All plots that are newer than the given plot can be destroyed with the `destroyto` command.

`destroyto [`*`plot_name`*`]`

If the *plot_name* is omitted, all plots newer than the current plot are destroyed. If the current plot is newer than the specified plot, the specified plot becomes the active plot after the `destroyto` command is finished. Otherwise, the current plot is not changed.

Renaming plots
The active plot can be renamed by means of the `nameplot` command.

`nameplot` *`newname`*

All plots, except for the `const` plot, can be renamed, provided that a plot with the same name as *newname* does not exist. Renaming of the active plot can also be achieved by writing the new name to the `curplot` variable.

Copying plots The `copyplot` command copies the contents of a plot into an existing or a new plot.

`copyplot` *`source`* *`destination`*

All vectors from the source plot are copied to the destination plot. If a vector exists in the destination plot, it is replaced. If the destination plot does not exist, it is created first.

5.3.3 Storing and Restoring Plots in Files

This section covers the `write` and `load` commands.

Storing plots in raw files
The contents of the active plot can be stored in a raw file with the help of the `write` command.

`write [`*`filename`*`]`

If the *filename* is omitted, the default value specified by the variable `rawfile` is used. A raw file can be either in ASCII or in binary format. ASCII files are more easily readable, whereas binary files are shorter. Switching between these two types can be achieved by setting the `filetype` variable to either `ascii` or `binary`.

A single raw file may contain multiple plots. Normally, writing to a raw file erases the old contents of the file. If one wants to append a plot to an existing raw file, the `appendwrite` variable must be set. This way the `write` command behaves differently if a file already exists. The old contents of the file are not discarded, and the plot that is being written is added at the end of the file.

```
* Set file type to binary.
set filetype=binary
* Go to plot op1 and write it to results.raw.
* This will erase the old contents of the file.
set op1
write results.raw
* Set append mode for write.
set appendwrite
* Append the op2 plot to results.raw.
setplot op2
write results.raw
```

Restoring plots from raw files

A plot (or multiple plots, if a raw file contains multiple plots) can be restored with the load command.

load [*filename*]

If the *filename* is omitted, the default name specified by the rawfile variable is used. The command loads all the plots that are stored in a file. The naming of the plots is done automatically. Therefore, it is recommended that the user rename the loaded plot(s) with the help of setplot previous and nameplot commands.

```
* File results.raw contains three plots.
* Load them and name the plots loaded1, loaded2, and loaded3.
* Load automatically recognises the file type and
* loads all of the plots stored in the file.
load results.raw
* Name the last plot loaded3.
nameplot loaded3
* Move to the second plot and name it loaded2.
setplot previous
nameplot loaded2
* Move to the first plot and name it loaded1.
setplot previous
nameplot loaded1
* Go to the last plot.
nameplot loaded3
```

5.3.4 Vectors

This section covers the let (regarding vectors and plots), setscale, display, settype, deftype, unlet, and print commands.

Accessing vectors in plots

A vector in the active plot can be accessed through its name. Take, for instance, frequency which accesses the vector named frequency in the active plot. To access vectors in other plots, use the following syntax:

plot_name . *vector_name*

Creating/changing a vector

A new vector in the current plot is created with the let command.

let *vector_name* = *expression*

A new vector can also be created in any other plot.

`let plot_name.vector_name = expression`

If the vector already exists, its value is replaced with the value of the *expression*. Otherwise, a new vector is created.

Changing individual vector components

`let vector_name[index] = expression`

If the expression is a single-valued vector, only a single component of vector *vector_name* is modified. If the expression evaluates to a vector with more than one component, its components are assigned to *vector_name*'s elements starting from *index*-th and leading up to the last element in *vector_name*.

```
* Create a vector
let a=(2;4;6;8;10)
* Change the third element to 9.
let a[2]=9
* Set the last two elements to 99.
let a[3]=(99;99)
* This affects only the last two elements.
* The length of a is not modified.
let a[3]=(98;98;98;98)
* a is now (2;4;9;98;98)
```

Default vector of a plot

Every nonempty plot has a default vector, also referred to as the default scale. In plots that are results of an analysis, this is usually a vector that contains swept values of a parameter, time, or frequency. In user-created plots the first vector added to a plot becomes the default scale.

The default scale of the current plot can be changed with the `setscale` command

`setscale [vector_name]`

Example:

```
* Create a new time scale, where time runs 2x slower
let mytime=time*0.5
setscale mytime
```

If the *vector_name* is omitted, the name of the default vector for the current plot is printed. The default vector of the current plot can also be changed by setting the `curplotscale` variable.

Displaying vector information

To display the type, length, and other properties of a vector, use the `display` command.

`display vector_name1 vector_name2 ...`

For every vector its name, type, domain (real or complex), and length are printed. If the vector is the default vector in the plot, a note (`default scale`) is printed.

To display the information on all vectors in the active plot, use the `display` command without arguments. This command also prints the short name (by which this plot is referred to in `setplot` and `destroy` commands), the full name, the title, and the date string of the current plot. A sample output of the `display` command for the `const` plot is in the next listing.

```
Here are the vectors currently active:

Title: Constant values
Name: const (constants)
Date: Sat Aug 16 10:55:15 PDT 1986

    boltz              : notype, real, 1 long
    c                  : notype, real, 1 long
    e                  : notype, real, 1 long
    echarge            : notype, real, 1 long
    false              : notype, real, 1 long
    i                  : notype, complex, 1 long
    kelvin             : notype, real, 1 long
    no                 : notype, real, 1 long
    pi                 : notype, real, 1 long
    planck             : notype, real, 1 long
    true               : notype, real, 1 long
    yes                : notype, real, 1 long [default scale]
```

Please do not be bothered by the fact that the date when the const plot was created is in 1986. Actually, this is true. First of all, the const plot is hardcoded into SPICE3, and second, that is just the time when SPICE3 was under development.

The short name of this plot is const, the full name is constants, the datestring is Sat Aug 16 10:55:15 PDT 1986, and the title of the plot is Constant values. The short name can be changed with the nameplot command or by writing to the curplot variable. The full name, the datestring, and the title can be changed by writing to the curplotname, curplotdate, and curplottitle variables.

Sorting of vectors by name can be turned off with the nosort variable.

Changing the type of a vector
The type of a vector can be changed with the settype command.
settype *type vec1 vec2* ...
    Example:

```
* Change the type of vectors a, b, and c to current.
settype current a b c
```

Defining new vector types
A type can be added to the list of types with the deftype command.
deftype [*name* [*abbreviation*]]
    If a type with name *name* already exists, it is modified. If no *abbreviation* is given, the abbreviation is not added (or removed if the type already exists).
    Example:

```
* Define new type named power with abbreviation W
deftype power W
* Change the abbreviation of current to Amps
deftype current Amps
```

Once a type is defined, it remains defined throughout the whole SPICE session and cannot be removed. It can be changed, however (i.e., the abbreviation can be changed).

The list of all defined types can be displayed by omitting the arguments to deftype. Some vector types are predefined in SPICE. The list of predefined types and their abbreviations is as follows:

```
SpiceOpus (c) 1 -> deftype
Name                   Abbreviation
----                   ------------
notype                 (no abbrev)
time                   s
frequency              Hz
voltage                V
current                A
onoise-spectrum        (V or A)^2/Hz
onoise-integrated      V or A
inoise-spectrum        (V or A)^2/Hz
inoise-integrated      V or A
pole                   1/s
zero                   1/s
s-param                (no abbrev)
```

If a type has no abbreviation, no abbrev is printed in place of the abbreviation. Note that if a predefined type (e.g., voltage) is changed, the change affects all vectors that are subsequently obtained as a result of some simulation.

Deleting a vector
A vector (or multiple vectors) can be deleted from the current plot with the unlet command.
unlet *vector_name1 vector_name2* ...

Printing a vector
The value contained by the vector can be printed with the print command.
print *vector_name1 vector_name2* ...
   If all vectors listed as print arguments are of length 1, the output looks like

```
vector_name1 = value1
vector_name2 = value2
...
```

For all other cases the output is in the form of a table. By default, the vector component index and the default vector are also printed. The default behavior of the print command can be changed with the set command. See Sect. 5.4 for details on noprintheader, noprintscale, noprintindex, nobreak, width, and height.

5.3.5 Expressions

Expressions appear on the right-hand side of the assignments in the let command. They also appear as arguments to the print, cursor, plot, fourier, and spec commands. Expressions that appear in the netlist in curly braces (parametrization of elements, models, and subcircuits) are also NUTMEG expressions.

NUTMEG expressions comprise constants, vector names, operators, and function invocations. Constants in NUTMEG expressions are written in the same manner as constants in the circuit description (e.g., 8.6k, 8.6e3, ...). Vectors were the subject of Sect. 5.3.4. Operators and functions are the topic of Sects. 5.3.6–5.3.8.

5.3.6 Operators in NUTMEG Expressions

There are one unary (sign change) and six binary (addition, subtraction, multiplication, division, power, and remainder) arithmetic operators available for NUTMEG expressions.

```
* Calculate -(-5).
print -(-5)
* Calculate 5*9+40-2.
print 5*9+40-2
* What is remainder when 18 is divided by 5?
print 18%5
* Cube root of 27.
print 27^(1/3)
```

Complex numbers can be constructed using the complex constructor operator (comma) by specifying the real and imaginary parts of the number separated by a comma.

```
* Multiply 2+3i by 4+5i.
print (2,3)*(4,5)
```

Relational operators (equal, not equal, less than, greater than, less than or equal to, greater than or equal to) compare two values and produce true (1) or false (0).

```
* Is 2*8 greater than 3*3?
print (2*8 gt 3*3)
```

Logical operators (negation, conjunction, disjunction) consider all operand values that are not zero as true. Zero is considered as false. The result is 0 (false) or 1 (true).

```
* not (1 and (0 or 1))
print (not (1 and (0 or 1)))
```

Two vectors can be joined with the vector join operator. This operator also enables you to construct a vector from its values.

```
* Construct vector x with components 2,3,4,0,9,5.
let x=(2;3;4;0;9;5)
* Join x with components 5,5.
let x=(x;5;5)
```

The vector component selection operator can be used for extracting components or subvectors from a vector. If the index is a real value, a single component is returned. For a complex index, the range of components starting at the real part of the index and ending at the imaginary part of the index is returned.

```
* Construct vector x with components 2,3,4,0,9,5.
let x=(2;3;4;0;9;5)
* Select the first component.
print x[0]
* Select the range of components from third to fifth component.
* Note that the index is complex.
print x[2,4]
```

The vector interpolation operator is used in conjunction with the cursor command. It returns the vector component that corresponds to a fractional index. Linear interpolation is used.

```
* Create a vector.
let x=(2;3;4;0;9;5)
* What value corresponds to the middlepoint between
*   the second and the third component (index 1.5)?
print x[%1.5]
```

The complete list of operators is given in Table 5.2. The operator precedence is in Table 5.3. Operator precedence can be changed by the use of parentheses.

All operators except for the vector join, vector component selection, and vector interpolation operate on vectors. The operation is performed on every element of a vector. For binary operators this means that both operands must be of the same length, or one of them must have exactly one element. In the latter case, the operand

Table 5.2 NUTMEG operators

Operator	Explanation	Example
unary $-$	Sign change	$-5 \rightarrow -5$
+	Addition	$3+5 \rightarrow 8$
$-$	Subtraction	$9 - 3 \rightarrow 6$
*	Multiplication	$3*4 \rightarrow 12$
/	Division	$9/4 \rightarrow 2.25$
^	Power	$9^0.5 \rightarrow 3$
%	Remainder	$8\%3 \rightarrow 2$
,	Complex constructor	$(2,3) \rightarrow 2 + 3i$
eq	Equal	$2 \text{ eq } 3 \rightarrow 0$
ne	Not equal	$2 \text{ ne } 3 \rightarrow 1$
gt	Greater than	$2 \text{ gt } 3 \rightarrow 0$
lt	Less than	$2 \text{ lt } 3 \rightarrow 1$
ge	Greater than or equal to	$2 \text{ ge } 3 \rightarrow 0$
le	Less than or equal to	$a \text{ le } b \rightarrow 1$
unary not	Logical negation	$\text{not } 1 \rightarrow 0$
and	Logical conjunction	$5 \text{ and } 0 \rightarrow 0$
or	Logical disjunction	$5 \text{ or } 0 \rightarrow 1$
;	Vector join	(a;b)
[]	Vector component selection	a[9]
[%]	Vector interpolation	a[%9.5]

Table 5.3 NUTMEG operator precedence from highest to lowest. Operators in the same row have the same precedence

Operator
[]
[%]
unary $-$
^ * / %
+ $-$
;
,
eq ne gt lt ge le
not
and
or

with one element is expanded to the length of the other operand. The elements of the expanded vector are all equal to the first element.

```
* Create two vectors.
let x=(5;8;12;99)
let y=(1;2;1;2)
* Multiply the two vectors resulting in (5;16;12;198).
print x*y
* Divide every element in x by 5.
print x/5
```

5.3.7 NUTMEG Built-in Functions

A complete list of NUTMEG built-in functions is given in Table 5.4.

Functions abs(x), mag(x), magnitude(x) return the magnitude of the components of a complex vector. For real vectors these functions return the absolute values of the components. Function db(x) returns the $20 \log(|x|)$ values of the components of a complex vector.

```
* Magnitude of 3+4i (equals 5).
print mag(3,4)
* dB value for 1000 (returns 60).
print db(1000)
```

Functions phase(x) and ph(x) return the arguments of the components of a complex vector. For real vectors the return value is 0. The angle is returned in radians by default. If the units variable is set to degrees the angle is returned in degrees. The angle wraps around at π radians. To unwrap it, use the unwrap(x) function. The argument must change in a sufficiently slow manner between individual components of a complex vector for the unwrap() function to detect a wraparound and correct it.

```
* Phase of 1+i (returns pi/2).
print phase(1,1)
* Switch units to degrees
set units=degrees
* This should return 45.
print phase(1,1)
* Switch back to radians.
set units=radians
* Suppose we calculate the phase of a and it is wrapped.
* This is how to fix it.
let unwrapped=unwrap(phase(a))
```

Functions re(x), real(x), im(x), and imag(x) return the real and imaginary parts of a complex vector. Function j(x) returns a complex vector multiplied by the imaginary unit.

```
* Print 3+4i
print (3,4)
* Print the real part of (3+4i)*i (returns -4).
print real((3,4)*(0,1))
* Multiply 4+5i by i (equals -5+4i).
print j(4,5)
```

Functions ln(x), log(x), and log10(x) return the natural and base 10 logarithms of a vector x. Function exp(x) returns e raised to the power specified by the

Table 5.4 NUTMEG built-in
functions

Function	Explanation
abs(x)	Magnitude of (x)
mag(x)	Same as abs(x)
magnitude(x)	Same as abs(x)
db(x)	$20\log(|x|)$
ph(x)	Argument of x
phase(x)	Same as (ph(x))
unwrap(x)	Unwrap phase
real(x)	Real part of (x)
re(x)	Same as (real(x))
imag(x)	Imaginary part of (x)
im(x)	Same as (imag(x))
j(x)	Multiplies (x) by i
ln(x)	Base e logarithm of x
log(x)	Base 10 logarithm of x
log10(x)	Same as log(x)
exp(x)	e^x
sqrt(x)	Square root of x
sin(x)	Sine of x
cos(x)	Cosine of x
tan(x)	Tangent of x
atan(x)	Inverse tangent of x
floor(x)	Largest integer below x
ceil(x)	Smallest integer above x
round(x)	Integer closest to x
length(x)	Length of x
mean(x)	Mean value of components of x
sum(x)	Sum of components of x
min(x)	Smallest value in x
max(x)	Largest value in x
vector(x)	Vector $0,1,2,\ldots,x$
unitvec(x)	Unit vector (all ones)
rnd(x)	Random integer $[0, \lfloor x \rfloor - 1]$
rndunif(x)	Uniformly distributed random number
rndgauss(x)	Normally distributed random number
interpolate(x)	Interpolate vectors from other plots
deriv(x)	Numeric derivative of a vector
integrate(x)	Numeric integral of a vector
timer(x)	CPU time in seconds from x
clock(x)	Wall clock time in seconds from x
area(x)	MOS transistor area

components of x. The sqrt(x) function returns the square root of the components of x. All of them take a complex or a real argument.

```
* Convert 85dB into voltage gain
print 10^(85/20)
* Now calculate the db value for voltage gain factor 3500
print 20*log10(3500)
```

Functions sin(x), cos(x), and tan(x) return the sine, cosine, and tangent of vector x. The atan(x) function returns the inverse tangent of the argument. The assumed unit is radians. If the `units` variable is set to `degrees`, the assumed unit is degrees.

```
* Sine at pi/2.
print sin(pi/2)
* Now switch to degrees.
set units=degrees
* Sine at 90 degrees.
print sin(90)
* Inverse tangent at 1
*    (returns 45 since units are set to degrees).
print atan(1)
* Switch back to radians.
set units=radians
```

Functions `floor(x)`, `ceil(x)`, and `round(x)` return integers associated with vector x. The first one returns the largest integer below x, the second one the smallest integer above x, and the last one the integer closest to x. For complex vectors these functions operate on the real and the imaginary parts independently.

```
* Floor and ceil of 1.5 (should be 1 and 2).
print floor(1.5) ceil(1.5)
* Rounding. 1.5 gets rounded to 2, whereas
*    1.4 gets rounded to 1.
print round(1.5) round(1.4)
```

Functions `length(x)`, `mean(x)`, and `sum(x)` return the number of components in x, the mean value of the components, and the sum of components in x, respectively. Functions `min(x)` and `max(x)` return the smallest and largest component in x. For complex vectors, these two functions return the component with the largest magnitude.

```
* Construct a vector.
let a=(9;4;5)
* Length is 3.
print length(a)
* Mean is 6 and sum is 18.
print mean(a) sum(a)
* Smallest component is 4 and the largest is 9.
print min(a) max(a)
```

Functions `vector(x)` and `unitvec(x)` return a vector with x components. The former function constructs a vector with integer components starting at 0 and growing to $x - 1$. The latter returns a vector whose components are all equal to 1. If x has more than one component, the first component is used for the length of the resulting vector. If x is complex, the magnitude of x is used.

```
* Construct a vector 0,1,2,...,9
let a=vector(10)
* Construct a vector consisting of 10 ones.
let b=unitvec(10)
```

The `rnd(x)` function returns a random integer from the interval $[0, \lfloor x \rfloor - 1]$. If x has multiple components, a separate range applies to every component of the resulting vector which has the same length as x. For complex arguments the real and the imaginary parts are treated as two independent range specifications for the real and the imaginary parts of the resulting vector.

```
* Uniformly distributed random integer from 1 to 10.
print rnd(10)+1
```

The rndunif(x) function returns a uniformly distributed real number from the interval $[a - d, a + d]$, where $d = \text{real}(x)$ and $a = \text{imag}(x)$. If x has multiple components, the result is a vector of the same length where every component is chosen randomly from a distribution specified by the component of x with the same index.

The rndgauss(x) function is similar to rndunif(x), except that the distribution is normal. The real part of x represents the standard deviation σ of the distribution. The distribution is truncated at $k\sigma$, where k is specified by writing to the gausstruncate variable. The truncation of the distribution forces a random number to be generated again if it falls outside $k\sigma$. The truncation factor must be positive. Initially, it is set to 10^{300}, which effectively means that no truncation is applied.

```
* Uniformly distributed number from [-5,5].
print rndunif(5)
* Uniformly distributed number from [4,6].
print rndunif(1,5)
* A set of 10 uniformly distributed numbers from [4,6].
print rndunif(unitvec(10)*(1,5))
* Normally distributed, sigma=5, mean=90.
print rndgauss(5,90)

* Enable truncation to 3 sigma.
set gausstruncate=3
* Normally distributed, sigma=5, mean=90.
*    Gets truncated at 3 sigma = 15.
*    All values should be between 75 and 105.
print rndgauss(5,90)
* Turn off truncation.
set gausstruncate=1e300
```

The initial state of the random generator when SPICE starts is always the same (-1). To reset the random generator to its initial state, write -1 to the rndinit variable. A positive value in the rndinit variable means that the random generator is no longer in its initial state. The initial state can be different than the default -1. In fact, any negative number can be written to rndinit. Note that the value of the rndinit variable does not reflect the state of the generator.

The random number generator is used by the rnd(x), the rndunif(x), and the rndgauss(x) functions. As the default initial state is always the same, the user can be assured that the same sequence of pseudo-random numbers is generated in all SPICE sessions.

The interpolate(x) function is used for interpolating vectors from other plots to the default scale of the current plot. It is best explained by the following example:

```
* Create a new plot, name it source.
setplot new
nameplot source
* Create scale (a) and vector (b).
* Note that the first vector created
*    in a new plot is the scale.
let a=(1;2;3;4;5)
let b=(10;20;30;40;50)
* Create the destination plot, name it dest.
setplot new
nameplot dest
* Create destination scale (x).
let x=(1.5;2.5;3.5;4;5)
```

```
* Interpolate vector b from plot source to
* current plot where scale x applies.
let y=interpolate(source.b)
* y is now (15;25;35;40;50)
```

Both scales must be strictly monotonic and real, and the source vector must be real. The degree of the interpolating polynomial can be set through the `polydegree` variable. The default value is 1 (linear interpolation).

The `deriv(x)` function returns the derivative of the argument. Interpolation is used for obtaining the derivative. The derivative is with respect to the scale of the plot from which x originates and is calculated at the respective scale points. The degree of the interpolating polynomial is specified by the `dpolydegree` variable. The same restrictions apply to the scale and the vector as in the `interpolate(x)` function.

```
* Create a new plot, name it source.
setplot new
nameplot source
set dpolydegree=2
* Create scale (a) and vector (b).
let a=(1;2;3;4;5)
let b=(10;20;30;20;10)
* Calculate derivative of b with respect to a.
let db=deriv(b)
* db is now (10;10;0;-10;-10)
```

The `integrate(x)` function integrates the argument with respect to its respective default scale vector. The resulting vector contains the values of the definite integral from the beginning of the default scale to the respective default scale component. The function uses interpolation for evaluating the integral. The degree of the interpolation polynomial is specified by the `polydegree` variable. The integral is evaluated at the respective scale points. The same restrictions apply to the scale and the vector as in the `interpolate(x)` function.

```
* Create a new plot, name it source.
setplot new
nameplot source
set polydegree=2
* Create scale (a) and vector (b).
let a=(1;2;3;4;5)
let b=(10;10;0;-10;-10)
* Calculate integral of b with respect to a.
let ib=integrate(b)
* ib is now (0;10;15;10;0)
```

Functions `timer(x)` and `clock(x)` return the CPU time, and the wall clock time in seconds decreased by the value of x. If x is complex, the real part is used. If x is a vector, only the first component is used.

```
* Remember initial CPU and wall clock time.
let cputime=timer(0)
let walltime=clock(0)
* Do something here
...
* Calculate elapsed times in seconds.
let cpuelapsed=timer(cputime)
let wallelapsed=clock(walltime)
```

The `area(x)` function returns the total area occupied by the MOS transistors. If x is 0, the total MOS area for the current circuit is returned (in square meters). The area occupied by a MOS transistor is calculated as M × (W × L + AD + AS). The channel dimensions are given by the W and the L MOS parameters. The drain and the source areas are given by the AD and the AS MOS parameters.

5.3.8 User-Defined Functions

This section covers the define and undefine commands.

New functions can be defined by means of the define command.

define *function_name*(*arg1*, *arg2*, ...) (*expression*)

The expression may contain arguments and vectors from the const plot. Example:

```
* Define it.
define limitlow(x,knee) ((x lt knee)*x+(x ge knee)*knee)
* Use it.
print limitlow(10,50) limitlow(90,50)
```

For printing the list of defined functions the define command is used without arguments.

undefine deletes a function definition.

undefine *function_name*

Example:

```
* Delete definition of limitlow.
undefine limitlow
```

5.3.9 Arithmetic Substitution

NUTMEG provides a mechanism that automatically converts the result of an expression into a string which in turn becomes a part of a NUTMEG command before that command is interpreted by NUTMEG. Expression substitution is invoked with curly braces.

{*expression*}

Here is an example of expression substitution:

```
echo Two plus five is {2+5}.
```

After NUTMEG finishes with expression substitution, the above command becomes

```
echo Two plus five is 7.
```

Finally, the command is interpreted, and the result is the following output:

```
Two plus five is 7.
```

Expression substitution is handy when the result of some expression needs to be passed as an argument to a command that does not take expressions for arguments. An example of such command is set.

```
* Set temperature to t1+50.
set temp={t1+50}
* This would try to set the temperature to a string
* 't1+50' which, of course, would fail.
set temp=t1+50
```

If an expression results in a vector, the substitution returns a NUTMEG expression that constructs that particular vector.

```
* This prints: (0;1;2;3;4;5;6;7;8;9)
echo {vector(10)}
```

5.4 Variables

Variables in NUTMEG are means for storing data and controlling NUTMEG and simulator behavior. This section explains the set, unset, and shift commands.

5.4.1 Variable Basics

There exists a single global namespace for variables. A variable has one of the following types: bool, integer, real, string, or list. integer and real variables are considered numeric.

The value of a variable can be set using the set command.

set *variable_name* [= *value*]

A variable is deleted from the variable namespace with the unset command.

unset *variable_name1 variable_name2* ...

One can set a bool variable by issuing a set command with no variable value. This automatically sets the variable's value to true. A bool variable is set to false by deleting it with the unset command.

```
* Set the appendwrite variable to true
set appendwrite
* Now reset it to false
unset appendwrite
```

Numeric variables require a value when they are created or changed with the set command.

```
* Set fourgridsize to 1024
set fourgridsize=1024
```

String variables are created or changed in a similar manner.

```
* Set filetype to ascii
set filetype=ascii
```

List variables are specified with parentheses.

set *variable_name* = (*value1 value2* ...)

Note the extra space after the open parenthesis and before the closed parenthesis. List variables are collections of numeric and string values.

```
* Set sourcepath to point to two directories
*   . and
*   /home/public/circuits
set sourcepath = ( . /home/public/circuits )
```

The shift command removes n elements from the beginning of a list variable.

shift *variable_name* [*n*]

n must be positive. If it is 0, no element is removed. If n is omitted, one element is removed. The variable must not be a read-only variable. An exception is the argv variable. By applying the shift command to the argv variable, the value stored in the argc variable decreases automatically so that it reflects the new state of argv.

```
* Create a list variable.
set a = ( one two three four )
* Remove the first two elements.
shift a 2
* a is now ( three four )
```

To print the complete list of variables and their values, use the set command without arguments. The printout looks something like this:

```
SpiceOpus (c) 9 -> set
+ abstol    1e-011
+ chgtol    1e-013
  color0    r255g255b255
  color1    r165g165b165
  color2    r127g063b000
  color3    r191g000b000
  color4    r223g159b000
* curplot   const
* curplotdate    Sat Aug 16 10:55:15 PDT 1986
* curplotname    constants
* curplottitle   Constant values
+ gmin  1e-011
  history   100
  linewidth 0
* plots ( const )
  plottype  normal
  plotwinheight 360
  plotwininfo
  plotwinwidth 360
  program   SpiceOpus (c)
  prompt    SpiceOpus (c) ! ->
  sourcepath    ( . c:/SpiceOpus/lib/scripts )
```

In the output the simulator parameters are marked with +. Special variables are marked with *. These variables depend on the state NUTMEG is in (i.e., active plot). Only those simulator parameters that were set with the .options netlist directive or that are the result of set commands issued on the current circuit will appear.

If you are bothered by the contents of the curplotdate variable (1986) in the above output, note that the output was produced when the const plot was active. For more details on the curious date stamp of the const plot, see Sect. 5.3.4.

5.4.2 NUTMEG Control Variables and Simulator Parameters

When a set command is encountered, the variable is first checked to see if its name matches the name of some simulator parameter. If it does, a check is performed to test if a circuit is loaded. If not, an error is reported. If there is a circuit loaded, the corresponding simulator parameter is set or changed for that circuit.

A similar check is performed when an unset command is encountered. Be careful with the unset command and simulator parameters. The value of an unset simulator parameter becomes 0.

```
* Set integration method to gear
set method=gear
* Set temperature to 50 degrees centigrade
set temp=50
* This sets the temperature to 0.
unset temp
* A better way would be:
set temp=0
```

In addition to simulator parameters, NUTMEG also recognizes several variables, as listed in Table 5.5. These variables set the behavior of various NUTMEG commands.

5.4.3   Variable Substitution and Text Input

Variable substitution provides the means for including the value of a variable in a NUTMEG command. Variable substitution is performed whenever the following pattern is encountered:

$(*descriptor*)

The simplest *descriptor* is the variable name (e.g., the one used with the set command). An example of variable substitution is given in the next listing. Note that the example works only if a circuit is loaded (since the method simulator parameter is being accessed).

```
echo Integration method before set: $(method)
set method=gear
echo Integration method after set: $(method)
```

The translation of the first echo command depends on the integration method that is set for the current circuit. The second echo command is translated to

```
echo Integration method after set: gear
```

which finally results in the following output:

```
Integration method after set: gear
```

In order to query whether a variable exists or not, a descriptor in the form of *?variable_name* can be used. It expands to 1 if the variable is defined; otherwise, the expansion is 0.

```
* Check if there is a temperature setting for the current circuit.
if $(?temp)
  echo Circuit temperature set to $(temp).
else
  echo Circuit has default temperature.
end
```

Table 5.5 NUTMEG special variables

Variable	Type	Meaning
aperiodic	Bool	Switches to aperiodic transform (spec)
appendwrite	Bool	Appends plot data if file exists (write)
argc	Integer	Read-only number of script Command line arguments
argv	List	Read-only list of script Command line arguments
badcktstop	Bool	Skip .control block execution if circuit Has errors (source)
badcktfree	Bool	Release the circuit from memory if it Contains errors (source)
color0	String	Background color for plot windows
color1	String	Grid color for plot windows
color$n$	String	Curve color for $n - 1$-th curve
circuits	String	Read-only list of circuits in memory
curcirc	String	Name of the current circuit Writing to this variable changes the Active circuit if a valid name is written.
curcirctitle	String	Title of the current circuit
curplot	String	Name of the current plot Writing to this variable changes the Active plot if a valid name is written.
curplotdate	String	Date string of the active plot
curplotname	String	Full name of the active plot
curplotscale	String	Default scale of the current plot
dpolydegree	Integer	Interpolating polynomial degree (deriv(x))
filetype	String	File type for write (ascii or binary)
forcerebuild	Bool	Forces circuit rebuild after change (netclass)
fourgridsize	Integer	Number of interpolated points (fourier)
gausstruncate	Real	Normal distribution truncation factor
history	Integer	Maximal number of commands in the history
iplottag	String	Plot window tag prefix for iplot
keepwindow	Bool	Keep the window in __specwindow For spec (spec)
length	Integer	Number of rows in a page for print
linewidth	Integer	Line width for the plot (0 is hairline)
magunits	String	Magnitude units for plot (normal, db20, or db10)
manualscktreparse	Bool	Turns on manual subcircuit reparsing
nfreqs	Integer	Number of harmonics for fourier
noaskquit	Bool	Don't ask user for confirmation (quit)
nobreak	Bool	Don't insert page breaks (print)
nogrid	Bool	Don't draw a grid in plot windows (plot)
nosort	Bool	Don't sort vectors (display)
noprintheader	Bool	Don't print header (print)
noprintindex	Bool	Don't print indices (print)

(continued)

Table 5.5 (continued)

Variable	Type	Meaning
noprintscale	Bool	Don't print default scale (print)
numdgt	Integer	Number of printed digits when printing numbers
plots	List	Read-only list of names of plots in memory
plottype	String	Plot type for plot (point, line, or comb)
pointtype	String	Point type for point-style plots (o, +, x, or d)
pointsize	Integer	Size of point for point-style plots
plotwinwidth	Integer	Width of plot window in pixels (plot)
plotwinheight	Integer	Height of plot window in pixels (plot)
plotwininfo	Bool	Show info frame in plot window (plot)
plotautoident	Bool	Turn on automatic vector identification (plot)
polydegree	Integer	Degree of interpolating polynomial (interpolate(x) and integrate(x))
prompt	String	Prompt string for the command window The ! character is replaced by the event number \ acts as an escape character
rndinit	Integer	Random generator seed
rawfile	String	Default filename for write
sourcepath	List	List of directories where spice looks for files (source, .include, and .lib directives
specwindow	String	Window type for spec
specwindoworder	Integer	Window order for spec
units	String	Unit for phase and angle (in expressions and plot) (degrees or radians)
width	Integer	Number of columns in a page for print
workdir	String	Current working directory (read-only)

If some variable is a list, the number of entries in the list can be obtained with the #*variable_name* descriptor. The substitution yields 0 if the variable is not found.

```
* How many directories are in the sourcepath?
echo There are $(#sourcepath) entries in the sourcepath list.
```

An individual entry can be retrieved from a list variable with the *variable_name* [*index*] descriptor. The first entry has index 0 and the last one $(#*variable_name*)−1. If the *index* exceeds this range the substitution results in an empty string.

```
* Set the sourcepath.
set sourcepath = ( . /home/public/circuits )
* Retrieve the second entry
echo Second sourcepath entry: $(sourcepath[1])
* This results in an empty string
echo Third sourcepath entry: $(sourcepath[2])
```

If the variable is not a list variable, indexing can still be used. Note, however, that the #*variable_name* descriptor results in 1 for such variables. So the only index that does not result in an empty string is 0.

Variable substitution can also be used to prompt the user for input. The −
descriptor prompts the user for a string. This string replaces the $(−) substitution
directive.

```
* Print a prompt message, do not jump to new line.
* Quotation marks are used to force the output of a trailing whitespace.
echo -n 'Enter a number: '
* Read and store the number in vector a.
let a=$(-)
* Print the result (arithmetic substitution is used).
echo You entered: {a}
```

Variable substitution takes place after arithmetic substitution. This way indexing
can be done with expressions.

```
* Print the (8-4-4)th plot name.
echo $(plots[{8-4-4}])
```

First the arithmetic substitution is performed. The expression is evaluated and
the result is 0. The echo command becomes

```
echo $(plots[0])
```

Next the variable substitution takes place and the echo command changes to

```
echo const
```

Finally the echo command is executed and the string const is printed.

Variable substitution makes it possible to read the environmental variables from
the environment of the running SPICE OPUS binary. If a variable is not recognized
as a SPICE variable, the environment is checked. If an environmental variable with
the given name is found, the substitution results in that variable's value.

```
* Print the operating system's path.
echo $(path)
```

Variable substitution for list variables omits the parentheses around the list.

```
set a = ( first second fifth )
echo $(a)
* ... will result in: first second fifth
* without parentheses
```

This option allows for appending entries to a list.

```
* Set a to: x y
set a = ( x y )
* Append z to a
set a = ( $(a) z )
* Prepend axis to a
set a = ( axis $(a) )
* a is now: axis x y z
```

5.5 Circuits

This section covers the source, setcirc, namecirc, delcirc, listing, show,
showmod, netclass, reset, and scktreparse commands.

5.5.1   Circuit and Script Input (source *Command*)

Circuits and scripts are stored in input files. An input file is read in with the source command.

source *filepath arg1 arg2* ...

SPICE first tries to open the file where *filepath* is assumed to be an absolute path to the file. If this fails, all directories specified by the sourcepath variable are tried.

When the file is read, the circuit is constructed from the circuit description found in the file. The circuit is given a name of the form ckt*n* where *n* is a unique number. The first line of the input file becomes the circuit's title.

The commands from the .control block are executed after circuit construction is finished. The additional arguments to the source command are stored in variable argv in the form of a list. The argc variable gives the number of additional arguments. Both argc and argv are read-only. An exception is the shift command which can be applied to argv. When argv is shifted, argc is adjusted accordingly.

If the badcktstop variable is set and an error occurs during the construction of the circuit, processing of the input file stops here. Otherwise, the commands in the .control block are executed.

If the badcktfree variable is set, the circuit is deleted from the memory if errors were encountered during the construction of the circuit.

Example:

```
This is a simple circuit

v1 (1 0) dc=10
r1 (1 0) r=1k

.control
* Perform an operating point analysis.
op
* Print the voltage at node 1.
print v(1)
.endc

.end
```

When the above example is stored in file sample.cir and the following command is issued in the command window:

```
source sample.cir
```

the file sample.cir is read. A circuit comprising a voltage source and a resistor is constructed. The circuit is named ckt1 (if this is the first circuit loaded in the current SPICE session) and becomes the current circuit. The title of the circuit is set to This is a simple circuit. Next the commands in the .control block are executed performing an operating point analysis and printing the resulting voltage.

After all of the commands from the .control block have been executed, the circuit remains in the computer memory and the user is presented with the prompt.

Input files that start with the # character are treated differently. They are interpreted as scripts. The first line of such a file is ignored. An example of such a file is given below.

```
# This is a script. This line is ignored.
* Print a message
echo hello world
* Print argc and argv
echo argc = $(argc)
echo argv = ( $(argv) )
```

The contents of such a file must follow the same rules as the contents of a
.control block.

5.5.2 Circuit Management

The list of circuits in computer memory can be obtained with the setcirc command. Example:

```
SpiceOpus (c) 9 -> setcirc
Current (ckt4) Demo circuit
        (ckt3) Inverter
        (ckt2) NAND gate
        (ckt1) amplifier
```

The current circuit is marked with Current. To change the current circuit the
setcirc command can be used.

setcirc *circuit_name*

The words previous and next switch to the older and the newer circuit, respectively. In the above example next results in an error message (because ckt4 is the
newest circuit) and previous switches the current circuit to ckt3. setcirc ckt2
switches to the NAND gate circuit.

The name of the current circuit is available through the curcirc variable. By
writing to this variable, the current circuit can be switched in the same manner as
with the setcirc command. The circuits variable is read-only and contains a list
of names for all circuits in computer memory. The current circuit's title is available
through the curcirctitle variable. If something is stored in this variable by using
a set command, the stored value becomes the title of the current circuit.

```
* Change the title of the current circuit
set curcirctitle='This is the new title'
```

The current circuit can be renamed with the namecirc command.

namecirc *new_name*

new_name must be unique.

```
* Rename ckt1 to amp1 and make it active.
setcirc ckt1
namecirc amp1
```

The delcirc command deletes a circuit.

delcirc *circuit_name1* *circuit_name2* ...

If no circuit names are specified, the current circuit is deleted. To delete all circuits from the memory, type delcirc all.

5.5.3 *Circuit Listing*

The listing command prints information about the current circuit. All examples
regarding the listing command will be based on the following netlist:

```
Demo netlist

.subckt vdiv in com out param: rtotal=1k fact=0.5
.param r1val=(1-fact)*rtotal
.param r2val=fact*rtotal
r1 (in out) r={r1val}
r2 (out com) r={r2val}
.ends

x1 (10 20 0) vdiv param: rtotal=10k
x2 (50 90 0) vdiv param: rtotal=10k fact=0.8

.end
```

listing logical prints only the lines that belong to the circuit description of
the current circuit (the .control block is omitted). listing physical prints the
complete input file. listing deck is the same as listing physical, except that
no line numbers are printed.

Subcircuit definitions in the current circuit can be printed with
listing subdef [*name1 name2* ...]

If no subcircuit definition names are given, the list of all subcircuit definition
names is printed. The main circuit is also regarded as a subcircuit with definition
name topdef_. For every definition the list of instances is printed, along with the
terminals, parameters and their default values, parametric expressions (.param),
and all elements (netlist lines) that constitute the definition.

Example:

```
SpiceOpus (c) 17 -> listing subdef
Active subcircuit definitions:
  vdiv
  topdef_
----
SpiceOpus (c) 18 -> listing subdef vdiv
Instances of vdiv :
  x2
  x1

Definition of vdiv :
  Terminals:
    in com out
  Parameters:
    rtotal = 1000
    fact = 0.5
  Parametric expressions:
    r1val = (1-fact)*rtotal
    r2val = fact*rtotal
  Elements:
    r1 (in out) r={r1val}
    r2 (out com) r={r2val}
----
```

The information on individual subcircuit instances in the current circuit can be
obtained with
listing sub [*name1 name2* ...]

If no names are given, the complete list of subcircuit instances for all subcircuit definitions is printed. The instantiation of the main circuit's definition is the main circuit itself. The name of this instantiation is xtopinst_. For every requested subcircuit instance, the place of its instantiation, connections, and parameter values are printed. If a parameter's value is taken from the subcircuit definition (the parameter is not specified at instantiation), its value is marked with (dfl).

Example:

```
SpiceOpus (c) 19 -> listing sub
Subcircuit instances of vdiv:
  x2
  x1

Subcircuit instances of topdef_:
  xtopinst_

SpiceOpus (c) 20 -> listing sub x1

Subcircuit instance x1 :
  Definition : vdiv
  Instantiated in top level circuit
  Connections (model -> instance) :
    in -> 10
    com -> 20
    out -> 0
  Parameters :
    rtotal = 10000
    fact = (dfl) 0.5
```

The hierarchy of inclusion for the current circuit (obtained with the .include and .lib input file directives) can be printed with the listing hierarchy command. listing global prints the list of global nodes (.global netlist directive) in the current circuit (the ground node is omitted).

listing activenc prints the active netclass blocks for all netclass groups. The consecutive number of the netclass group and active block is printed in parentheses.

```
SpiceOpus (c) 27 -> listing activenc
Active netclasses (2):
  (0) r1 : (0) normalr
  (1) c1 : (0) normalc
```

In the above example the circuit has two netclass groups named r1 and c1. Both of them are set to the first block in the respective group named normalr and normalc, respectively.

5.5.4 Printing Parameter Values of SPICE Primitives

The values of SPICE primitive instances and models can be printed with the show and showmod commands.

The show command displays the selected instance parameters for selected instances.

show key1 key2 ... [: par1 par2 ...]

Instances are specified with keys. A key is either an instance name, a model name prefixed with #, or the word all. If a model name is specified, all instances that are using the specified model are printed. If the word `all` is encountered, all instances are printed.

If no parameters are specified, only the most important ones are printed. The word `all` expands to all instance parameters.

Example:

```
* This prints the most important instance parameters of r1.
show r1
* Print instance parameters r and tc1 for r1.
show r1 : r tc1
* Print all instance parameters for r1.
show r1 : all
* Print all instance parameters for all resistors
* using model named rmod.
show #rmod : all
* Print all instance parameters for all instances.
print all : all
```

The `showmod` command prints model parameters.

`showmod key1 key2 ... [: par1 par2 ...]`

Here the key specifies the instance for which the corresponding model's parameters are printed. By prepending the name with a # character the name represents a model name. The word `all` has the same meaning as in `show`.

Example:

```
* Most important model parameters of r1's model.
showmod r1
* All model parameters of r1's model.
showmod r1 : all
* All model parameters of model rmod.
showmod #rmod : all
* All model parameters of all models.
showmod all : all
```

5.5.5 Selecting the Topology and Rebuilding the Circuit

SPICE offers limited facilities for changing the circuit without reloading it from an input file. In an input file blocks of circuit description can be defined. A block starts with

`.netclass group_name block_name`

and ends with

`.endn`

Of all the blocks that belong to the same group (i.e., have the same *group_name*), only one block can be active. Active blocks are the basis for building the circuit in computer memory. Lines that are outside `.netclass` blocks are always active. When the input file is read and the circuit is built, the active block in every group is the one that appears first in the input file. The active block can be switched without changing the input file and reloading it. The command that switches the active block of a group is the `netclass` command.

`netclass select group::block`

The group can be specified with the name or the number. Groups are enumerated in the same order as they appear in the circuit description. Blocks in a group are also enumerated in order of appearance. The enumeration of groups and block starts with 0.

The set of the circuit's active blocks can be reset to its initial state with the

```
netclass reset
```

command. Whenever the set of active blocks changes, the circuit in memory does not correspond to the state described by this set. Suppose that the open block in group diode is active. After the

```
netclass select diode::short
```

command is issued, the circuit in the memory becomes outdated. Checking whether the circuit is outdated can be done with the

```
netclass uptodate [vecname]
```

command. If the circuit is up to date, the vector *vecname* is set to 0; otherwise, it is set to 1. If the *vecname* is not provided, status is assumed.

The circuit can be brought up to date with the

```
netclass rebuild
```

command. The old circuit structures are released from the memory, and a new circuit that corresponds to the set of active blocks is constructed. All circuit parameters are reset to the values specified in the netlist. If the variable forcerebuild is set, the circuit is rebuilt even if it is up to date. Otherwise, the rebuild is performed only if the circuit is not up to date.

The reset command has the same effect as netclass rebuild with the forcerebuild variable set.

Example:

```
* A sample netlist
v1 (1 0) dc=0 pulse=(0 1 1m 1u)

.netclass r1 normal
r1 (1 2) r=1k
.endn

.netclass r1 open
r1 (1 20) r=1k
copen_r1 (20 2) c=10p
.endn

r2 (2 0) r=100k
c1 (2 0) c=1u

.end
```

After the circuit is loaded, normal block from the group r1 is active. To switch to the open block (r1 fails and becomes an open circuit), type the following commands in the command window:

```
netclass select r1::open
netclass rebuild
```

If another `netclass rebuild` is issued, the circuit is not rebuilt, unless the `forcerebuild` variable is set or the set of active blocks changes since the last rebuild. Probing for the circuit's state can be done in the following manner:

```
netclass uptodate needsrebuild if needsrebuild
  echo Circuit needs to be rebuilt.
else
  echo Circuit is up to date.
end
```

5.5.6 Changing Nodesets and Initial Conditions

Nodesets
The initial operating point solution for the Newton–Raphson algorithm can be set with the `.nodeset` netlist directive and also through the use of the NUTMEG `nodeset` command.

Nodesets can be printed with the

```
nodeset
```

command. Individual nodesets can be given as arguments in the same manner as with the `.nodeset` netlist directive.

Example:

```
* Set nodesets for nodes 10 (to 8.0) and 11 (to 9.0).
nodeset v(10)=8.0 v(11)=9.9
```

If a plot name is specified as the argument, the contents of the plot are interpreted as an operating point solution and the nodesets are set according to these results. If vectors have more than one component, the last component is used. This is useful, if the nodesets are to be set to the circuit's solution at the last time point of a transient analysis. In such a case you simply issue a `nodeset` command with one argument: the name of the plot containing the results of the transient analysis.

Example:

```
* Transient analysis with the current initial conditions.
tran 0.1u 100u 0u 0.1u uic
* Set the nodesets to the last time point of the transient analysis.
nodeset $(curplot)
```

Initial conditions
Initial conditions can also be manipulated from NUTMEG with the `ic` command. The syntax of the `ic` command is the same as that of the `nodeset` command.

Example:

```
* Do a long transient analysis (100 periods).
tran 0.01m 100m
* Set the initial conditions to the last transient point.
ic $(curplot)
* Analyze one period with the given initial conditions.
tran 0.01m 1m 0m 0.01m uic
```

5.5.7 Changing Instance, Model, and Subcircuit Parameters

The let command can also be used for changing instance, model, and subcircuit parameters.

let *parameter_specification* = *expression*

The following parameter specifications are available.

Instance parameter

@*inst* [*par*]

Using this specification you can access instance parameter *par* of instance *inst*. If the parameter is a vector parameter, the complete vector is accessed.

Example:

```
* Set the resistance of r1 to 9k.
let @r1[r]=9k
* Set the transient waveform of v1 to sine
* with 0V offset, 0.1V amplitude and 1kHz.
let @v1[sin]=(0;0.1;1k)
```

If *inst* is a subcircuit instance, the corresponding subcircuit parameter for that instance is changed.

```
* Set the fact subcircuit parameter for subcircuit xdiv1 to 0.95.
let @xdiv1[fact]=0.95
```

Instance parameter's component

@*inst* [*par*] [*comp*]

This specification accesses component *comp* of vector parameter *par* of instance *inst*.

Example:

```
* Change the amplitude of the sinusoidal transient voltage at v1 to 0.2.
let @v1[sin][1]=0.2
```

Model parameter

@@*mod* [*par*]

This specification accesses model parameter *par* of model *mod*.

Example:

```
* Change the tc1 parameter of model rmod1 to 0.
let @@rmod1[tc1]=0
```

If the model name is a subcircuit definition name, the default value of parameter *par* of subcircuit definition *mod* is changed. The values of the given parameter are also written to all subcircuit instances.

Example:

```
* Suppose there is a subcircuit definition named vdiv that is used
* by subcircuits x1 and x2.
* This changes the default value of parameter fact for subcircuit definition
* vdiv to 0.2. It also sets fact to 0.2 for instances x1 and x2.
let @@vdiv[fact]=0.2
```

Model parameter's component

@@mod [par] [comp]

This is similar to accessing a component of a vector parameter of some instance, except that this time it applies to a vector parameter of model mod.

All subinstances in a particular subcircuit definition

@inst :@def [par]

or

@inst :@def [par] [comp]

This specification accesses parameter par of all instances named inst in all subcircuit instances of definition def.

Example:

```
* Suppose subcircuit definition attn contains resistor r1.
* There are 3 instances of attn in the circuit (x1, x2, and x3).
* This changes r1 in x1, x2, and x3 to 10k.
let @r1:@attn[r]=10k
```

5.5.8 Parameter Value Propagation in Subcircuits

When a parameter value is changed for some subcircuit instance, the change propagates through the subcircuit hierarchy to deeper levels, until it reaches all SPICE primitive instances and models that are hierarchically within the subcircuit instance where the parameter was changed by the user. Only then is the circuit affected by the parameter change.

The process of parameter propagation can be time consuming. As the propagation occurs every time a subcircuit instance parameter is changed, considerable amounts of time can be spent on this task. In the following example the propagation occurs three times (once for every instance parameter).

```
let @x1[rtotal]=1k
let @x1[fact]=0.9
let @x1[tol]=0.01
```

In the next example the propagation occurs as many times as there are instances of subcircuit definition vdiv in the circuit.

```
let @@vdiv[rtotal]=1k
```

The propagation is triggered automatically whenever a subcircuit instance parameter changes. This can be prevented by setting the manualscktreparse variable. A propagation can then be triggered manually after all instance parameters are written.

The propagation is triggered with the scktreparse command.

scktreparse spec1 spec2 ...

The subcircuit specification (*specn*) can either be a subcircuit instance name or a subcircuit definition name prefixed with @. In the latter case the propagation occurs in all instances of the specified subcircuit definition.

Now consider the following circuit description:

```
.subckt vdiv in out com param: rtotal=10k fact=0.5
r1 (in out) r={(1-fact)*rtotal}
r2 (out com) r={fact*rtotal}
.ends

x1 (1 2 0) param: rtotal=1k fact=0.9
```

What happens in various steps of parameter propagation is explained by the following script:

```
* After the circuit is loaded rtotal for x1 is 10k.
* Therefore resistances of r1:x1 and r2:x1 are 1k and 9k, respectively.
* Now turn off automatic parameter propagation.
set manualscktreparse
* Change rtotal to 1k
let @x1[rtotal]=1k
* Resistances of r1:x1 and r2:x1 are still unchanged (1k and 9k).
* Now propagate parameter values.
scktreparse x1
* Now the values of r1:x1 and r2:x1 resistance reflect the
* change in rtotal at x1.
* They are now 0.1k and 0.9k, respectively.
* Now turn off manual propagation.
unset manualscktreparse
```

Here is an example of propagation starting at a particular subcircuit instance.

```
* Turn off automatic parameter propagation.
set manualscktreparse
* Set the parameters for instance x1:xamp
* (instance x1 within instance xamp).
let @x1:xamp[leak]=1p
let @x1:xamp[rout]=0.1k
let @x1:xamp[rin]=100k
let @x1:xamp[gain]=10000
* Now propagate the changes down the hierarchy.
scktreparse x1:xamp
* Only one propagation was performed instead of four.
```

This time the propagation is invoked for all instances of the given subcircuit definition.

```
* Turn off automatic parameter propagation.
set manualscktreparse
* Set the parameter rout for all instances of definition attn.
let @@attn[rout]=1k
* Now propagate the changes down the hierarchy.
scktreparse @attn
```

When accessing instances with a given name across all instances of a particular definition, the manual propagation makes sense only if it is invoked for all instances with the given name within all instances of the given definition.

```
set manualscktreparse
let @x1:@attn[tol]=0.1
* Now propagation must be invoked manually for all instances named x1
* within all instantiations of definition attn.
```

5.5.9 Changing and Accessing Simulator Parameters for the Current Circuit

The simulator parameters can be accessed as variables through the set command.
set *option_name* = *value*
> Example:

```
* Increase the circuit temperature to 100 degrees centigrade.
set temp = 100
```

The value must be a constant. If the value is an expression, expression substitution must be used (curly braces).

```
* Change the temperature according to the step we're at.
set temp = {step*10+20}
```

The value of a simulator parameter can be accessed through variable substitution.

```
* Print the temperature in kelvin.
* kelvin is a constant from the const plot (-273.15).
let curtemp = $(temp)-kelvin
echo Current circuit temperature is {curtemp}K.
```

The current value of some circuit parameter cannot be used in an expression that is part of arithmetic substitution (in curly braces) because arithmetic substitution takes place before variable substitution.

```
* Increase the temperature by 10 degrees.
* This is wrong:
*    set temp = {$(temp)+10}
* This is the right way to do it.
let curtemp = $(temp)
set temp = {curtemp+10}
```

5.5.10 Accessing Instance, Model, and Subcircuit Parameter Values

All of the instance, model, and subcircuit parameters that can be accessed with the let command can also be used in NUTMEG expressions, as long as there is no vector with the same name in the current plot. Table 5.6 lists the syntax for accessing parameter values.

Table 5.6 Accessing instance, model, and subcircuit parameters in NUTMEG expressions

Syntax	Explanation	Example
@*inst*[*par*]	Instance parameter	@v1[dc]
@@*mod*[*par*]	Model parameter	@@nmos0u5[vto]
@*subinst*[*par*]	subcirc. parameter	@x1[fact]
@@*subdef*[*par*]	subcirc. parameter default	@@vdiv[fact]
@*inst*:@*def*[*par*]	Instance parameter in subcirc.	@r1:@vdiv[r]

The first two rows in Table 5.6 refer to primitive instance and model parameters. If a parameter is a vector, a vector is returned. The third row refers to an instance parameter of a subcircuit instance.

The fourth row refers to the default value of a subcircuit parameter. This is the value specified at the subcircuit definition.

The syntax in the last row is for accessing primitive and subcircuit instance parameter values if the instance is within another subcircuit. The latter subcircuit is referred to through its definition name. The returned value is taken from an instance named *inst* that is one of the instantiations of subcircuit definition *def*. If different instantiations of definition *def* contain instances *inst* with different values of parameter *par*, the return value is not defined.

Examples:

```
* Set a to the resistance of r1.
let a=@r1[r]
* Set b to the threshold voltage of MOS model nmos0u5.
let b=@@nmos0u5[vto]
* Retrieve rtotal from subcircuit instance x1.
let c=@x1[rtotal]
* Retrieve the resistance of r1 from instantiations
* of subcircuit definition vdiv.
let d=@r1:@vdiv[r]
```

5.6 Simulation

This section explains the save, iplot, trace, delete, and status commands. The commands that invoke a simulation were explained in Chap. 4. Here only brief summaries are given.

5.6.1 Choosing What the Simulator Stores in Vectors (Saving)

By default, only node voltages and currents through voltage sources and inductances are saved during simulation. This behavior can be changed with the save command.
save *spec1 spec2 ...*

The save command instructs the simulator to save the specified device properties (parameters) and simulation results (node voltages and branch currents) to vectors. The effect of the save command becomes visible in subsequent simulations. The specification can be one of the items discussed below.

Node voltage
v(*name*)
or
name
Example:

```
* Instruct simulator to store node voltages for nodes 1, 10, and 99.
save v(1) v(10) v(99)
```

Current flowing through a voltage source or an inductance

i(name)

or

name#branch

In fact, the values of the current are stored in a vector named name#branch.
Example:

```
* Instruct simulator to store branch current for voltage sources
* v1, v2 and v9.
save i(v1) i(v2) i(v9)
```

Instance property calculated by the simulator

@inst [par]

The values of the instance property are stored in a vector named @inst [par].

```
* Instruct simulator to store resistor power for r1.
* v1, v2 and v9.
save @r1[p]
```

All other properties

name

Example:

```
* Save only the output noise spectrum and the output integrated noise.
save onoise_spectrum onoise_total
```

The default vector is always saved, regardless of the save directives issued by
the user. Once at least one save directive is present, the simulator starts to store
only those properties that were specified with the save command. To specify all
node voltages and all currents flowing through voltage sources and inductances, the
keyword all can be used.

```
* Save the power dissipation at r1 and everything else
* that normally gets saved.
save @r1[p] all
```

If the save all directive is issued, all directives for node voltage and branch
current are removed because they become unnecessary. Only directives for storing
instance properties calculated by the simulator remain unaffected.

Assuming that the following input file is loaded into the simulator:

```
Simple circuit
v1 (1 0) dc=9
r1 (1 0) r=1k
.control
* Set the save directives
save @r1[p]
save i(v1)
* Run a dc sweep of v1 (0V to 10V, 1V step).
dc v1 0 10 1
* Display the list of vectors.
display
* Print the power dissipated at r1.
print @r1[p]
.endc
.end
```

the following output is obtained in the command window:

```
Here are the vectors currently active:

Title: Simple circuit
Name: dc1 (DC transfer characteristic)
Date: Mon Feb 20 11:47:42  2006

    @r1[p]              : notype, real, 11 long
    sweep               : voltage, real, 11 long [default scale]
    v1#branch           : current, real, 11 long
                          Simple circuit
                          DC transfer characteristic  Mon Feb 20 11:47:42  2006
--------------------------------------------------------------------------
Index   sweep          @r1[p]
--------------------------------------------------------------------------
0    0.000000e+000   0.000000e+000
1    1.000000e+000   1.000000e-003
2    2.000000e+000   4.000000e-003
3    3.000000e+000   9.000000e-003
4    4.000000e+000   1.600000e-002
5    5.000000e+000   2.500000e-002
6    6.000000e+000   3.600000e-002
7    7.000000e+000   4.900000e-002
8    8.000000e+000   6.400000e-002
9    9.000000e+000   8.100000e-002
10   1.000000e+001   1.000000e-001
```

The vectors that contain the values of an instance property calculated by the simulator (the vectors starting with an @) have the type set to notype.

The list of all save (and also iplot and trace) directives can be printed with the status
command. Every directive is marked with a number. The status command produces the following printout after the last example is loaded:

```
1     save @r1[p]
2     save all
```

The numbers enable the user to remove individual directives with the delete command.

delete n1 n2 ...

All directives specified with the save command (and also iplot and trace commands) can be deleted with the delete all command. After delete all is issued, the enumeration of directives starts again with 1.

The delete all command is often issued in the beginning of a script to remove all previously set save, iplot, and trace directives.

Example:

```
.control
* Remove old directives.
delete all
* Remove old results
destroy all
...
```

5.6.2   Tracing Simulation Results During Simulation

Any node voltage or branch current that is normally saved by the simulator can be printed during simulation. The printing (tracing) can be turned on with the `trace` command.

trace *spec1* *spec2* ...

A `trace` specification has the same form as a `save` specification. `i(`*name*`)` and *name*`#branch` stand for branch currents. `v(`*name*`)` and *name* stand for node voltages.

The keyword `all` turns on tracing of all node voltages and all currents flowing through voltage sources and inductances. All other trace directives are removed from the list of trace directives.

Example (the circuit from the `save` example is used):

```
* Remove all save, trace, and iplot directives.
delete all
* Trace node voltage at node 1 and current through v1.
trace v(1) i(v1)
* Start a dc analysis
dc v1 0 10 1
```

The following printout is obtained after the DC analysis is started:

```
Execution trace (remove with the "delete" command).
Simple circuit
DC transfer characteristic   Mon Feb 20 12:31:38  2006
-----------------------------------------------
  sweep          V(1)              v1#branch
-----------------------------------------------
0.000000e+000  0.000000e+000   0.000000e+000
1.000000e+000  1.000000e+000  -1.000000e-003
2.000000e+000  2.000000e+000  -2.000000e-003
3.000000e+000  3.000000e+000  -3.000000e-003
4.000000e+000  4.000000e+000  -4.000000e-003
5.000000e+000  5.000000e+000  -5.000000e-003
6.000000e+000  6.000000e+000  -6.000000e-003
7.000000e+000  7.000000e+000  -7.000000e-003
8.000000e+000  8.000000e+000  -8.000000e-003
9.000000e+000  9.000000e+000  -9.000000e-003
1.000000e+001  1.000000e+001  -1.000000e-002
```

The traced device properties, simulation results, and the value of the default scale are printed during analysis. The `trace` directives can be displayed with the `status` command and deleted with the `delete` command (see `save` for details and an example).

Example (status printout for the above listed `trace` commands):

```
SpiceOpus (c) 2 -> status
1    trace v(1)
2    trace i(v1)
```

Tracing is a feature intended for debugging. It considerably slows down the simulator and should not be used normally. Use the `print` command after the simulation is finished to print the results.

5.6.3 Plotting Simulation Results During Simulation (Iplotting)

Any simulation result that can be traced can also be plotted during simulation. The plotting is turned on with the `iplot` command.

iplot *spec1 spec2 ...*

`iplot` specifications have the same format as the `trace` specifications. The `all` directive instructs the simulator to plot all node voltages and branch currents that are the result of the simulation.

Every `iplot` command instructs the simulator to create a new plot window and plot the listed simulation results as they are calculated. Plotting does not take place if SPICE is started in the console mode.

The plot window receives a tag of the form *tagn* where the *tag* is specified by the `iplottag` variable (default is `iplot`). The number *n* is chosen by the simulator. The plot windows that are opened during an analysis as a result of `iplot` directives are numbered consecutively starting with 0.

The created plot windows are affected by variables. Some plot window properties are affected at window creation; others are affected every time a curve is added to the plot window.

Example: two separate graphs, one with the node voltage at node 1 and one with the current flowing through v1.

```
* Delete all directives.
delete all
* Set line width to 2.
set linewidth=2
* Set up iplot.
* First iplot plots v(1) (tag is ipl0),
*   the second one plots i(v1) (tag is ipl1).
set iplottag=ipl
iplot v(1)
iplot i(v1)
* During this analysis two plots are created.
dc v1 0 10 1
* Now we have two plot windows tagged ipl0 and ipl1.
* ipl0 is displaying the node voltage at node 1.
* ipl1 is displaying the current flowing through v1.
```

Example: One graph displaying the power dissipated at r1 and the current flowing through v1.

```
* Now delete all directives.
delete all
* Set default iplot tag prefix (iplot).
unset iplottag
* We need to save @r1[p] so that we can plot it.
* We also add the keyword all to save node voltages and other
*   stuff that normally gets saved.
save @r1[p] all
* Plot @r1[p] and i(v1) in the same graph (tagged with iplot0).
iplot @r1[p] i(v1)
* Start the analysis and do the plotting.
dc v1 0 10 1
* Now we have one plot window tagged iplot0 displaying
*   the power dissipated at r1 and the current flowing through v1.
```

Iplotting is a feature intended for debugging. It considerably slows down the simulator. Normally the simulation results are plotted with the `plot` command after the simulation is finished.

5.6.4 Invoking Simulation

How various analyses can be invoked is the subject of Chap. 4. This is only a brief summary of commands that start various analyses. The subject of the analysis is the current circuit. Every analysis produces one or more plots as its result. The title of these plots is set to the title of the circuit used in the analysis (the current circuit).

Operating point analysis

The operating point analysis is invoked with the

op

command. The result is a plot named op$n$ where n is uniquely chosen by the simulator. All vectors in a plot are real with a single component. They represent the operating point of the circuit. Although one vector is the default scale of the plot, it has no special meaning.

Operating point sweep analysis

The operating point can be swept in many different ways. The result is a plot with real vectors where every component represents the operating point at one value of the swept parameter. The plot is named dc$n$ where n is uniquely chosen by the simulator. The `sweep` vector is the default scale and contains the swept values of the parameter. All primitive instance and model parameters can be swept along with the circuit temperature and global parameter measurement temperature in kelvins or degrees centigrade.

A linear sweep is obtained with

dc *parameter start stop step*

Example:

```
* Sweep the dc parameter of v1 from 0V to 10V with 0.1V step.
dc v1 0 10 0.1
* Sweep the dc parameter of i1 from 0A to 10mA with 1mA step.
dc i1 0 10m 1m
* Sweep the resistance of r1 from 1k to 10k in 1k steps.
dc @r1[r] 1k 10k 1k
* Sweep the threshold voltage of nmos model
*    from 0.5V to 0.9V in 0.01V steps.
dc @@nmos[vto] 0.5 0.9 0.01
* Sweep the circuit temperature from -50 to 150 degrees centigrade
*    in 10 degree steps.
dc @@@temp -50 150 10
* Sweep the circuit temperature from 100K to 350K in 10K steps.
* Don't write any units as they are interpreted as the kilo prefix (K).
dc @@@tempk 100 350 10
* Sweep global parameter measurement temperature
*    from 20 to 30 degrees centigrade in 1 degree steps.
dc @@@tnom 20 30 1
```

Another form for the linear sweep is

dc *parameter* *start* *stop* lin *nsteps*

Example:

```
* Sweep dc parameter of v1 from 0V to 10V with 20 steps.
dc v1 0 10 lin 20
```

A logarithmic sweep is similar to the linear sweep mentioned above, except that the keyword lin is replaced by dec or oct. *nsteps* is the number of steps per decade or octave. The default scale is marked with the flag log meaning that the x-axis is by default logarithmic when a vector from the resulting plot is plotted with the plot command.

Example:

```
* Sweep resistance of r1 from 1k to 100k, 10 points per decade.
dc @r1[r] 1k 100k dec 10
* Same as before, but with 12 points per octave.
*    i.e., 1 octave is the range between x and 2x.
dc @r1[r] 1k 100k oct 12
```

Small signal DC transfer function

The analysis evaluates the small signal DC transfer function and the input and output impedances. The result is a plot with three vectors: input_impedance, output_impedance, and transfer_function. The default scale of the resulting plot has no special meaning. The plot is named tf*n* where *n* is uniquely chosen by the simulator.

tf *output* *input*

The output can be a node voltage (v(*name*)), a differential voltage (v(*node1*, *node2*)), or a branch current (i(*name*)). The input is either an independent voltage source (v*name*) or an independent current source (i*name*). The output impedance is correct only if the output is a voltage. More details are available in Sect. 4.3.

Example:

```
* Transfer function from input at vin
*    to output between nodes 10 and 11.
* Transfer function is of type voltage/voltage.
tf v(10,11) vin
* Transfer function from input at iin
*    to output between nodes 10 and 11.
* Transfer function is of type voltage/current.
tf v(10,11) iin
```

Small signal (frequency-domain, AC) transfer function

The frequency-domain complex small signal transfer function is evaluated. If the keepopinfo simulator parameter is set, the operating point is stored in a separate plot named op*n*.

The plot with the frequency domain analysis results is named ac*n*. All vectors in the plot are complex. The default scale vector is named frequency. Its imaginary part is zero. Every component of the resulting vectors represents the complex small signal response of the circuit (linearized around the operating point) at one frequency. In order to obtain nonzero results, at least one independent source in the circuit must have its ac (or acmag) parameter set to a nonzero value.

ac *sweeptype* nsteps start stop
sweeptype can be one of lin, dec, or oct. If dec or oct is chosen, the default scale is marked with the log flag meaning that the x-axis is by default logarithmic when a vector from the resulting plot is plotted with the plot command.

Example:

```
* 10 points per decade from 1Hz to 1MHz.
ac dec 10 1 1meg
* 12 points per octave from 2kHz to 8kHz.
ac oct 12 2k 8k
```

Small signal (AC) transfer function poles and zeros

The poles and zeros of the small signal transfer function are evaluated. If the keepopinfo simulator parameter is set, the operating point is dumped to a separate plot named op*n*. The poles and zeros are dumped to a plot named pz*n*.

The poles can be found in the complex vector named pole and the zeros are in a complex vector named zero. Not all entries in the pole vector are valid poles. Valid entries are marked with a nonzero value in the polevalid vector. Valid zeros are marked in the same way with entries in the zerovalid vector. The default scale vector has no special meaning.

The syntax of the command that invokes a pole-zero analysis is

pz *in+ in- out+ out- type results*

The input is specified with nodes *in+* and *in-*. The output is determined by nodes *out+* and *out-*. The *type* parameter is either vol or cur. In the former case the transfer function is of type output voltage divided by input voltage. In the latter case the transfer function is of the type output voltage divided by input current. The *results* parameter specifies what results are desired. pol evaluates only poles, zer evaluates only zeros, and pz evaluates both. More details on the pole-zero analysis can be found in Sect. 4.5.

Example:

```
* Poles and zeros, voltage/voltage transfer function.
*  Input between nodes 1 and 2. Output is between 5 and 9.
pz 1 2 5 9 vol pz
* The same, except that only zeros are evaluated.
pz 1 2 5 9 vol zer
```

Small signal frequency-domain noise

This analysis evaluates the noise at the output of the circuit and the equivalent noise at the circuit's input. The sources of noise are the circuit components. The shot noise and the $1/f$ noise generated by a component depend on the operating point, whereas the thermal noise does not. See Sect. 1.10 for details.

Two plots are created. The first one is named noise*n* and contains the noise spectra. The default scale vector is frequency and is real. The second one is named noise*n+1* and contains integrated noise. The default scale vector in this plot has no special meaning. The syntax of the noise command is

noise *output input sweeptype nsteps start stop [ptssum]*

The first argument specifies the output at which the noise is evaluated. It can either be a node voltage (v(*name*)) or a differential voltage (v(*node1*,*node2*)). The input which the equivalent input noise refers to can be either an independent voltage source (*vname*) or an independent current source (*vname*). In order to obtain the correct value of the equivalent input noise, the ac parameter of the source must be set to 1.

The *sweeptype*, *nsteps*, *start*, and *stop* parameters have the same meaning as for the ac command. If the sweep type is logarithmic, the default scale vector of the first resulting plot is marked with the log flag, meaning that the x-axis is by default logarithmic when a vector from the respective plot is plotted with the plot command.

The last argument is optional. If it is omitted, the results are less detailed (only the input and output noise are available in the plots; no individual device contributions are saved). If the last argument is specified, it represents the number of points in the vectors of the second noise plot (integrated noise). Usually it is set to 1, which means that the whole frequency range from start to stop is integrated into a single value.

Example:

```
* Output is between nodes 9 and 5. Input is at v1.
* Frequency range is between 1Hz and 1MHz with 10 pts per decade.
* One point per summary is requested.
noise v(9,5) v1 dec 10 1 1meg 1
* The same as the last analysis, except that the results are
* limited to the input and output noise. No individual device
* contributions are stored in the plots.
noise v(9,5) v1 dec 10 1 1meg
* Now go to the plot with the noise spectra.
setplot previous
* Plot the output noise spectrum.
plot onoise_spectrum
* Return back to the plot with the integrated noise.
setplot next
* Print the integrated output noise.
print onoise_total
```

Transient (time-domain) analysis
The transient analysis evaluates the circuit's behavior in the time domain. A single plot is produced with the name tran*n*. The syntax is
tran *step stop* [*start* [*maxstep* [uic]]]
The last three arguments are optional. *step* and *stop* define the initial step and the time at which the analysis is stopped. *start* defines the time at which the results start to be recorded. *maxstep* limits the time step from above. If the uic argument is specified, the initial point is taken from the initial conditions set by the instance parameters (e.g., ic parameter of a capacitor) and .ic netlist directives (which in turn can be changed from NUTMEG with the ic command). If uic is specified, the initial operating point analysis is skipped.

Example:

```
* Initial step 1us, end time 10ms.
tran 1u 10m
* Same as before, start recording at 5ms.
tran 1u 10m 5m
* Same as the last one, except that the step is
* bound from above to 2us.
tran 1u 10m 5m 2u
* Same as the last one, except that the initial operating
```

```
* point analysis is skipped and the initial conditions
* are taken from the .ic netlist directives and
* instance parameter values (e.g., ic parameter of capacitor).
tran 1u 10m 5m 2u uic
```

5.7 Analyzing the Results

This section explains how the results that are obtained with circuit analyses can be further analyzed. The linearize, cursor, fourier, and spec commands are explained.

5.7.1 Linearizing the Scale of a Plot

Often the result of an analysis has a nonuniform scale vector (e.g., the transient analysis varies the time step). The linearize command can be used for changing the scale of a plot to a linear scale. The syntax of the command is the following:
linearize *step* [*vec1 vec2* ...]
A new plot is created named linn. The default scale of the plot has the same name as the default scale of the original plot, except that this time the points of this scale are equidistant (the distance is *step*). The transformed vectors also have the same names. The points in the transformed vectors correspond to the transformed scale points. The full name of the linearized plot is the same as the full name of the original plot, except that the word (linearized) is added. The title of the plot is copied from the original plot.

The *step* must be smaller than the span of the original scale. The original scale and all original vectors must be real. The original scale must also be monotonic. If no vectors are specified, the complete set of vectors from the current plot is transformed to the new scale. Linear interpolation is used for transforming the vectors to a new scale.

Example:

```
* Create a new plot and name it source.
setplot new
nameplot source
* Create a scale (a) and two vectors (b and c). The scale
* is not linear. b and c can be expressed as b=2*a and c=5*a.
let a=(0;1;2;5;10;20;50;100)
let b=2*a
let c=5*a
* Now linearize the scale to step 10. This will result in
* vectors of length 11.
linearize 10
* Print the vectors.
print a b c
```

The following printout is obtained:

```
                              Anonymous
                              unknown (linearized)  Tue Feb 21 15:34:21  2006
-------------------------------------------------------------------------------
Index   a                b                  c
-------------------------------------------------------------------------------
0    0.000000e+000    0.000000e+000     0.000000e+000
1    1.000000e+001    2.000000e+001     5.000000e+001
2    2.000000e+001    4.000000e+001     1.000000e+002
3    3.000000e+001    6.000000e+001     1.500000e+002
4    4.000000e+001    8.000000e+001     2.000000e+002
5    5.000000e+001    1.000000e+002     2.500000e+002
6    6.000000e+001    1.200000e+002     3.000000e+002
7    7.000000e+001    1.400000e+002     3.500000e+002
8    8.000000e+001    1.600000e+002     4.000000e+002
9    9.000000e+001    1.800000e+002     4.500000e+002
10   1.000000e+002    2.000000e+002     5.000000e+002
```

If in the above listing the linearize 5 command is replaced by

```
linearize 5 b
```

the resulting plot contains only two vectors: a (the scale) and b (the vector listed as the argument to linearize).

5.7.2 Performing Measurements on Vectors

A vector can be interpreted as a table of values. Often a value is needed from this table that corresponds to a position between two listed values (two vector components). Such a value can be extracted with the help of the [%] operator. Linear interpolation is used for determining the corresponding vector value.

Example:

```
* Create a vector.
let a=(1;2;5;10)
* Get the value half way between index 1 and 2
* (between vector component values 2 and 5).
print a[%1.5]
* The result is 3.5.

* This returns the value that lies one quarter of the way
* between the third and the fourth component (note that
* indexing starts at 0).
print a[%2.25]
* The result is 6.25.
```

In conjunction with the cursor command the [%] operator is a powerful tool. The cursor command can be used for positioning a fractional index to various positions on a vector that depend on the vector's components. The syntax of the cursor command is the following:

```
cursor cvec dir expr [level [intercept edge]]
```

A cursor is a fractional index into a vector. The cvec argument specifies the name of the single-valued vector that holds the cursor position (fractional index). The dir argument specifies the direction in which the cursor should move. The direction can be either left (moving toward lower index values) or right (moving toward higher index values).

The *expr* must evaluate to a real vector. This is the vector that is referred to by the cursor *cur*. The most simple form of the `cursor` command requires only these three arguments. Depending on the value of the *dir* argument the cursor moves to the lowest or the highest index of the vector (obtained by evaluating *expr*).

Example:

```
* Create a vector.
let a=(0;1;2;5;10;5;2;1;0)
* Create a cursor named c and set it to index 5.
let c=5
* Move the cursor to the leftmost position on a (lowest index).
cursor c left a
* Print the cursor position.
print c
* The result is 0.

* Move the cursor to the rightmost position on a (highest index).
cursor c right a
* Print the cursor position.
print c
* The result is length(a)-1 (8).
```

The *level* argument is an expression that must evaluate to a single-valued real vector. This value represents the value of expression *expr* sought be the cursor. The search is performed in the direction specified by *dir*. If the target level is not found, the cursor (*c*) becomes negative, thereby indicating an error. If the cursor is already at the specified level in *expr* it is not moved. Linear interpolation is used for determining the fractional index corresponding to the target level.

Example:

```
* Create a vector.
let a=(0;1;2;5;10;5;2;1;0)
* Create a cursor named c. Position it at index 0.
let c=0
* Move cursor towards right (higher index values) until
*  the value 8 is reached in vector a.
cursor c right a 8
* Print the cursor position.
print c
* The result is 3.6.
* This is 3/5 of the way between indices 3 and 4
*  (i.e., values 5 and 10).

* Now try to move it again. and print position.
cursor c right a 8
print c
* The cursor remains at the same index (3.6).

* Move it to level 2 toward higher index values.
cursor c right a 2
print c
* It is now precisely on index 6.
```

When the same level occurs multiple times in the value of an *expr* the user may want the cursor to stop after a certain number of level crossings occur. This number is specified by the *intercept* argument, which is an expression that must evaluate to a single-valued real vector. Negative values of *intercept* reverse the direction of search. The default value of this argument is 1.

The edge argument specifies the type of the edges that are taken into account.
falling, rising, and any stand for falling, rising, and all edges, respectively.
Default is any.

Example:

```
* Create a vector.
let a=(0;1;2;5;10;5;2;1;0)
* Create a cursor named c. Position it at index 0.
let c=0
* Move it to the position where the vector value reaches
*   8.0 for the second time. All crossings count.
cursor c right a 8.0 2 any
print c
* The cursor is now at index 4.4 (2/5 of the way between
*   values 10 and 5 corresponding to index values 4 and 5).

* Now try to move it left (toward lower index values)
*   until the first crossing with 8.0. All edges count.
cursor c left a 8.0 1 any
print c
* This time the cursor's position remains unchanged.
* We are already at level 8.0.

* Now move it to the second crossing with 8.0 toward left.
cursor c left a 8.0 2 any
print c
* This time we land at the first crossing with 8.0 at
*   index 3.6. The crossing at which the cursor was
*   initially positioned was the first crossing!

* Now position cursor at index 0 and try to find the
*   first crossing with 8.0 towards right. This time only
*   falling edges count.
let c=0
cursor c right a 8.0 1 falling
print c
* This time we land at index 4.4.
```

Advanced measurements can be performed by moving the cursor to a particular
level in one vector (with the cursor command) and then reading the corresponding
value from another vector (with the [%] operator).

Example (measuring the rise time of a signal at the second pulse):

```
* Suppose the signal is stored in vector out and
*   the time is stored in vector time.
* Detect the top and the bottom level of the signal.
let sigmax=max(out)
let sigmin=min(out)

* Determine the 10% and the 90% signal level.
let sig10=sigmin+(sigmax-sigmin)*0.1
let sig90=sigmin+(sigmax-sigmin)*0.9

* Create a cursor and move it to the second rising
*   edge at the 10% level.
let c1=0
cursor c1 right out sig10 2 rising
* Create another cursor at c1's position.
let c2=c1

* Move it to the 90% level toward higher index values.
cursor c2 right out sig90

* Calculate the time difference corresponding
*   to cursors c1 and c2.
let trise=time[%c2]-time[%c1]
```

The last example calculates the rise time of the second pulse correctly if all pulses in the signal have the same top and bottom levels. Automatic level determination is changed to user-selected levels if the first two let commands assign constants to vectors sigmax and sigmin.

5.7.3 Fourier Analysis of Vectors

The fourier command performs a frequency-domain analysis of a signal. The syntax of the command is
fourier *fund expr1 expr2* ...
fund is the fundamental frequency of the signal components (it must be greater than zero). The signals are obtained by evaluating the expressions exprn. The time scale for the signals is the default scale of the current plot. It must be real and monotonic.

If the polydegree variable is set to a nonzero value, the part of the vector corresponding to the last $1/f_{fund}$ (f_{fund} is the fundamental frequency) seconds is interpolated to an equidistant grid. The polydegree variable specifies the degree of the interpolating polynomial (1 by default). The number of gridpoints is by default 200 and can be changed with the fourgridsize variable.

If polydegree is zero, no interpolation is performed. The number of gridpoints equals the number of points in the vector. The time scale of the vector is ignored, and the vector is treated as if it represents equidistantly distributed points from one period of the fundamental frequency.

Next the signals are decomposed in harmonics. If $y(t)$ denotes the signal, the following decomposition is used:

$$y(t) = \sum_{n=0}^{n_{freq}-1} A_n \cos(2\pi f_{fund} n t + \varphi_n) + \epsilon(t),$$

where $\epsilon(t)$ is the residual error due to Fourier series truncation. The n_{freq} parameter is specified by the nfreqs variable (10 by default).

For every signal the harmonic component index (n), the component frequency ($n f_{fund}$), the magnitude (A_n), the phase (φ_n), the normalized magnitude (A_n/A_1), and the normalized phase ($\varphi_n - \varphi_1$) are printed. The normalized magnitude and phase for the DC component ($n = 0$) are not calculated. The printed value of these two quantities is always 0.

Besides individual harmonic components, the total harmonic distortion (THD) is also calculated and printed,

$$THD = \frac{1}{A_1} \left(\sum_{n=2}^{n_{freq}-1} A_n^2 \right)^{1/2}.$$

Example (Fourier analysis of the results of a transient analysis):

```
* Transient analysis.
tran 1u 10m
* Analyse voltage at node 9. Base frequency in 1kHz.
fourier 1k v(9)
```

Example (Fourier analysis of constructed vectors):

```
* Create a plot and scale from 0 to 1, 1001 points.
setplot new
nameplot freq
let t=vector(1001)/1000
* Fourier analysis of the following signal:
*    0.1+1.2*cos(2*pi*10*t+pi/2)+0.2*sin(2*pi*10*8*t-pi/4)
* Fundamental frequency is 10Hz.
* There are two harmonics and a DC component in this signal.
*    0Hz    A=0.1
*    10Hz   A=1.2    phi= 1/2 pi (90 degrees)
*    80Hz   A=0.2    phi=-3/4 pi (-135 degrees)
* Do the Fourier analysis.
fourier 10 0.1+1.2*cos(2*pi*10*t+pi/2)+0.2*sin(2*pi*10*8*t-pi/4)
```

The following output is obtained for the last example:

```
Fourier analysis for 0.1+1.2*cos(2*pi*10*t+pi/2)+0.2*sin(2*pi*10*8*t-pi/4):
  No. Harmonics: 10, THD: 16.4089 %, Gridsize: 200, Interpolation Degree: 1
```

Harmonic	Frequency	Magnitude	Phase	Norm. Mag	Norm. Phase
0	0	0	0	0	0
1	10	1.1997	90	1	0
2	20	1.8402e-015	63.0845	1.53388e-015	-26.915
3	30	1.56783e-015	56.6654	1.30685e-015	-33.335
4	40	2.57458e-015	63.5734	2.14601e-015	-26.427
5	50	8.0379e-016	30.5094	6.6999e-016	-59.491
6	60	1.0823e-015	159.925	9.0214e-016	69.9255
7	70	1.35893e-015	79.6397	1.13272e-015	-10.36
8	80	0.196858	-135	0.164089	-225
9	90	7.72173e-016	-32.809	6.43636e-016	-122.81

Note that the phase is irrelevant for the harmonics with insignificant magnitude.

5.7.4 Transforming Vectors to Frequency Domain

The fourier command only prints the frequency content of vectors. If one wants to further analyze it, the vectors must first be transformed to the frequency domain with the spec command.

spec fstart fstop fstep vec1 vec2 ...

Listed vectors must be real and have the same length as the default scale of the current plot which is used as the time scale of the vectors. Let N denote the number of points in a vector. The index of the vector components starts at 0 and ends at $N - 1$. w_j and t_j denote the j-th point of the window function and the time scale, respectively. Δ denotes the time span of the scale ($\Delta = t_{N-1} - t_0$).

The input vectors are windowed during the calculation of the frequency-domain coefficients. The window type is determined by the specwindow variable (default is hanning). The window functions listed below are available.

No windowing

```
set specwindow=none
```
or
```
set specwindow=rectangular
```

$$w_j = 1 \tag{5.1}$$

Hann window

```
set specwindow=hanning
```
or
```
set specwindow=cosine
```

$$w_j = \beta(0.5 - 0.5\cos(2\pi\frac{t_j - t_0}{\Delta})), \quad \beta = 2 \tag{5.2}$$

Hamming window

```
set specwindow=hamming
```

$$w_j = \beta(0.54 - 0.46\cos(2\pi\frac{t_j - t_0}{\Delta})), \quad \beta = 1/0.54 \tag{5.3}$$

Triangle (Bartlett) window

```
set specwindow=triangle
```
or
```
set specwindow=bartlet
```

$$w_j = \beta(1 - |1 - 2\frac{t_j - t_0}{\Delta}|), \quad \beta = 2 \tag{5.4}$$

Blackman window

```
set specwindow=blackman
```

$$w_j = \beta(0.42 - 0.5\cos(2\pi\frac{t_j - t_0}{\Delta}) + 0.08\cos(4\pi\frac{t_j - t_0}{\Delta})), \quad \beta = 1/0.42 \tag{5.5}$$

Gaussian window

```
set specwindow=gaussian
```
The order of the window is specified by the `specwindoworder` variable (default is 2) and is denoted by α. If the order is set to less than 2, 2 is used.

$$w_j = \beta e^{-0.5\alpha(1-2\frac{t_j - t_0}{\Delta})^2}, \quad \beta = \sqrt{\frac{\pi}{2\alpha}}\mathrm{erf}(\sqrt{\frac{\alpha}{2}}) \tag{5.6}$$

fstop must not exceed *fstart*. *fstart* must not be negative. *fstep* must be greater than 0. *fstep* must not be smaller than $1/\Delta$. Both the start and the stop frequencies are truncated to a multiple of *fstep*.

The results are stored in a new plot named spn. The title of the plot is the same as the title of the plot from which the vectors were obtained. The full name of the plot is Spectrum. The resulting plot contains a real frequency scale (named frequency) with frequency values between *fstart* and *fstop*. The frequency step equals *fstep*. If the keepwindow variable is set, the vector __specwindow contains the window used for windowing the signal. Note that the scale of the window is the default scale vector of the original plot.

The remaining vectors are complex and have the same names as the vectors listed as arguments to spec. The number of components in these vectors equals the length of the frequency vector. Let $y(t_j)$ denote the j-th component of the original vector and $Y(f_k)$ the k-th component of the resulting complex vector. k denotes the index of a component of the resulting complex vector and f_k is the corresponding frequency from the frequency vector. $Y(f_k)$ is calculated as

$$Y(f_k) = \frac{2}{M} \sum_{j=0}^{M-1} y(t_j)e^{-2\pi f_k t_j}, \quad k = 0, 1, \ldots, \frac{f_{stop} - f_{start}}{f_{step}}. \quad (5.7)$$

The value of M equals N if the variable aperiodic is set. Otherwise, M equals $N - 1$. Coefficients $Y(f_k)$ are meaningful if the time points t_j are equally spaced. If the time points are not equally spaced, the linearize command can be used to preprocess the vectors for spec.

Periodic signals

If the aperiodic variable is not set and *fstep* is a multiple of $1/\Delta$, a coefficient $Y(f_k)$ represents the frequency content of a periodic signal with period Δ at frequency f_k. Note that the first and the last points of the signal ($y(t_0)$ and $y(t_{M-1})$) represent the first points of two consecutive periods. The last point of the signal is not used for calculating $Y(f_k)$ (since $M = N - 1$). Due to aliasing effects the coefficients for frequencies above $\frac{M}{2\Delta}$ are complements of coefficients below that frequency.

Assuming that all the frequencies from the set $\mathcal{F} = \{k f_{step}; k = 0, 1, \lfloor M/2 \rfloor\}$ are represented in the frequency vector and that no windowing is used for obtaining coefficients $Y(f_k)$, the perfect reconstruction of $y(t_j)$ for time points $t_0, \ldots, t_{M-1}$ can be written as

$y(t_j) = \sum_{f_k \in \mathcal{F}} u(f_k)|Y(f_k)| \cos(2\pi f_k t_j + \arg(Y(f_k))) \quad j = 0, 1, \ldots, M$
$u(f_k)$ is $1/2$ if $f_k = \frac{M}{2\Delta}$ (Nyquist frequency) and 1 otherwise.

Aperiodic signals

On the other hand, the results can also be interpreted as the frequency content estimation of an aperiodic signal. In this case the aperiodic variable should be set. This includes the last time point in the calculation of $Y(f_k)$. Again, the time points must be equally spaced to obtain meaningful results.

Example (Fourier analysis of a periodic signal):

```
* Create a new plot.
setplot new
nameplot td
* We analyze a periodic signal with period equal to 0.1s.
* The analyzed vectors must contain exactly one period of the signal.
* We have exactly 101 points. The Nyquist frequency is at
* 1/2*100/0.1s = 500Hz
let t=0.1*vector(101)/100
* We will be using trigonometric functions.
* Make sure the correct units are set.
set units=radians
* Create the signal. It contains
*   a DC component (0.1) and frequencies
*   100Hz (0.2, 90deg), 200Hz (0.1, 120deg), 250Hz (0.05, 0deg), and
*   500Hz (0.4, 0deg; Nyquist frequency)
let a=0.1+0.2*cos(2*pi*100*t+pi/2)+
+         0.1*cos(2*pi*200*t+2*pi/3)+
+         0.05*cos(2*pi*300*t)+
+         0.4*cos(2*pi*500*t)
* Disable windowing.
set specwindow=none
* Set periodic transform.
unset aperiodic
* Analyze it from 0 to 500Hz in 100Hz steps.
spec 0 500 100 a
* We have a new plot. Print the frequency scale,
*   the magnitude and the phase.
set units=degrees
print frequency abs(a) phase(a)
```

The following output is printed:

```
                                     Anonymous
                                     Spectrum  Thu Feb 23 15:15:27  2006
-----------------------------------------------------------------------------

Index   frequency        abs(a)           phase(a)
-----------------------------------------------------------------------------

0     0.000000e+000   1.000000e-001    0.000000e+000
1     1.000000e+002   2.000000e-001    9.000000e+001
2     2.000000e+002   1.000000e-001    1.200000e+002
3     3.000000e+002   5.000000e-002   -7.442498e-012
4     4.000000e+002   1.305320e-014   -1.695562e+002
5     5.000000e+002   8.000000e-001    1.570284e-012
```

Note that the phase of the components with very small magnitude is insignificant. The magnitude at the Nyquist frequency is doubled. This explains the $u(f_k) = 1/2$ in the perfect reconstruction formula at the Nyquist frequency.

Example (distortion analysis of transient results):

```
* The signal is periodic with period equal to 1ms.
* Transient analysis.
tran 1u 1m
* We're analyzing v(9). Time step is not constant.
* Linearize the step to 10us.
linearize 10u v(9)
* Transform to frequency domain. 0 Hz to 10kHz with 1kHz step.
* Note that the step is a multiple of 1/period=1kHz.
unset aperiodic
set specwindow=none
spec 0 10k 1k v(9)
* Print the first and the second harmonic magnitude.
print abs(v(9))[1] abs(v(9))[2]
```

5.8   Plotting the Results

Vectors can be displayed in plot windows graphically with the `plot` command. In the following subsections the plotting features available through the `plot` command are described. The `plot` command is not supported in the console mode.

5.8.1   The Plot Window

Results are plotted in plot windows. Every plot window has a tag. This tag is either user supplied or assigned by SPICE. In the latter case the tag has the form `plot`*n*. Tags are the means for referring to plot windows. If two plots have the same tag, the older one is not accessible from NUTMEG. The plot tag of the plot in Fig. 5.1 is `plot55`. The initial size of newly created plot windows (in pixels) can be changed by setting the `plotwinwidth` and `plotwinheight` variables.

A plot window can have a frame displaying various information based on mouse cursor position (the info frame). A plot window with an info frame is shown in Fig. 5.2. The info frame can be turned on or off by pressing `Ctrl+H`. If the `plotwininfo` variable is set, the info frame of a newly created plot is enabled.

Every plot window displays a collection of curves. Every curve consists of two sets of values: curve and scale values. From these two sets the x–y coordinates of the displayed points are obtained.

When the cursor moves across a plot window, the coordinates are displayed in the lower part of the window. By pressing the `SPACE` key, the nearest curve can be identified (Fig. 5.3, left). The curve is marked with a green line to the cursor, and the name of the curve is displayed in the lower part of the window. The curve name and the green line disappear when the cursor is moved. Curve identification can be invoked automatically when the cursor does not move for about 1 s. The automatic

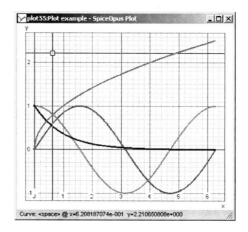

Fig. 5.1 Plot window

Fig. 5.2 Plot window with an
info frame

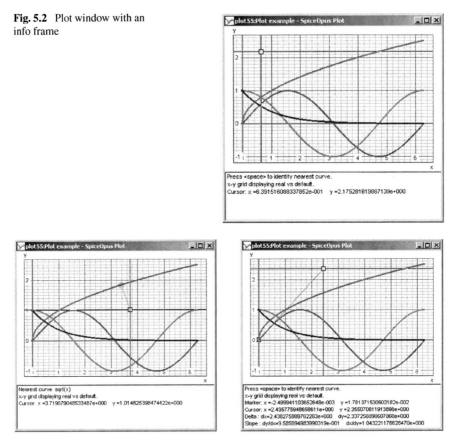

Fig. 5.3 Plot window. *Left*: identifying the nearest curve. *Right*: performing measurements with
cursor and marker

identification mode can be toggled by pressing Ctrl+I. Automatic identification
is turned off by default when the plot window is created. This can be changed by
setting the plotautoident variable.

Usually only the cursor position is displayed (Fig. 5.1). If, however, the info
frame is enabled (Ctrl+I), more advanced measurements are possible. The info
frame displays the grid mode (i.e., x-y grid) and the vector interpretation mode
(i.e., real vs. default). Below that the cursor coordinates are displayed. The
coordinate display depends on the type of the grid and on the active vector interpretation mode.

A marker can be placed with a left mouse button click (Fig. 5.3, right). When a
marker is placed, the marker coordinates are displayed below the cursor coordinates.
Besides marker coordinates, the coordinate difference between the marker and the
cursor (dx and dy), and the slope of the line connecting the marker and the cursor (dy/dx and dx/dy) are displayed. The marker can be removed by pressing the
Esc key.

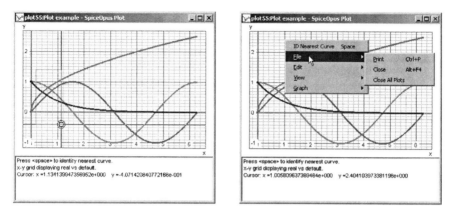

Fig. 5.4 *Left*: box zoom. *Right*: plot window menu

The contents of the info frame change as the cursor moves. The info frame can be frozen by pressing `Ctrl+F`. Now the contents can be copied to the clipboard by pressing `Ctrl+C`. Freeze mode for the info frame is disabled if `Ctrl+F` is pressed again.

The plot window can be box-zoomed by pressing the Z key upon which the cursor changes its form (Fig. 5.4, left). The box is selected by moving the cursor to one corner of the box, pressing the left mouse button and holding it, moving the cursor to the diagonally opposite box corner, and then releasing the mouse button. Box-zoom can be canceled by pressing the Z key again. Two additional zoom functions are available. By pressing the − (+) key, the plot is zoomed out (in) around the mouse cursor. By pressing `Ctrl+R` the aspect ratio can be corrected to 1:1 so that circles are circles and not ellipses. The plot window can be autoscaled so that all curves are fully visible by pressing `Ctrl+A`.

Every plot window has a menu. The plot window menu can be opened by clicking with the right mouse button on the contents of the plot window (Fig. 5.4, right). The contents of the menu are as follows: **ID Nearest Curve** option (invokes automatic identification) and the **File**, **Edit**, **View**, and **Graph** submenus.

The **File** submenu contains the **Print** option (shortcut `Ctrl+P`) for printing the contents of the plot window, the **Close** option (shortcut `Alt+F4`) for closing the plot window, and the **Close All Plots** option for closing all open plot windows.

The **Edit** submenu contains the **Copy** option (shortcut `Ctrl+C`) for copying the contents of the info frame to the clipboard, the **Freeze Messages** (shortcut `Ctrl+F`) option for toggling the frozen mode of the info frame, and the **Clear Marker** option (shortcut `Esc`) for clearing the marker.

Zoom options (**Box Zoom, Center Zoom In**, and **Center Zoom Out**) can be found in the **View** submenu (shortcuts Z, +, and −). The **Autoscale** (shortcut `Ctrl+A`), **Aspect Ratio 1:1** (shortcut `Ctrl+R`), **Show Info Frame** (shortcut `Ctrl+H`), and **Automatic Identification** (shortcut `Ctrl+I`) options are also located in this submenu.

The **Graph** submenu has numerous options. The bottom 12 options select the vector interpretation mode. Besides changing the vector interpretation mode, these options also change the grid. The first eight of them (**Real (real)**, **Imaginary (imag)**, **Magnitude (mag)**, **Phase (phase)**, **R (realz)**, **X (imagz)**, **G (realy)**, and **B (imagy)**) switch the grid to x–y mode. The **Complex (cx), x–y grid** option switches to the cx vector interpretation mode and x–y grid. The last three options (**Complex (cx), polar grid**, **Complex (cx), Smith Z grid**, **Complex (cx), Smith Y grid**) all activate the cx vector interpretation mode. They also switch to the polar grid, Smith impedance grid, and Smith admittance grid, respectively. The vector interpretation mode specifies the way the points of the plotted vectors are interpreted. All these modes (except for the real mode) make sense only when the vectors are complex. The active vector interpretation mode applies to all curves in the plot window.

The **Graph/Curve Representation** submenu enables one to change the curve style, line width, and point size for all curves simultaneously. In the **Graph/Units** submenu the units for the magnitude and phase displayed in the plot window can be changed. The setting is valid when the vector interpretation mode is set to mag or phase. The units setting also affects the cursor coordinate units when the grid is set to polar. Units can also be set with the units and magunits variables. The options in the **Graph/Grid** submenu switch to a different grid (x–y, polar, Smith impedance, and Smith admittance grid). The active vector interpretation is not changed. The logarithmic setting for the x and y axes, available in this menu, is valid only when the x–y grid is selected.

5.8.2 Curve Display Styles

There are several ways the resulting pairs of (x–y) coordinates can be displayed. Figure 5.5 (left) displays the curve styles. The curve style is determined by the

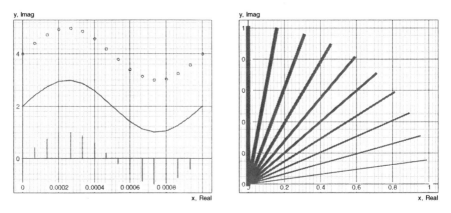

Fig. 5.5 *Left*: different curve styles (plottype variable set to point, line, and comb). *Right*: different line widths (linewidth set to values from 0 to 10). The axis labels are the SPICE OPUS default labels and have no special meaning for the plots in the figure

contents of the `plottype` variable. Three styles are available (from top to bottom): `point`, `line`, and `comb`. The last style (`comb`) displays the points as lines connecting the respective point with the x-axis. The default style is `line`.

The line width is determined by the contents of the `linewidth` variable. Figure 5.5 (right) displays different line widths. The default line width is 0. If the curve style is `point` additional settings are available. The `pointsize` variable specifies the point size (starting at 0). The default point size is 2. The type of the points plotted is selected with the `pointtype` variable. Its value can be one of o (circles), x (crosses), + (plus signs), q (squares), d (diamonds), or t (triangles). The default is circles (o). Figure 5.6 displays different combinations of line width, point size, and point type.

The colors used in the plot window can be set with the `color`n variables. `color0` and `color1` represent the background and the grid color. They apply only to newly created plots. Colors `color2`, `color3`, etc. represent curve colors. The first curve plotted by a `plot` command is plotted in `color2`, the second one in `color3`, etc. The format for defining colors is the following: `r`xxx`g`xxx`b`xxx where individual color components are given as decimal numbers between 0 and 255.

Example (setting up the colors):

```
* Default background color (white).
set color0  = r255g255b255
* Default grid color (bluish).
set color1  = r80g80b255
* Default curve colors.
set color2  = r220g000b000
set color3  = r000g180b000
set color4  = r000g000b200
set color5  = r210g060b210
set color6  = r000g000b000
set color7  = r150g180b020
set color8  = r255g000b128
set color9  = r000g128b128
set color10 = r100g000b230
```

5.8.3 Plot Window Grids

Different grids are available in the plot window (see Fig. 5.7). Depending on the grid type, the various coordinates of the cursor are displayed (in the info frame and in the bottom of the window). The x–y grid has four variations (linear, x-log, y-log, and log-log). In Fig. 5.7 the function $y = x^n$ ($n \in \{-2, 1, 0, 1, 2\}$) is displayed on the four variations of the x–y grid.

In bottom two plots in Fig. 5.8 the complex function $z(t) = 4\cos(t)e^{-0.2t} + 4i\sin(t)e^{-0.2t}$ is plotted for $t \in [0, 8\pi]$. The left plot has a polar grid, whereas the right plot has a Smith grid. In the polar grid mode the info frame prints the r and the φ coordinates. The r and φ coordinates displayed in the info frame are affected by the magnitude (`absolute`, `db20`, or `db10`) and angle units (`radians` or `degrees`) set in the **Graph/Units** submenu (or through `units` and `magunits`

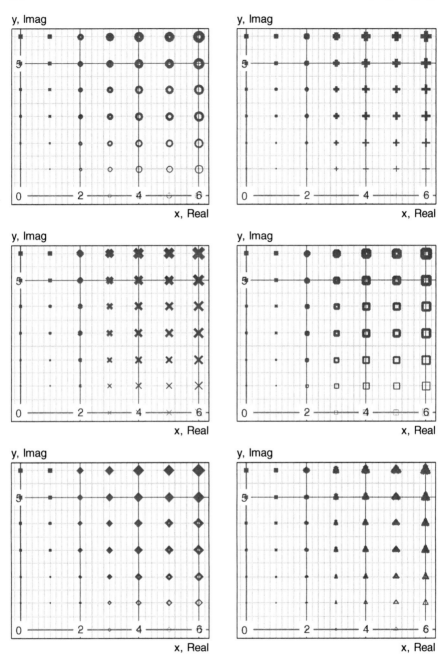

Fig. 5.6 Various settings for `linewidth` (*rows*), `pointsize` (*columns*), and `pointtype` (o, +, x, q, d, and t from left to right and top to bottom). The axis labels are the SPICE OPUS default labels and have no special meaning for the plots in the figure

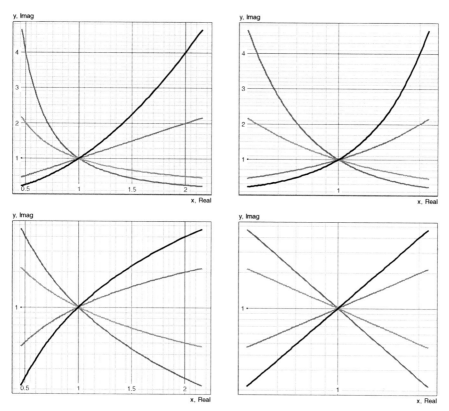

Fig. 5.7 Function $y = x^n$ on variations of the x–y grid (linear, x-log, y-log, and log-log). The axis labels are the SPICE OPUS default labels and have no special meaning for the plots in the figure

variables). `absolute` stands for absolute gain ($|gain|$), `db20` is for gain decibels ($20 \log |gain|$), and `db10` is for power gain decibels ($10 \log |gain|$). The default is `absolute` and `degrees`. The units setting does not affect polar grid labeling. Angle and magnitude in the polar grid are always labeled in degrees and absolute magnitude.

In the right plot in Fig. 5.8 the Smith grid is displayed. There are two variants of the Smith grid available (although both look the same): Smith impedance grid (Smith Z grid, `smithz`) and Smith admittance grid (Smith Y grid, `smithy`). In the Smith grid mode the cursor coordinates are displayed as normalized load impedances or normalized load admittances that belong to individual (x, y) points in the graph representing complex reflectance values. Assuming that the complex reflectance is denoted by $\Gamma = x + iy$, the normalized load impedance (Z) and admittance (Y) are calculated as

$$Z(\Gamma) = R + iX = \frac{1 + \Gamma}{1 - \Gamma} \tag{5.8}$$

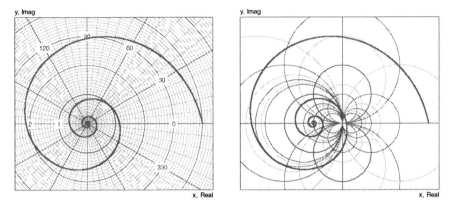

Fig. 5.8 Complex function $z(t) = 4\cos(t)e^{-0.2t} + 4i\sin(t)e^{-0.2t}$ for $t \in [0, 8\pi]$ on polar grid (*left*) and Smith grid (*right*). The axis labels are the SPICE OPUS default labels and have no special meaning for the plots in the figure

and

$$Y(\Gamma) = G + iB = \frac{1 - \Gamma}{1 + \Gamma}. \tag{5.9}$$

5.8.4 Vector Interpretation Modes

As mentioned before, every curve consists of two sets of values: curve and scale values. The way (x, y) coordinates are constructed from these two sets depends on the vector interpretation mode. The vector interpretation mode is common to all vectors in a plot window. All vector interpretation modes (except for the `real` mode) make sense only if the plotted vectors are complex.

The four basic modes are `real`, `imag`, `mag`, and `phase`. The default is `real`. These modes are the first four in the **Graph** menu. The corresponding options in the `Graph` menu (besides switching to respective vector interpretation modes) also switch the grid to x–y mode. Table 5.7 shows the way (x, y) pairs are constructed from curve and scale values.

The units are set to absolute magnitude and degrees (third and fifth row in Table 5.7) when the plot window is created. This behavior can be changed by setting the `units` and `magunits` variables. Note that in the `phase` mode the phase is not unwrapped. In Fig. 5.9 the complex function $z(t) = 0.5\cos(t)e^{-0.2t} + 0.5i\sin(t)e^{-0.2t}$ for $t \in [0, 8\pi]$ is displayed in the `real`, `imag`, `mag`, and `phase` modes.

The `realz`, `imagz`, `realy`, and `imagy` modes interpret the curve values as complex reflectances (Γ). The corresponding normalized load impedance (Z) or admittance (Y) is calculated according to (5.8) and (5.9). Then the real or imaginary

Table 5.7 Vector interpretation for the `real`, `imag`, `mag`, and `phase` vector interpretation modes and various **Graph/Units** settings

Mode	x coordinate	y coordinate
`real`	Re(*curve*)	Re(*scale*)
`imag`	Im(*curve*)	Re(*scale*)
`mag` (`absolute`)	|*curve*|	Re(*scale*)
`mag` (`db20`)	$20\log(\|curve\|)$	Re(*scale*)
`mag` (`db10`)	$10\log(\|curve\|)$	Re(*scale*)
`phase` (`radians`)	arg(*curve*)	Re(*scale*)
`phase` (`degrees`)	arg(*curve*) $\cdot\, 180/\pi$	Re(*scale*)

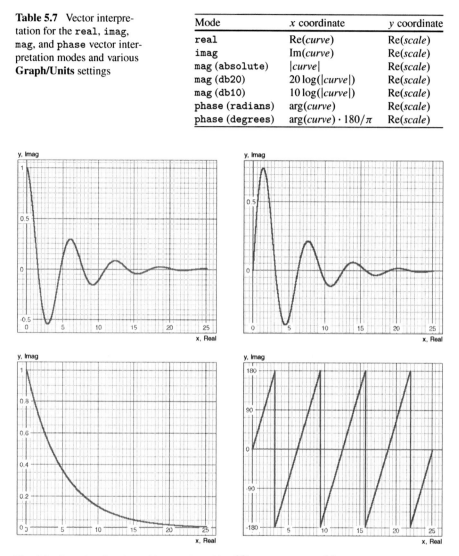

Fig. 5.9 Complex function $z(t) = 0.5\cos(t)e^{-0.2t} + 0.5i\sin(t)e^{-0.2t}$ displayed in the `real`, `imag`, `mag`, and `phase` modes. Units are set to absolute magnitude and degrees. The axis labels are the SPICE OPUS default labels and have no special meaning for the plots in the figure

part of the normalized load impedance or admittance is plotted. See Table 5.8 on how the (x, y) pairs are constructed.

The complex function $z(t) = 0.5\cos(t)e^{-0.2t} + 0.5i\sin(t)e^{-0.2t}$ is plotted in the aforementioned four modes in Fig. 5.10. Switching to one of these four modes in the **Graph** submenu also switches the grid to the x–y grid.

Table 5.8 Vector interpretation for the `realz`, `imagz`, `realy`, and `imagy` vector interpretation modes

Mode	x coordinate	y coordinate
`realz`	$\mathrm{Re}(Z(curve))$	$\mathrm{Re}(scale)$
`imagz`	$\mathrm{Im}(Z(curve))$	$\mathrm{Re}(scale)$
`realy`	$\mathrm{Re}(Y(curve))$	$\mathrm{Re}(scale)$
`imagy`	$\mathrm{Im}(Y(curve))$	$\mathrm{Re}(scale)$

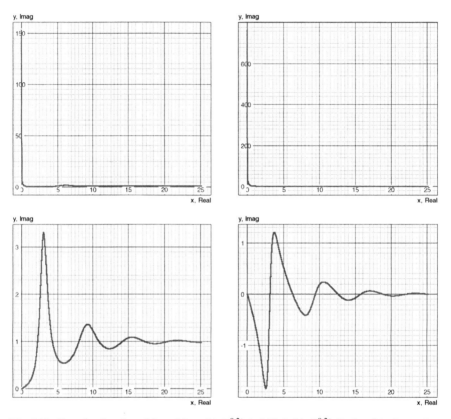

Fig. 5.10 Complex function $z(t) = 0.5\cos(t)e^{-0.2t} + 0.5i\,\sin(t)e^{-0.2t}$ displayed in the `realz`, `imagz`, `realy`, and `imagy` modes. The axis labels are the SPICE OPUS default labels and have no special meaning for the plots in the figure

The last four options in the **Graph** submenu are for the `cx` vector interpretation mode. Besides switching to the `cx` mode these options also switch the grid to one of the following four grid types: x–y, polar, Smith impedance, or Smith admittance grid. (x, y) pairs are constructed from the real and the imaginary parts of the curve values. In this mode the scale values are ignored.

The complex function $z(t) = 0.5\cos(t)e^{-0.2t} + 0.5i\,\sin(t)e^{-0.2t}$ is plotted in the complex `cx` vector interpretation mode on the x–y, the `polar`, and the `smithz` (or the `smithy`) grids in Fig. 5.11. The coordinates displayed in the info frame and in the

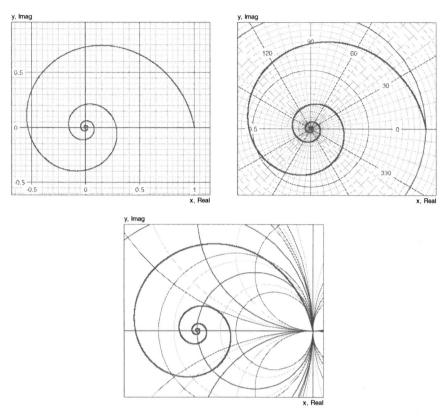

Fig. 5.11 Complex function $z(t) = 0.5 \cos(t)e^{-0.2t} + 0.5i \sin(t)e^{-0.2t}$ displayed in the complex cx mode on x–y, polar, Smith impedance, and Smith admittance grids. Both Smith grids look the same. The difference is only in the coordinates displayed in the info frame. For the smithz (smithy) grid the real and imaginary parts of the normalized impedance (admittance) are denoted by R and X (G and B). The axis labels are the SPICE OPUS default labels and have no special meaning for the plots in the figure

bottom of the plot window depend on the type of the grid and the units chosen in the **Graph/Grid** and **Graph/Units** submenus (or by units and magunits variables).

5.8.5 The plot Command

The syntax of the plot command is
plot [*tagspec*] [*options*] [*vecspec1 vecspec2* ...]

Tag specification
The following tag specifications (*tagspec*) are available.

`create` *tag*
Creates a plot window with the specified tag. Multiple plot windows with the same tag can be created. Of all the windows with the same tag, only the one that was created as last can be accessed by referring to it through its tag.

`append` *tag*
This forces the plot command to refer to the plot window tagged with `tag`. If no such plot window exists, a new window tagged by `tag` is created. All options and vector specifications refer to the specified window. After all options are applied and all vectors are added, the plot window is refreshed.

`quickappend` *tag*
Same as `append`, except that it does not refresh the plot window. For refreshing the window manually, use the `append` tag specification.

`destroy` *tag*
Closes the plot window with the specified tag. If no such window exists, nothing happens. All options and vector specifications following the tag specification are ignored.

`autoscale` *tag*
Scales the specified window so that all vectors are fully visible. If no such window exists, nothing happens. All options and vector specifications following the tag specification are ignored.

The tag specification can also be omitted. In such case a new plot window is created with a tag of the form `plot`*n* where the number *n* is chosen by NUTMEG.

Options
The tag specification is followed by one or more options. If multiple conflicting options are given, the last option is the one that is valid.

`xygrid`
Selects x–y grid. The type of scale (linear or logarithmic) depends on the displayed vectors and on the previous state of the plot window. This is the default grid type.

`xlin`
Switches to x–y grid. Forces a linear scale on the x-axis. Whether or not y-axis will use a logarithmic scale depends on the displayed vectors and on the previous state of the plot window.

`ylin`
Switches to x–y grid. Forces a linear scale on the y-axis. Whether or not the x-axis will use a logarithmic scale depends on the displayed vectors and on the previous state of the plot window.

`xlog`
Forces logarithmic scale on the x-axis.

`ylog`
Forces logarithmic scale on the y-axis.

`polar`
Switches to polar grid.

`smithz` or `smith`
Switches to Smith impedance grid.

`smithy`
Switches to Smith admittance grid.

`xlimit` *low high* or
`xl` *low high*
Forces the given scale range on the x-axis.

`ylimit` *low high* or
`yl` *low high*
Forces the given scaling on the y-axis.

`xindices` *low high* or
`xind` *low high*
Sets the range of indices that are plotted. The setting applies to all vectors plotted in
a plot window.

`xdelta` *step* or
`xdel` *step*
Sets the tick step for the x-axis.

`ydelta` *step* or
`ydel` *step*
Sets the tick step for the y-axis.

`aspect`
Sets the plot window aspect ratio to 1:1.

`xlabel` *text*
Labels the x-axis with the given text. If there are spaces in `text` it should be placed
in single quotes. The default x-axis label is `x, Real`.

`ylabel` *text*
Labels the y-axis with the given text. If there are spaces in `text` it should be placed
in single quotes. The default y-axis label is `y, Imag`.

`title` *text*
Sets the title of the plot window (displayed in the titlebar of the plot window). If
there are spaces in `text` it should be placed in single quotes.

`mode` *mode_name*
Switches the plot window to the given vector interpretation mode. The mode name
is one of the following: `real`, `imag`, `mag`, `phase`, `realz`, `imagz`, `realy`, `imagy`, or
`cx`. The default is `real`.

Vector group specification

A vector group specification (*vecspec*) has the following form:

vec1 vec2 ... [vs scale]

The vectors *vecn* are the basis for the curves that are plotted. If the *scale* vector is not specified, the default scale vector of the current plot is used. There can be multiple vector group specifications in a single plot command.

If any of the vectors has the xlog, ylog, or loglog grid flag set (i.e., default scale vectors of DC, AC, and NOISE analysis results when logarithmic sweep is used) the scale for the respective axis is set to logarithmic. The grid flag is displayed when the vector information is printed with the display command.

The number of plotted points depends on the length of the shortest vector in the vector specification group. The current plot's default scale length is also considered, if the *scale* vector is not specified.

Variables that affect plot windows

Besides arguments to the plot command, the style of the plot window can be set through several variables.

color0 and color1

specify the background and the grid color. They affect the plot window only when it is created. The format for color specification is r*xxx*g*xxx*b*xxx* where individual color components are specified as decimal integers between 0 and 255. The defaults for the background and the grid color are r255g255b255 and r80g80b255.

color2, color3, etc.

specify the colors for the vectors in the vector group specification. If there are more vectors in a vector group specification than there are colors defined for vectors, the defined colors are used in a cyclic manner.

If no colors are defined, there are nine predefined colors available. If there are more than nine vectors in a vector group specification, the colors for additional vectors are selected randomly. The predefined nine colors are listed in the example in Sect. 5.8.2.

plottype

specifies the curve style. It can be point, line, or comb. The default is line. The variable affects the curves when they are added to a plot window.

pointtype

sets the point style when the plottype variable is set to point. pointtype can be o (circles), x (crosses), + (plus signs), q (squares), d (diamonds), or t (triangles). The default is circles (o). The variable affects the curves when they are added to a plot window.

linewidth

specifies the line width. The default is 0 (hairline). The variable affects the curves when they are added to a plot window.

`pointsize`
specifies the point size when the `plottype` variable is set to `point`. The default is 2. The variable affects the curves when they are added to a plot window.

`plotwinwidth` and `plotwinheight`
specify the width and height of a plot window. The two variables affect only newly created plots. The default plot window size is 280×280. The plot window cannot be smaller than 200×200 pixels.

`plotwininfo`
turns on the info frame of a newly created window.

`plotautoident`
turns on automatic identification of curves for newly created windows.

`nogrid`
turns off the grid. Affects the plot window when there is no tag specification or when the `create` or append tag specification is used.

`units`
sets the units for the angle (cursor position for polar grid and phase calculation in the `angle` vector interpretation mode). Possible values are `degrees` and `radians`. The default is `degrees`. Affects the plot window when there is no tag specification or when the `create` or append tag specification is used.

`magunits`
sets the units for the angle (cursor position for polar grid and magnitude calculation in the `mag` vector interpretation mode). Possible values are `absolute`, `db10`, and `db20`. The default is `absolute`. Affects the plot window when there is no tag specification or when the `create` or append tag specification is used.

5.8.6 Examples

The plots in Figs. 5.1–5.3 have been obtained with the following script:

```
destroy all
set plotwinwidth=410
set plotwinheight=360

setplot new
nameplot pl1
set units=radians
set linewidth=2
set plottype=line

* Uncomment this if you want an info frame.
* set plotwininfo

let x=vector(1001)/1000*2*pi
plot sin(x) cos(x) exp(-x) sqrt(x) xlabel 'x' ylabel 'y' title 'Plot example'
```

Figure 5.5 (left):

```
destroy all
set plotwinwidth=250
set plotwinheight=250

setplot new
nameplot pl1
set units=radians
set pointsize=3
set pointtype=o
set linewidth=1

let x=vector(16)/15*1m
let y=sin(2*pi*1k*x)

plot create plottypes
set plottype=comb
plot append plottypes y
set plottype=line
plot append plottypes y+2
set plottype=point
plot append plottypes y+4
```

Figure 5.5 (right):

```
destroy all
set plotwinwidth=250
set plotwinheight=250

setplot new
nameplot pl1
* x is the default scale.
let x=(0;1)
let y=(0;0)

let cnt=0
let degstep=9
let deg=0
set units=degrees
unset plotwininfo
set plottype=line
plot create lines
while cnt le 10
  let x[1]=cos(degstep*cnt)
  let y[1]=sin(degstep*cnt)
  set linewidth={cnt}
  * The default scale is used.
  plot append lines y
  let cnt=cnt+1
end
plot append lines title 'Line demo'
```

Figure 5.7:

```
destroy all
set plotwinwidth=250
set plotwinheight=250

setplot new
nameplot pl1
set plottype=line
set pointstyle=o
```

```
set linewidth=0
set pointsize=0

let x=(vector(101)/100)*2-1
let x=10^(x/3)
plot x^(-2) x^(-1) 1 x x^2 vs x
plot xlog x^(-2) x^(-1) 1 x x^2 vs x
plot ylog x^(-2) x^(-1) 1 x x^2 vs x
plot xlog ylog x^(-2) x^(-1) 1 x x^2 vs x
```

Figure 5.8:

```
destroy all
set plotwinwidth=250
set plotwinheight=250

setplot new
nameplot pl1
set units=radians
let t=vector(10001)/10000*8*pi
let x=4*cos(t)*exp(-0.2*t)
let y=4*sin(t)*exp(-0.2*t)

plot xygrid aspect y vs x
plot polar y vs x

* These two look the same.
* The difference is only in cursor coordinates.
plot smithz y vs x
plot smithy y vs x
```

Figures 5.9–5.11:

```
destroy all
set plotwinwidth=250
set plotwinheight=250

setplot new
nameplot pl1
set units=radians
let t=vector(10001)/10000*8*pi
let x=cos(t)*exp(-0.2*t)
let y=sin(t)*exp(-0.2*t)

set units=degrees
set magunits=absolute

* Create complex vector
let z=(x,y)

plot mode real  z vs t
plot mode imag  z vs t
plot mode mag   z vs t
plot mode phase z vs t

plot mode realz z vs t
plot mode imagz z vs t
plot mode realy z vs t
plot mode imagy z vs t

plot mode cx xygrid aspect z vs t
plot mode cx polar  z vs t
plot mode cx smithz z vs t
plot mode cx smithy z vs t
```

5.9   Control Structures

Control structures control the flow of a NUTMEG script. The basic units of every control structure are blocks of statements. A block of statements consists of statements and `break` and `continue` control lines.

The flow control often depends on *conditions*. A *condition* is a NUTMEG expression whose result is interpreted either as true or as false. If the result of an expression is a vector with all components equal to zero, the interpretation is false. The interpretation is also false if an error occurs at evaluation. In all other cases the interpretation is true.

5.9.1   *if-elseif-else* **Block**

The `if-elseif-else` statement conditionally executes various blocks of statements. There are several forms of this statement available.

```
if condition
      statement_block
end
```

This form executes the *statement_block* if the *condition* is true. The following form has two statement blocks. The second one is executed if the condition is false and the first one if the condition is true.

```
if condition
      statement_block1
else
      statement_block2
end
```

To specify a chain of *conditions* one can use the `elseif` control line.

```
if condition1
      statement_block1
elseif condition2
      statement_block2
      .
      .
      .
else
      statement_blockn
end
```

The `else` control line and *statement_blockn* can be omitted. If *conditionx* is true *statement_blockx* is executed. If none of the specified conditions is true, *statement_blockn* is executed (if specified).

Example:

```
* The user types a color
echo "Enter a color code."
echo "(1=red, 2=greeen, 3=blue)."
```

```
let color=$(-)
if color eq 1
  echo red
elseif color eq 2
  echo green
elseif color eq 3
  echo blue
else
  echo Unknown color code.
end
```

Expression and variable substitution can be used in *conditions*.

5.9.2 *while*, *dowhile*, *and* *repeat* **Loops**

A while loop repeats a *statement_block* while the specified *condition* is true.
The *condition* is checked before every execution of the *statement_block*.
while *condition*
       statement_block
end
   Example:

```
* Print the square roots of numbers from 1 to 10.
let n=1
while n le 10
  echo sqrt({n})={sqrt(n)}
  let n=n+1
end
```

The dowhile loop is similar to the while loop, except that the condition is checked after every execution of the *statement_block*. This way the *statement_block* is always executed at least once.
dowhile *condition*
       statement_block
end
   Expression and variable substitution can be used in *conditions*. Substitutions occur at every iteration of the loop.
   The repeat loop repeats the *statement_block* a specified number of times.
repeat *count*
       statement_block
end
   The *count* must be a numeric constant. If the *count* is to be obtained from a vector, a variable, or an expression, corresponding substitution must be used. Substitutions occur only once per loop execution.
   Example:

```
* Were you ever punished like this?
repeat 100
  echo "Class clown is not a paid position."
end
```

5.9.3 Breaking and Continuing a Loop

Loops can be broken or continued just as in C. To break out of a loop, use the `break` control line.

`break [n]`

The optional argument tells how many loops to break out of. If it is omitted, one is assumed. If n exceeds the number of loop levels available, all of the loops are broken.

Example:

```
* This will print numbers from 1 to 9.
let n=0
while n lt 100
  echo {n}
  if n eq 9
    break
  end
  let n=n+1
end
```

Continuing a loop means that the current iteration is interrupted and a new iteration of a loop is started. To continue a loop, use the `continue` control line.

`continue [n]`

The optional argument tells which loop to continue. By default, it is one (the innermost loop is continued). Larger values of n stand for loop levels outside the innermost loop. If the specified number exceeds the number of loop levels available, the topmost loop is continued.

Example:

```
* This will print two messages for numbers from 0 to 4
* and only a single message for numbers 5 to 9.
let n=0
while n lt 10
  echo msg1 {n}
  let n=n+1
  if n-1 gt 4
    continue
  end
  echo msg2 {n-1}
end
```

5.10 Miscellaneous

5.10.1 Directory and File Management

This section covers the `cd`, `copy`, `move`, and `remove` commands.

Changing the current directory
The current directory can be changed with the `cd` command.

`cd [dir_name]`

In Windows the slash character (/) must be used as the path separator. The backslash character works too, but it must be escaped. It is recommended that the slash character be used as the path separator in Windows. The backslash character does not work in Linux.

Example:

```
* Change current directory to c:\work
cd c:/work

* This works in windows only.
* Note how the backslash is escaped.
cd c:\\work
* This doesn't work in windows.
* The backslash escapes the 'w' character.
cd c:\work
```

The current directory can be retrieved from the read-only variable `workdir`.

Copying, moving, and deleting files

A file can be copied with the `copy` command.

```
copy source destination
```

The source file is copied to the destination file. If the destination file exists, it is overwritten.

To move or rename a file, use the `move` command.

```
move source destination
```

The original file ceases to exist. If the destination file exists, it is overwritten.

To delete a file, use the `remove` command.

```
remove name
```

The command removes a file.

Testing if a file exists

To test whether a file exists, use the `test` command.

```
test exist filename vector
```

If the specified file exists, the vector is set to 1; otherwise, it is set to 0.

Example:

```
* Test if a file exists and delete it.
test exist junk.txt flag
if flag
  remove junk.txt
else
  echo File junk.txt not found and therefore cannot be deleted.
end
```

5.10.2 Obtaining Statistics from the Simulator

To print the version of the SPICE OPUS binary, use the `version` command.

```
version
```

The `rusage` command with no arguments prints the elapsed time since the last call to `rusage`, the total elapsed time, and memory usage information.

Table 5.9 Arguments to rusage

Argument	Meaning
cputime	The amount of time elapsed since the Last rusage cputime call
totalcputime	The total amount of elapsed CPU time
space	Memory usage statistics
temp	Current circuit temperature
tnom	Parameter measurement temperature
equations	Number of circuit equations
time	Total analysis time
totiter	Total iterations count
accept	Accepted time point count
rejected	Rejected time point count
loadtime	Matrix and equation right-hand side load time
reordertime	Matrix reordering time
lutime	LU decomposition time
solvetime	Matrix solve time
trantime	Transient analysis time
tranpoints	Transient time points count
traniter	Transient iterations count
trancuriters	Transient iterations count for last time point
tranlutime	Transient LU decomposition time
transolvetime	Transient matrix solve time
all	All of the above

If an argument is given to rusage, various statistics collected during simulation can be obtained.

rusage *arg1 arg2 ...*

Possible arguments to rusage are listed in Table 5.9.

5.10.3 The *spinit* File

The spinit file is read by SPICE OPUS at startup. It contains a NUTMEG script that customizes the environment for the user. It does not have to start with a # character because it is always treated as a script file. The simulator searches for the spinit file in the scripts subdirectory of the lib directory (see Sect. 2.1).

Example:

```
* Standard spice and nutmeg init file

* Some aliases and variables for setting up the frontend
alias exit quit
alias acct rusage all

* Setting colors in plots
* Background (white)
set color0 = r255g255b255
* Grid (blue)
set color1 = r80g80b255
```

```
* Trace 1 (red)
set color2 = r220g000b000
* Trace 2 (green)
set color3 = r000g180b000
* Trace 3 (blue)
set color4 = r000g000b200
* Trace 4 (violet)
set color5 = r210g060b210
* Trace 5 (black)
set color6 = r000g000b000
* Trace 6 (yellow)
set color7 = r150g180b020
* Trace 7 (ligt red)
set color8 = r255g000b128
* Trace 8 (light green)
set color9 = r000g128b128
* Trace 9 (light violet)
set color10 = r100g000b230


* Plot window size
set plotwinwidth=360
set plotwinheight=360
* Comment this out if you don't want the info box to be shown on plot windows
set plotwininfo
* Uncomment this if you want automatic instead of manual
* vector identification (right-click) in plot windows
* set plotautoident

* Default plot type
set plottype=normal
* set plottype=point
* set plottype=comb

* Default line weight
set linewidth=1

* Uncomment this if you want Spice Opus to automatically remove circuits with
* errors from memory.
* set badcktfree

* Uncomment this if you don't want to execute the commands in the .control
* block if the circuit is bad. This makes spice crash less often.
* You must comment out badcktfree too.
* set badcktstop

* This is the space-separated path for .include, .lib and source.
* Add your own directories here.
* set sourcepath = ( $sourcepath directory1 directory2 )
```

Chapter 6
Mathematical Background

This chapter attempts to explain the mechanisms of SPICE OPUS in formal terms. You can certainly use a circuit simulator without understanding its mathematical background, just as you can drive a car without knowing what is going on under the hood. This may sound like a good excuse to skip this chapter, in which case many phenomena will just have to go unexplained. For instance, why does SPICE only compute nodal voltages and currents through independent voltage sources? Or why is it that sometimes a simple operating point analysis will last longer than a complex transient? However, advanced users will find this chapter very useful when working with tricky circuits that require manual adjustments of the many simulator parameters available in SPICE OPUS. Questions like the following are tackled.

- When does one switch from the default trapezoidal integration algorithm to Gear integration?
- What exactly is the meaning of all those cryptic warnings and error messages?
- Where is the source of convergence problems and how can one avoid them?
- How accurate are the obtained simulation results?

In order to be as comprehensive as possible, we will proceed systematically from very simple linear resistive networks and later add nonlinear and dynamic behavior.

6.1 Linear Resistive Networks

We do have to start with some 101-course basics, but we will move very fast from there.

Let us observe an arbitrary network consisting of branches numbered $1, 2, \ldots, b$ with two terminals each, interconnected through $n + 1$ nodes numbered $0, 1, \ldots, n$. We know that voltages cannot be considered absolute, so we need to name exactly one reference node, also referred to as the ground node. Let the ground node be the one numbered 0. So we have n nodal voltages across each respective node to the ground node $v_{n_1}, v_{n_2}, \ldots, v_{n_n}$. Each of the b branches is conducting a branch current $i_{b_1}, i_{b_2}, \ldots, i_{b_b}$, which is flowing from the respective positive branch terminal t_+ to

T. Tuma and Á. Bűrmen, *Circuit Simulation with SPICE OPUS*, Modeling and
Simulation in Science, Engineering and Technology, DOI 10.1007/978-0-8176-4867-1_6,
© Birkhäuser Boston, a part of Springer Science+Business Media, LLC 2009

the negative one t_-. We also formulate voltages across each branch $v_{b_1}, v_{b_2}, \ldots, v_{b_b}$, that is, from the respective positive branch terminal t_+ to the negative t_-.

We have barely started and we already have a lot of indices in our expressions, so let us simplify the notation by introducing vectors. We will refer to column vectors $\mathbf{v}_n$, $\mathbf{v}_b$, and $\mathbf{i}_b$ rather than their individual components,

$$\mathbf{v}_n = \begin{bmatrix} v_{n_1} \\ v_{n_2} \\ \vdots \\ v_{n_n} \end{bmatrix}, \quad \mathbf{v}_b = \begin{bmatrix} v_{b_1} \\ v_{b_2} \\ \vdots \\ v_{b_n} \end{bmatrix}, \quad \mathbf{i}_b = \begin{bmatrix} i_{b_1} \\ i_{b_2} \\ \vdots \\ i_{b_b} \end{bmatrix}. \tag{6.1}$$

Next, we need to determine the node numbers of the positive and negative terminals for each branch. The configuration of the branches is also called the circuit topology and is usually given as a full incidence matrix $\mathbf{A}_f$ with $n + 1$ rows corresponding to nodes and b columns corresponding to branches,

$$\mathbf{A}_f = \begin{bmatrix} a_{0,1} & a_{0,2} & \ldots & a_{0,b} \\ a_{1,1} & a_{1,2} & \ldots & a_{1,b} \\ \vdots & \vdots & \ddots & \vdots \\ a_{n,1} & a_{n,2} & \ldots & a_{n,b} \end{bmatrix}. \tag{6.2}$$

The positive terminal of each branch is marked by $+1$ in the respective row, while the negative terminal is denoted by -1. All other elements in $\mathbf{A}_f$ have zero values. So branch number k in Fig. 6.1 with its positive terminal at node l and its negative terminal at node m would have zero elements in the k-th column of $\mathbf{A}_f$ except for $a_{l,k} = +1$ and $a_{m,k} = -1$. Each column in $\mathbf{A}_f$ thus has exactly two nonzero elements and each row has one nonzero element for each branch connected to it.

Note that, by definition, a positive branch current i_{b_k} is flowing from t_+ to t_- through the branch regardless of the branch type. This seems a bit unnatural when the branch contains an independent voltage source – we expect supply voltage sources to drive supply current through the network, rather than the other way around. This is why you normally see negative supply currents in SPICE OPUS output listings.

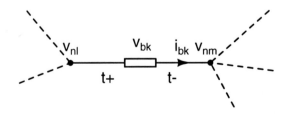

Fig. 6.1 Arbitrary branch in the network

Table 6.1 The vectors and matrices used in the mathematical description of the circuit topology

Notation	Dimension	Description
$n + 1$	–	Number of nodes + one reference node (ground)
b	–	Number of branches
$\mathbf{v}_n$	$n \times 1$	Column vector of nodal voltages
$\mathbf{v}_b$	$b \times 1$	Column vector of branch voltages
$\mathbf{i}_b$	$b \times 1$	Column vector of branch currents
$\mathbf{A}_f$	$n + 1 \times b$	Full incidence matrix of the network
$\mathbf{A}$	$n \times b$	Reduced incidence matrix of the network

Besides the full incidence matrix $\mathbf{A}_f$, we also need the reduced incidence matrix $\mathbf{A}$, which actually is $\mathbf{A}_f$ minus the first row, describing the connections to the ground node. By reducing the matrix in this way, we are not losing any information because we know that each column in $\mathbf{A}$ with a single nonlinear element has to be connected to ground. In fact, by reducing $\mathbf{A}_f$ to $\mathbf{A}$ we are removing redundant information. We summarize the notation we have so far in Table 6.1.

At this point we are ready to write down Kirchhoff's voltage law applied to each branch,

$$\mathbf{A}^\mathrm{T}\mathbf{v}_n - \mathbf{v}_b = \mathbf{0}, \tag{6.3}$$

and its twin, Kirchhoff's current law applied to each node,

$$\mathbf{A}\mathbf{i}_b = \mathbf{0}. \tag{6.4}$$

In a very formal but elegant way, (6.3) is actually saying that the sum of voltage drops along the loop from ground to positive branch terminal, negative branch terminal and back to ground is equal to zero. This holds for all branches. Equation (6.4) is stating that the sum of all branch currents entering or leaving each node equals zero.

Let us now focus on the devices in the branches, where the functional dependency between the branch voltages $\mathbf{v}_b$ and the respective branch currents $\mathbf{i}_b$ is determined. Because we are discussing only linear resistive networks in this section, we can state the branch equations as a linear function of the branch variables,

$$\mathbf{Y}\mathbf{v}_b + \mathbf{Z}\mathbf{i}_b - \mathbf{e} = \mathbf{0}. \tag{6.5}$$

$\mathbf{Y}$ and $\mathbf{Z}$ are b-dimensional square matrices containing admittance and impedance. In most cases these matrices will have nonzero elements only on their respective diagonal, as most branch relations involve only their own branch voltage and current. However, sometimes we need extra-diagonal nonzero elements to accommodate different kinds of controlled sources. We also need to consider independent voltage and current sources, as they will obviously cause zero rows in $\mathbf{Z}$ and $\mathbf{Y}$, respectively. The additions to the notation are listed in Table 6.2.

Notation	Dimension	Description
Table 6.2 The vector and the matrices used in the mathematical description of the circuit's branches		
$\mathbf{e}$	$b \times 1$	Column vector of excitation
$\mathbf{Y}$	$b \times b$	Admittance coefficient matrix
$\mathbf{Z}$	$b \times b$	Impedance coefficient matrix

6.1.1 Circuit Tableau

Equation (6.5) can be viewed as an extension to the well-known Ohm's law, so we can set up a complete equation set for any linear resistive network. To this end we simply combine (6.3), (6.4), and (6.5) in matrix notation,

$$\begin{bmatrix} \mathbf{A}^{\mathrm{T}} & -\mathbf{I} & \mathbf{0} \\ \mathbf{0} & \mathbf{0} & \mathbf{A} \\ \mathbf{0} & \mathbf{Y} & \mathbf{Z} \end{bmatrix} \begin{bmatrix} \mathbf{v}_n \\ \mathbf{v}_b \\ \mathbf{i}_b \end{bmatrix} - \begin{bmatrix} \mathbf{0} \\ \mathbf{0} \\ \mathbf{e} \end{bmatrix} = \mathbf{0}. \qquad (6.6)$$

The equation set (6.6) is also known as the circuit tableau, where $\mathbf{A}^{\mathrm{T}}$ denotes the transposition of the reduced incidence matrix and $\mathbf{I}$ is a diagonal unity matrix. In principle, we could solve (6.6) directly and obtain all circuit variables. The equation set, however, is very large as it involves $2b + n$ equations with $2b + n$ unknowns. Due to the topological part, the equation set includes a large trivial portion with integer coefficients. A direct solution would thus be rather inefficient.

But before we continue, it is time we illustrated the circuit tableau with a simple example. Consider the resistive circuit in Fig. 6.2. Besides the ground node, it only has two more nodes and a total of four branches.

First we need to capture the circuit topology by setting up the reduced incidence matrix. To this end we need the branch numbers, the node numbers, and the branch orientation, that is, we need to decide which is the positive and the negative terminal of each branch. With some devices, like the independent current source, we have no choice, because their functionality is terminal dependent, but with simple resistors, we can choose any orientation we want. So let us set up the incidence matrix according to branch current directions as indicated in Fig. 6.2,

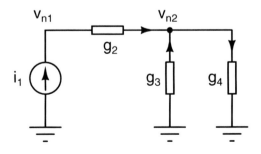

Fig. 6.2 A simple resistive network

$$\mathbf{A} = \begin{bmatrix} -1 & 1 & 0 & 0 \\ 0 & -1 & -1 & 1 \end{bmatrix}. \tag{6.7}$$

With the incidence matrix in place, we can immediately formulate Kirchhoff's voltage law using (6.3),

$$\begin{bmatrix} -1 & 0 \\ 1 & -1 \\ 0 & -1 \\ 0 & 1 \end{bmatrix} \begin{bmatrix} v_{n_1} \\ v_{n_2} \end{bmatrix} - \begin{bmatrix} v_{b_1} \\ v_{b_2} \\ v_{b_3} \\ v_{b_4} \end{bmatrix} = 0, \tag{6.8}$$

as well as Kirchhoff's current law from (6.4),

$$\begin{bmatrix} -1 & 1 & 0 & 0 \\ 0 & -1 & -1 & 1 \end{bmatrix} \begin{bmatrix} i_{b_1} \\ i_{b_2} \\ i_{b_3} \\ i_{b_4} \end{bmatrix} = 0. \tag{6.9}$$

When setting up branch relations, we again have to make some choices. Simple resistors can be set up either as admittances in $\mathbf{Y}$ or as impedances in $\mathbf{Z}$. As we will see later, the admittance form is much more efficient. Therefore, we would like to express branch currents with branch voltages. In this case the impedance matrix is $\mathbf{Z} = -\mathbf{I}$ and (6.5) becomes

$$\begin{bmatrix} 0 & 0 & 0 & 0 \\ 0 & g_2 & 0 & 0 \\ 0 & 0 & g_3 & 0 \\ 0 & 0 & 0 & g_4 \end{bmatrix} \begin{bmatrix} v_{b_1} \\ v_{b_2} \\ v_{b_3} \\ v_{b_4} \end{bmatrix} + \begin{bmatrix} -1 & 0 & 0 & 0 \\ 0 & -1 & 0 & 0 \\ 0 & 0 & -1 & 0 \\ 0 & 0 & 0 & -1 \end{bmatrix} \begin{bmatrix} i_{b_1} \\ i_{b_2} \\ i_{b_3} \\ i_{b_4} \end{bmatrix} - \begin{bmatrix} -i_1 \\ 0 \\ 0 \\ 0 \end{bmatrix} = 0. \tag{6.10}$$

The complete circuit tableau (6.6) for our simple resistive circuit thus looks like this:

$$\begin{bmatrix} -1 & 0 & -1 & 0 & 0 & 0 & 0 & 0 & 0 & 0 \\ 1 & -1 & 0 & -1 & 0 & 0 & 0 & 0 & 0 & 0 \\ 0 & -1 & 0 & 0 & -1 & 0 & 0 & 0 & 0 & 0 \\ 0 & 1 & 0 & 0 & 0 & -1 & 0 & 0 & 0 & 0 \\ 0 & 0 & 0 & 0 & 0 & 0 & -1 & 1 & 0 & 0 \\ 0 & 0 & 0 & 0 & 0 & 0 & 0 & -1 & -1 & 1 \\ 0 & 0 & 0 & 0 & 0 & 0 & -1 & 0 & 0 & 0 \\ 0 & 0 & 0 & g_2 & 0 & 0 & 0 & -1 & 0 & 0 \\ 0 & 0 & 0 & 0 & g_3 & 0 & 0 & 0 & -1 & 0 \\ 0 & 0 & 0 & 0 & 0 & g_4 & 0 & 0 & 0 & -1 \end{bmatrix} \begin{bmatrix} v_{n_1} \\ v_{n_2} \\ v_{b_1} \\ v_{b_2} \\ v_{b_3} \\ v_{b_4} \\ i_{b_1} \\ i_{b_2} \\ i_{b_3} \\ i_{b_4} \end{bmatrix} - \begin{bmatrix} 0 \\ 0 \\ 0 \\ 0 \\ 0 \\ 0 \\ -i_1 \\ 0 \\ 0 \\ 0 \end{bmatrix} = 0. \tag{6.11}$$

As expected, we are facing a 10 by 10 equation set, which can be solved directly. However, in order to improve computational efficiency, let us try to compact this equation set.

6.1.2 Basic Nodal Equations

With some simplifications we can considerably reduce the circuit tableau. For easier handling we simplify our notation (6.6) by appending the excitation vector to the coefficient matrix,

$$
\begin{bmatrix}
\mathbf{A}^\mathsf{T} & -\mathbf{I} & \mathbf{0} & \mathbf{0} \\
\mathbf{0} & \mathbf{0} & \mathbf{A} & \mathbf{0} \\
\mathbf{0} & \mathbf{Y} & \mathbf{Z} & \mathbf{e}
\end{bmatrix}
\begin{bmatrix}
\mathbf{v}_n \\
\mathbf{v}_b \\
\mathbf{i}_b \\
-1
\end{bmatrix}
= \mathbf{0}.
\tag{6.12}
$$

Then we rearrange the order of variables as well as the order of equations. In terms of matrices, we are permuting columns and rows. Specifically, we move column 1 in (6.12) between columns 3 and 4. At the same time, we exchange rows 2 and 3, thus getting

$$
\begin{bmatrix}
-\mathbf{I} & \mathbf{0} & \mathbf{A}^\mathsf{T} & \mathbf{0} \\
\mathbf{Y} & \mathbf{Z} & \mathbf{0} & \mathbf{e} \\
\mathbf{0} & \mathbf{A} & \mathbf{0} & \mathbf{0}
\end{bmatrix}
\begin{bmatrix}
\mathbf{v}_b \\
\mathbf{i}_b \\
\mathbf{v}_n \\
-1
\end{bmatrix}
= \mathbf{0}.
\tag{6.13}
$$

As the right-hand side of the equation is zero, we are allowed to multiply the left-hand side with any nonsingular matrix expression as long as the dimensions of the matrix match. We choose a unity matrix with the addition of $\mathbf{Y}$ in the first column of the second row. It is not difficult to see that the inverse of such a matrix always exists, thus guaranteeing nonsingularity. After the multiplication is carried out, we get

$$
\begin{bmatrix}
\mathbf{I} & \mathbf{0} & \mathbf{0} \\
\mathbf{Y} & \mathbf{I} & \mathbf{0} \\
\mathbf{0} & \mathbf{0} & \mathbf{I}
\end{bmatrix}
\begin{bmatrix}
-\mathbf{I} & \mathbf{0} & \mathbf{A}^\mathsf{T} & \mathbf{0} \\
\mathbf{Y} & \mathbf{Z} & \mathbf{0} & \mathbf{e} \\
\mathbf{0} & \mathbf{A} & \mathbf{0} & \mathbf{0}
\end{bmatrix}
\begin{bmatrix}
\mathbf{v}_b \\
\mathbf{i}_b \\
\mathbf{v}_n \\
-1
\end{bmatrix}
=
\begin{bmatrix}
-\mathbf{I} & \mathbf{0} & \mathbf{A}^\mathsf{T} & \mathbf{0} \\
\mathbf{0} & \mathbf{Z} & \mathbf{Y}\mathbf{A}^\mathsf{T} & \mathbf{e} \\
\mathbf{0} & \mathbf{A} & \mathbf{0} & \mathbf{0}
\end{bmatrix}
\begin{bmatrix}
\mathbf{v}_b \\
\mathbf{i}_b \\
\mathbf{v}_n \\
-1
\end{bmatrix}
= \mathbf{0}.
\tag{6.14}
$$

By this transformation we have eliminated branch voltages $\mathbf{v}_b$ from the equation set, much as one step of a Gaussian elimination would eliminate one variable. The only difference is that we are working with entire submatrices instead of real numbers. Instead of the $2b + n$ dimensional equation set, we now have only $b + n$ independent remaining equations,

$$
\begin{bmatrix}
\mathbf{Z} & \mathbf{Y}\mathbf{A}^\mathsf{T} & \mathbf{e} \\
\mathbf{A} & \mathbf{0} & \mathbf{0}
\end{bmatrix}
\begin{bmatrix}
\mathbf{i}_b \\
\mathbf{v}_n \\
-1
\end{bmatrix}
= \mathbf{0},
\tag{6.15}
$$

which can be solved directly. The solution ($\mathbf{i}_b$ and $\mathbf{v}_n$) would then be substituted back into (6.14) to obtain $\mathbf{v}_b$. The trick was to arrange for a zero column below the diagonal element $-\mathbf{I}$ in (6.14). If we repeat this strategy once more, we can eliminate $\mathbf{i}_b$ as well. So this time we multiply the system by a unity matrix with the addition of $-\mathbf{A}\mathbf{Z}^{-1}$ in the second column of the third row. Of course, we are assuming that the inverse of the impedance matrix $\mathbf{Z}^{-1}$ exists, which is not always the case, but we will return to this question later. Right now we get

$$
\begin{bmatrix} \mathbf{I} & \mathbf{0} & \mathbf{0} \\ \mathbf{0} & \mathbf{I} & \mathbf{0} \\ \mathbf{0} & -\mathbf{AZ}^{-1} & \mathbf{I} \end{bmatrix} \begin{bmatrix} -\mathbf{I} & \mathbf{0} & \mathbf{A}^{\mathrm{T}} & \mathbf{0} \\ \mathbf{0} & \mathbf{Z} & \mathbf{YA}^{\mathrm{T}} & \mathbf{e} \\ \mathbf{0} & \mathbf{A} & \mathbf{0} & \mathbf{0} \end{bmatrix} \begin{bmatrix} \mathbf{v}_b \\ \mathbf{i}_b \\ \mathbf{v}_n \\ -1 \end{bmatrix} =
$$

$$
\begin{bmatrix} -\mathbf{I} & \mathbf{0} & \mathbf{A}^{\mathrm{T}} & \mathbf{0} \\ \mathbf{0} & \mathbf{Z} & \mathbf{YA}^{\mathrm{T}} & \mathbf{e} \\ \mathbf{0} & \mathbf{0} & -\mathbf{AZ}^{-1}\mathbf{YA}^{\mathrm{T}} & -\mathbf{AZ}^{-1}\mathbf{e} \end{bmatrix} \begin{bmatrix} \mathbf{v}_b \\ \mathbf{i}_b \\ \mathbf{v}_n \\ -1 \end{bmatrix} = \mathbf{0}. \qquad (6.16)
$$

The final reduced equation set only has n independent variables and is the well-known circuit nodal equation set,

$$
-\mathbf{AZ}^{-1}\mathbf{YA}^{\mathrm{T}}\mathbf{v}_n + \mathbf{AZ}^{-1}\mathbf{e} = \mathbf{0}. \qquad (6.17)
$$

We can directly solve (6.17) for the nodal voltage vector $\mathbf{v}_n$. All other circuit variables can be obtained by substitution back in (6.16). Note that we need not solve the equation sets, just evaluate the matrix expressions. As expected, the branch voltages are trivial,

$$
\mathbf{v}_b = \mathbf{A}^{\mathrm{T}}\mathbf{v}_n. \qquad (6.18)
$$

The branch currents, on the other hand, involve the inverse of the impedance coefficient matrix,

$$
\mathbf{i}_b = -\mathbf{Z}^{-1}\mathbf{YA}^{\mathrm{T}}\mathbf{v}_n + \mathbf{Z}^{-1}\mathbf{e}. \qquad (6.19)
$$

We can postpone the discussion about the existence of $\mathbf{Z}^{-1}$ for just a little longer: fortunately, the example circuit in Fig. 6.2 does have a very convenient impedance matrix $\mathbf{Z} = -\mathbf{I}$. The inverse is trivial, so nodal equations (6.17) are greatly simplified, $\mathbf{AYA}^{\mathrm{T}}\mathbf{v}_n - \mathbf{Ae} = \mathbf{0}$. By taking $\mathbf{A}$ from (6.7) and $\mathbf{Y}$ and $\mathbf{e}$ from (6.10), we can directly formulate the nodal equation set for the example circuit,

$$
\begin{bmatrix} g_2 & -g_2 \\ -g_2 & g_2 + g_3 + b_4 \end{bmatrix} \begin{bmatrix} v_{n_1} \\ v_{n_2} \end{bmatrix} - \begin{bmatrix} i_1 \\ 0 \end{bmatrix} = 0. \qquad (6.20)
$$

This equation set is much easier to solve than the entire circuit tableau (6.11), but the trick is to have a simple impedance coefficient matrix $\mathbf{Z}$ which is easily inverted. With resistive devices in branches we always have the choice of expressing the relation in admittance or impedance form. By choosing the admittance formulation, we automatically get -1 on the respective diagonal of $\mathbf{Z}$. Independent current sources will give us a -1 diagonal as well. Independent voltage sources, on the other hand, will necessarily cause a zero column and a zero row, thus rendering a singular $\mathbf{Z}$ without an inverse.

In order to solve this kind of problem, we need to modify the nodal equations.

6.1.3 Modified Nodal Equations

As we have seen in the previous section, any nodal approach is greatly simplified if we can provide a unity impedance coefficient matrix $\mathbf{Z} = -\mathbf{I}$. So let us separate the voltage sources that cannot be expressed using $\mathbf{Z} = -\mathbf{I}$ right from the beginning by splitting the relational equations (6.5) into two parts,

$$\begin{bmatrix} \mathbf{Y}_1 & \mathbf{0} \\ \mathbf{0} & \mathbf{Y}_2 \end{bmatrix}\begin{bmatrix} \mathbf{v}_{b1} \\ \mathbf{v}_{b2} \end{bmatrix} + \begin{bmatrix} \mathbf{Z}_1 & \mathbf{0} \\ \mathbf{0} & \mathbf{Z}_2 \end{bmatrix}\begin{bmatrix} \mathbf{i}_{b1} \\ \mathbf{i}_{b2} \end{bmatrix} - \begin{bmatrix} \mathbf{e}_1 \\ \mathbf{e}_2 \end{bmatrix} = \mathbf{0}. \tag{6.21}$$

Consider all independent voltage sources indexed 1 and everything else indexed 2. Therefore, considering $\mathbf{Z}_1 = \mathbf{0}$, in the complete circuit tableau (6.12) we have

$$\begin{bmatrix} \mathbf{A}_1^\mathrm{T} & -\mathbf{I} & \mathbf{0} & \mathbf{0} & \mathbf{0} & \mathbf{0} \\ \mathbf{A}_2^\mathrm{T} & \mathbf{0} & -\mathbf{I} & \mathbf{0} & \mathbf{0} & \mathbf{0} \\ \mathbf{0} & \mathbf{0} & \mathbf{0} & \mathbf{A}_1 & \mathbf{A}_2 & \mathbf{0} \\ \mathbf{0} & \mathbf{Y}_1 & \mathbf{0} & \mathbf{0} & \mathbf{0} & \mathbf{e}_1 \\ \mathbf{0} & \mathbf{0} & \mathbf{Y}_2 & \mathbf{0} & \mathbf{Z}_2 & \mathbf{e}_2 \end{bmatrix}\begin{bmatrix} \mathbf{v}_n \\ \mathbf{v}_{b1} \\ \mathbf{v}_{b2} \\ \mathbf{i}_{b1} \\ \mathbf{i}_{b2} \\ -1 \end{bmatrix} = \mathbf{0}. \tag{6.22}$$

Similar to the manipulations in the previous section, we permute column 1 between columns 5 and 6 and row 3 to the bottom. The equation set is then multiplied by a nonsingular matrix in order to eliminate branch voltages $\mathbf{v}_{b1}$ and $\mathbf{v}_{b2}$,

$$\begin{bmatrix} \mathbf{I} & \mathbf{0} & \mathbf{0} & \mathbf{0} & \mathbf{0} \\ \mathbf{0} & \mathbf{I} & \mathbf{0} & \mathbf{0} & \mathbf{0} \\ \mathbf{Y}_1 & \mathbf{0} & \mathbf{I} & \mathbf{0} & \mathbf{0} \\ \mathbf{0} & \mathbf{Y}_2 & \mathbf{0} & \mathbf{I} & \mathbf{0} \\ \mathbf{0} & \mathbf{0} & \mathbf{0} & \mathbf{0} & \mathbf{I} \end{bmatrix}\begin{bmatrix} -\mathbf{I} & \mathbf{0} & \mathbf{0} & \mathbf{0} & \mathbf{A}_1^\mathrm{T} & \mathbf{0} \\ \mathbf{0} & -\mathbf{I} & \mathbf{0} & \mathbf{0} & \mathbf{A}_2^\mathrm{T} & \mathbf{0} \\ \mathbf{Y}_1 & \mathbf{0} & \mathbf{0} & \mathbf{0} & \mathbf{0} & \mathbf{e}_1 \\ \mathbf{0} & \mathbf{Y}_2 & \mathbf{0} & \mathbf{Z}_2 & \mathbf{0} & \mathbf{e}_2 \\ \mathbf{0} & \mathbf{0} & \mathbf{A}_1 & \mathbf{A}_2 & \mathbf{0} & \mathbf{0} \end{bmatrix}\begin{bmatrix} \mathbf{v}_{b1} \\ \mathbf{v}_{b2} \\ \mathbf{i}_{b1} \\ \mathbf{i}_{b2} \\ \mathbf{v}_n \\ -1 \end{bmatrix} = \mathbf{0}. \tag{6.23}$$

So far the result is not much different in comparison to the elimination step (6.14) in the previous section,

$$\begin{bmatrix} -\mathbf{I} & \mathbf{0} & \mathbf{0} & \mathbf{0} & \mathbf{A}_1^\mathrm{T} & \mathbf{0} \\ \mathbf{0} & -\mathbf{I} & \mathbf{0} & \mathbf{0} & \mathbf{A}_2^\mathrm{T} & \mathbf{0} \\ \mathbf{0} & \mathbf{0} & \mathbf{0} & \mathbf{0} & \mathbf{Y}_1\mathbf{A}_1^\mathrm{T} & \mathbf{e}_1 \\ \mathbf{0} & \mathbf{0} & \mathbf{0} & \mathbf{Z}_2 & \mathbf{Y}_2\mathbf{A}_2^\mathrm{T} & \mathbf{e}_2 \\ \mathbf{0} & \mathbf{0} & \mathbf{A}_1 & \mathbf{A}_2 & \mathbf{0} & \mathbf{0} \end{bmatrix}\begin{bmatrix} \mathbf{v}_{b1} \\ \mathbf{v}_{b2} \\ \mathbf{i}_{b1} \\ \mathbf{i}_{b2} \\ \mathbf{v}_n \\ -1 \end{bmatrix} = \mathbf{0}. \tag{6.24}$$

As $\mathbf{Z}_1 = \mathbf{0}$ we now have a missing pivot in row 3. To avoid singularity we exchange columns 3 and 4 and move row number 3 to the bottom to obtain

$$
\begin{bmatrix}
-\mathbf{I} & 0 & 0 & 0 & \mathbf{A}_1^{\mathrm{T}} & 0 \\
0 & -\mathbf{I} & 0 & 0 & \mathbf{A}_2^{\mathrm{T}} & 0 \\
0 & 0 & \mathbf{Z}_2 & 0 & \mathbf{Y}_2\mathbf{A}_2^{\mathrm{T}} & \mathbf{e}_2 \\
0 & 0 & \mathbf{A}_2 & \mathbf{A}_1 & 0 & 0 \\
0 & 0 & 0 & 0 & \mathbf{Y}_1\mathbf{A}_1^{\mathrm{T}} & \mathbf{e}_1
\end{bmatrix}
\begin{bmatrix}
\mathbf{v}_{b1} \\
\mathbf{v}_{b2} \\
\mathbf{i}_{b2} \\
\mathbf{i}_{b1} \\
\mathbf{v}_n \\
-1
\end{bmatrix}
= \mathbf{0}. \qquad (6.25)
$$

We cannot eliminate the entire branch current vector anymore; hence, we settle for $\mathbf{i}_{b2}$ only,

$$
\begin{bmatrix}
\mathbf{I} & 0 & 0 & 0 & 0 \\
0 & \mathbf{I} & 0 & 0 & 0 \\
0 & 0 & \mathbf{I} & 0 & 0 \\
0 & 0 & -\mathbf{A}_2\mathbf{Z}_2^{-1} & \mathbf{I} & 0 \\
0 & 0 & 0 & 0 & \mathbf{I}
\end{bmatrix}
\begin{bmatrix}
-\mathbf{I} & 0 & 0 & 0 & \mathbf{A}_1^{\mathrm{T}} & 0 \\
0 & -\mathbf{I} & 0 & 0 & \mathbf{A}_2^{\mathrm{T}} & 0 \\
0 & 0 & \mathbf{Z}_2 & 0 & \mathbf{Y}_2\mathbf{A}_2^{\mathrm{T}} & \mathbf{e}_2 \\
0 & 0 & \mathbf{A}_2 & \mathbf{A}_1 & 0 & 0 \\
0 & 0 & 0 & 0 & \mathbf{Y}_1\mathbf{A}_1^{\mathrm{T}} & \mathbf{e}_1
\end{bmatrix}
\begin{bmatrix}
\mathbf{v}_{b1} \\
\mathbf{v}_{b2} \\
\mathbf{i}_{b2} \\
\mathbf{i}_{b1} \\
\mathbf{v}_n \\
-1
\end{bmatrix}
= \mathbf{0}, \quad (6.26)
$$

to obtain

$$
\begin{bmatrix}
-\mathbf{I} & 0 & 0 & 0 & \mathbf{A}_1^{\mathrm{T}} & 0 \\
0 & -\mathbf{I} & 0 & 0 & \mathbf{A}_2^{\mathrm{T}} & 0 \\
0 & 0 & \mathbf{Z}_2 & 0 & \mathbf{Y}_2\mathbf{A}_2^{\mathrm{T}} & \mathbf{e}_2 \\
0 & 0 & 0 & \mathbf{A}_1 & -\mathbf{A}_2\mathbf{Z}_2^{-1}\mathbf{Y}_2\mathbf{A}_2^{\mathrm{T}} & -\mathbf{A}_2\mathbf{Z}_2^{-1}\mathbf{e}_2 \\
0 & 0 & 0 & 0 & \mathbf{Y}_1\mathbf{A}_1^{\mathrm{T}} & \mathbf{e}_1
\end{bmatrix}
\begin{bmatrix}
\mathbf{v}_{b1} \\
\mathbf{v}_{b2} \\
\mathbf{i}_{b2} \\
\mathbf{i}_{b1} \\
\mathbf{v}_n \\
-1
\end{bmatrix}
= \mathbf{0}. \qquad (6.27)
$$

The two bottom lines actually represent an independent equation set, which is often referred to as the first modification of nodal equations. Besides the nodal voltages $\mathbf{v}_n$, it also involves branch currents of independent voltage sources $\mathbf{i}_{b1}$. Assuming all devices, apart from independent voltage sources, may have a clean admittance form, we can set $\mathbf{Z}_2 = -\mathbf{I}$, yielding

$$
\begin{bmatrix}
\mathbf{A}_1 & \mathbf{A}_2\mathbf{Y}_2\mathbf{A}_2^{\mathrm{T}} \\
0 & \mathbf{Y}_1\mathbf{A}_1^{\mathrm{T}}
\end{bmatrix}
\begin{bmatrix}
\mathbf{i}_{b1} \\
\mathbf{v}_n
\end{bmatrix}
-
\begin{bmatrix}
\mathbf{A}_2\mathbf{e}_2 \\
\mathbf{e}_1
\end{bmatrix}
= \mathbf{0}. \qquad (6.28)
$$

One might be tempted to interpret the zero below pivot $\mathbf{A}_1$ as an elimination of $\mathbf{i}_{b1}$, but note that $\mathbf{A}_1$ is not a square matrix. All unknowns are therefore independent and must be solved for simultaneously.

We have arrived at an equation set of the order $n + b_1$, where b_1 is the number of independent voltage sources in the circuit. The basic nodal equations are easily recognized on the top right-hand side of the matrix in (6.28). They are now framed by b_1 additional unknowns and supplemented with $b1$ equations. Normally a circuit will only have a few voltage sources, so the order of the set of equations is only insignificantly larger than the basic nodal equations (6.17). Looking at any default output of SPICE OPUS, you will notice a listing of all nodal voltages plus currents through independent voltage sources.

So far we have covered both kinds of independent sources as well as arbitrary resistive branches. Next we need to consider controlled sources as well.

Voltage-controlled current sources will have -1 on the diagonal of $\mathbf{Z}$ and an extra-diagonal nonzero coefficient in $\mathbf{Y}$, so they would be included in $\mathbf{Y}_2$ in the modification of (6.28). Voltage-controlled voltage sources will cause an empty row in $\mathbf{Z}$ and an additional nonzero coefficient beside the diagonal 1 in the respective row of $\mathbf{Y}$. Similar to independent voltage sources, they belong in $\mathbf{Y}_1$. Current-controlled current sources will have -1 on the diagonal of $\mathbf{Z}$ in addition to a nonzero element in the respective row. This does not cause singularity, but the inverse of $\mathbf{Z}$ is not that trivial anymore. Basically, they can be included in $\mathbf{Y}_2$. But the problem of current-controlled voltage sources still exists. They have some nonzero coefficient in the respective row of $\mathbf{Z}$, but not on the diagonal. They may or may not cause singularities, depending on the circuit topology. To solve this problem, we would need to introduce yet another modification of the basic tableau. The procedure is not difficult, but it is rather complicated because the relational equations now need to be split four ways. In order to include all types of controlled sources, SPICE OPUS is actually based on a more complex modification than (6.28).

However, the order of the set of equations and all its properties are heavily dominated by the basic nodal equations, so we will focus on these equations in the following section. Remember that all the procedures described next will equally apply to any modification.

6.2 Solving the Set of Circuit Equations

In this section we will look at ways to solve sets of linear equations. In general, a linear equation set may be formulated as n linear expressions with n unknowns. Assuming a square coefficient matrix $\mathbf{G}$, a constant vector $\mathbf{c}$, and a vector of unknowns $\mathbf{x}$, a linear equation set is formulated as $\mathbf{Gx} = \mathbf{c}$. The explicit notation of individual components thus is

$$g_{1,1}\, x_1 + g_{1,2}\, x_2 + \cdots + g_{1,n}\, x_n = c_1$$
$$g_{2,1}\, x_1 + g_{2,2}\, x_2 + \cdots + g_{2,n}\, x_n = c_2$$
$$\vdots$$
$$g_{n,1}\, x_1 + g_{n,2}\, x_2 + \cdots + g_{n,n}\, x_n = c_n. \tag{6.29}$$

As we have established in the previous section, SPICE OPUS is actually based on a high order modification of the basic nodal equation set, but all equation set properties are heavily dominated by the basic nodal equations, so we will assume our general equation set (6.29) to be formulated according to (6.17). Assuming $\mathbf{Z} = -\mathbf{I}$, the coefficient matrix becomes $\mathbf{G} = \mathbf{AYA}^\mathrm{T}$, the constant vector $\mathbf{c} = \mathbf{Ae}$, and the unknowns $\mathbf{x} = \mathbf{v}_n$.

6.2.1 Sparse Matrices

A nonsingular circuit must necessarily have at least one branch more than there are nodes $b > n$; moreover, real-life circuits will have something like two branches for each node $b \approx 2n$. According to (6.2) each row of incidence matrix $\mathbf{A}_f$ will thus have an average of about four nonzero elements. Especially with large circuits, the reduced incidence matrix $\mathbf{A}$ will yield an insignificantly small lower row element average. The majority of elements in $\mathbf{A}$ will be zero. Matrices with mostly zero elements are referred to as sparse matrices.

Just looking at the definition (6.5) of the admittance matrix, we observe that $\mathbf{Y}$ is sparse as well. So both components of the coefficient matrix $\mathbf{G} = \mathbf{AYA}^{\mathsf{T}}$ are sparse. This should make $\mathbf{G}$ sparse too.

Consider the arbitrary admittance branch y_k in Fig. 6.1. It causes only two nonzero elements $a_{l,k} = +1$ and $a_{m,k} = -1$ in the kth column of $\mathbf{A}$. It also places a single nonzero coefficient y_k on the diagonal of $\mathbf{Y}$. It is not difficult to see that product $\mathbf{AY}$ will yield only two nonzero elements in the kth column, namely $+y_k$ in row l and $-y_k$ in row m. Just as the multiplication by $\mathbf{A}$ has propagated the admittance y_k column-wise, the subsequent multiplication by its transposition $(\mathbf{AY})\mathbf{A}^{\mathsf{T}}$ will additionally propagate y_k row-wise. Therefore, an arbitrary admittance y_k between nodes l and m only contributes to four coefficients in G, namely to the diagonal $g_{l,l} = +y_k$, $g_{m,m} = +y_k$ and to the anti-diagonal $g_{l,m} = -y_k$, $g_{m,l} = -y_k$.

With approximately four branch connections to each node, we have the sum of all four admittances on the diagonal in addition to four nondiagonal nonzero elements in each row and column of $\mathbf{G}$. So we have typically something like five nonzero elements per row and column in $\mathbf{G}$ regardless of the circuit size. A circuit with 100 nodes will give us a 100 by 100 coefficient matrix $\mathbf{G}$, but of the 10,000 coefficients only about 500 will be nonzero. $\mathbf{G}$ definitely qualifies as a sparse matrix.

Consequently, a majority of coefficients in equation set (6.29) are equal to zero, and we can take advantage of this fact when attempting to solve (6.29).

6.2.2 LU Decomposition

One way of solving linear equation sets is by decomposing the square coefficient matrix $\mathbf{G}$ into the product of a lower triangular matrix $\mathbf{L}$ and an upper triangular matrix $\mathbf{U}$,

$$\mathbf{G} = \mathbf{LU}. \tag{6.30}$$

By definition, all elements above the diagonal of a lower triangular matrix are equal to zero, $l_{i,j} = 0\ \forall i < j$. Similarly, all elements below the diagonal of an upper triangular matrix are zero, $u_{i,j} = 0\ \forall i > j$. Consequently, the components of the product in (6.30) are

$$
\begin{bmatrix}
g_{1,1} & g_{1,2} & \cdots & g_{1,n} \\
g_{2,1} & g_{2,2} & \cdots & g_{2,n} \\
\vdots & \vdots & \ddots & \vdots \\
g_{n,1} & g_{n,2} & \cdots & g_{n,n}
\end{bmatrix}
=
\begin{bmatrix}
l_{1,1} & 0 & \cdots & 0 \\
l_{2,1} & l_{2,2} & \cdots & 0 \\
\vdots & \vdots & \ddots & \vdots \\
l_{n,1} & l_{n,2} & \cdots & l_{n,n}
\end{bmatrix}
\begin{bmatrix}
u_{1,1} & u_{1,2} & \cdots & u_{1,n} \\
0 & u_{2,2} & \cdots & u_{2,n} \\
\vdots & \vdots & \ddots & \vdots \\
0 & 0 & \cdots & u_{n,n}
\end{bmatrix}.
$$

$$(6.31)$$

After the multiplications we get n^2 individual equations with $n^2 + n$ unknowns,

$$
g_{i,j} = \sum_{k=1}^{\min(i,j)} l_{i,k} u_{k,j} \quad \forall i, j = 1, 2, \ldots, n. \tag{6.32}
$$

Obviously, the decomposition is not unique, because we have more unknowns than equations. We can freely choose the values for n variables. Usually the diagonal of **L** is set to 1,

$$
l_{i,i} = 1 \quad \forall i = 1, 2, \ldots, n. \tag{6.33}
$$

We can now formally resolve (6.32) for the individual components of **L** and **U** as

$$
l_{i,j} = \frac{1}{u_{j,j}} \left(g_{i,j} - \sum_{k=1}^{j-1} l_{i,k} u_{k,j} \right) \quad \forall i = 1, 2, \ldots, n, \quad j = 1, 2, \ldots, i-1 \tag{6.34}
$$

and

$$
u_{i,j} = g_{i,j} - \sum_{k=1}^{i-1} l_{i,k} u_{k,j} \quad \forall i = 1, 2, \ldots n, \quad j = i, i+1, \ldots, n. \tag{6.35}
$$

These expressions look complex, but if we compute them in the right order, then each equation holds only one unknown. All $l_{i,j}$ and $u_{i,j}$ are computed as a difference between the corresponding element $g_{i,j}$ and a sum of products of previously computed $l_{i,k}$ and $u_{k,j}$, that is, elements with lower indices $k < i, j$. This is very convenient, because we can store the components of **L** and **U** in the original matrix **G** as we proceed. Note that the unit diagonal of **L** does not need storing. The decomposition can proceed row-wise or column-wise, but in our case we prefer an alternating pattern between rows and columns,

$$
\begin{bmatrix}
u_{1,1} & \cdots & & \vdots & & \vdots & \\
\vdots & \ddots & & u_{k,i} & & u_{k,j} & \\
& & & \vdots & & \vdots & \\
\cdots & l_{i,k} & \cdots & g_{i,i} & \cdots & g_{i,j} & \cdots \\
& & & \vdots & \ddots & & \\
\cdots & l_{j,k} & \cdots & g_{j,i} & & &
\end{bmatrix}
\mapsto
\begin{bmatrix}
u_{1,1} & \cdots & & \vdots & & \vdots & \\
\vdots & \ddots & & u_{k,i} & & u_{k,j} & \\
& & & \vdots & & \vdots & \\
\cdots & l_{i,k} & \cdots & u_{i,i} & \cdots & u_{i,j} & \cdots \\
& & & \vdots & \ddots & & \\
\cdots & l_{j,k} & \cdots & l_{j,i} & & &
\end{bmatrix}.
$$

$$(6.36)$$

In each step we first finish the row, mapping $g_{i,i} \ldots g_{i,n}$ to $u_{i,i} \ldots u_{i,n}$ and then the respective column $g_{i+1,i} \ldots g_{n,i}$ to $l_{i+1,i} \ldots l_{n,i}$. This pattern is very important because **G** is a sparse matrix and needs special treatment, as we shall see in the next section.

The decomposition of **G** in itself does not solve our equation set directly. Two more steps are needed: a forward and a backward substitution. Our equation set (6.30) can now be formulated as $(\mathbf{LU})\mathbf{x} = \mathbf{c}$ or $\mathbf{L}(\mathbf{Ux}) = \mathbf{c}$. By introducing a new vector of unknowns $\mathbf{Ux} = \mathbf{y}$, we first solve $\mathbf{Ly} = \mathbf{c}$ for $\mathbf{y}$. As **L** is lower triangular, only a forward substitution is needed,

$$y_i = c_i - \sum_{j=1}^{i-1} l_{i,j} y_j \quad \forall i = 1, 2, \ldots, n. \tag{6.37}$$

Knowing y, we now solve $\mathbf{Ux} = \mathbf{y}$ for $\mathbf{x}$. This time the coefficient matrix is upper triangular, so we employ a back substitution,

$$x_i = \frac{1}{u_{i,i}} \left(y_i - \sum_{j=i+1}^{n} u_{i,j} x_j \right) \quad \forall i = n, n-1, \ldots, 1. \tag{6.38}$$

Let us summarize the total computational effort. With a little arithmetic we can establish that the LU decomposition in expressions (6.34) and (6.35) demands $\frac{n^3}{3} + \frac{n^2}{2} - \frac{n}{3}$ multiplications and just as many additions. The forward (6.37) and the backward substitution (6.38) each need another $\frac{n^2}{2} + \frac{n}{2}$ operations. So we have a total of $\frac{n^3}{3} + \frac{3n^2}{2} + \frac{2n}{3}$ multiplications and additions to solve the equation set $\mathbf{Gx} = \mathbf{c}$. With large circuits (e.g., $n > 100$) the n^3 term is predominant, so we can state that the computational effort is approximately $\frac{n^3}{3}$.

So far we have discussed the process of LU decomposition from a general point of view. However from the previous section we know that our equation set will only yield approximately five nonzero elements in each row in the coefficient matrix **G**. This is potentially dangerous because we have divisions by $u_{i,i}$ in the decomposition (6.34) as well as the back substitution (6.38). Another consideration is the total number of arithmetic operations $\frac{n^3}{3}$ needed for the solution. This number should be much lower if we employ an efficient pivoting technique as explained in the next section.

6.2.3 Successful Pivoting Techniques

For several reasons, the key elements in the LU decomposition are the pivots, that is, the diagonal elements $u_{i,i}$ and $l_{i,i}$. The latter are not really a problem because we have wisely chosen unit pivots in (6.33). However, the pivots in **U** are used for divisions in (6.34) and (6.38). Obviously, all pivots need to be nonzero; moreover, the pivot values must not be too small compared to all other coefficients as this can cause fatal numerical errors, as will be seen in the following section.

The way to influence the pivots is by matrix permutation. Consider the component notation of $\mathbf{Gx} = \mathbf{c}$ in (6.29). We have n individual equations with n unknowns. The order in which we write the equations surely cannot have any effect on the results and neither can the order of the unknowns. We are allowed to change the order of rows and columns in $\mathbf{G}$ as long as we reflect the changes in $\mathbf{x}$ and $\mathbf{c}$. Specifically, the order of elements in $\mathbf{x}$ must follow the order of columns in $\mathbf{G}$. Similarly, the order of $\mathbf{c}$ must accompany the order of rows in $\mathbf{G}$. Changing the order of rows and columns in a matrix or vector is termed permutation.

So let us consider the birth of pivots (6.35, 6.36) from the permutation point of view. Each pivot is computed from the respective coefficient $g_{i,i}$ and the sum of products of all ls and gs above and to the left of it. If we ensured large, nonzero diagonals $g_{i,i}$ in advance, we should get nonzero pivots $u_{i,i}$ which are large enough not to cause fatal numerical errors. In some unlucky situations, however, the sum of products may be equal or very close to $g_{i,i}$, in which case we get pivots which are either zero or too small. These rare occasions cannot be foreseen.

Nevertheless, the first step is a permutation of $\mathbf{G}$ aiming to bring the largest coefficients to the diagonal. This is not difficult as most diagonal elements are nonzero and relatively large already. Consider the fact explained in Sect. 6.2.1 that all resistive branches contribute to the diagonal. Thus, only a few permutations will be required to prepare the diagonal in $\mathbf{G}$ for solid pivots during the LU decomposition.

But there is another aspect to permutation. The number of arithmetical operations during the decomposition heavily depends on the pattern of nonzero elements. Consider the following example of LU decomposition where the sparsity is lost in the process of elimination (l denotes a nonzero element of the $\mathbf{L}$ matrix and u denotes a nonzero element of the $\mathbf{U}$ matrix):

$$
\begin{bmatrix} g & g & g & g \\ g & g & 0 & 0 \\ g & 0 & g & 0 \\ g & 0 & 0 & g \end{bmatrix} \mapsto \begin{bmatrix} u & u & u & u \\ l & u & u & u \\ l & l & u & u \\ l & l & l & u \end{bmatrix}. \tag{6.39}
$$

Regardless of the coefficient values, we can see that the decomposition algorithm (6.34, 6.35) creates a lot of fill-in. Zero elements become nonzero and the sparsity of $\mathbf{G}$ is lost during the decomposition. The number of operations is in the $\frac{n^3}{3}$ range. Now consider the same example where row 1 is permuted to the bottom and column 1 is permuted to the extreme right,

$$
\begin{bmatrix} g & 0 & 0 & g \\ 0 & g & 0 & g \\ 0 & 0 & g & g \\ g & g & g & g \end{bmatrix} \mapsto \begin{bmatrix} u & 0 & 0 & u \\ 0 & u & 0 & u \\ 0 & 0 & u & u \\ l & l & l & u \end{bmatrix}. \tag{6.40}
$$

From (6.34), (6.35) we see that most *lu* products are zero; thus, they do not require computation and are not causing any fill-in at all. Of course, this simple example is specially designed to show the effect of permutation on the fill-in, but in real situations fill-in control is much less efficient.

The exact pivoting algorithm as it is implemented in SPICE OPUS would exceed the scope of this book. So we present the basic algorithm only.

1. Scan all nonzero coefficients in **G** and tag the ones that are not too small to become pivot. These are the pivot candidates.
2. Let $k = 1$.
3. Loop through all pivot candidates in $g_{i,j} \; \forall i, j \geq k$ and count the number of fill-ins each would generate if selected as pivot. Select the one which causes the least number of fill-ins. In case of a tie, select the larger one.
4. Permute the pivot to position $g_{i,i}$ and add all fill-ins caused to $g_{i,j} \; \forall i, j > k$.
5. Let $k = k + 1$ and loop back to step 3 until $k = n$.

Note that the above pivoting algorithm is performed before the actual decomposition. It simulates the decomposition fill-in only. There is no information regarding the actual coefficient values, except for the initial **G** scan. Consequently, the fill-ins are not pivot candidates. Nevertheless, the algorithm would perform sufficiently in the majority of real cases.

The weak spot of our algorithm might pop up during the subsequent decomposition if some pivot candidate is reduced too much or some fill-in becomes too large compared with its pivot.

6.2.4 Numerical Error Control

In the previous section we kept stressing the fact that pivots should not be too small compared to all other coefficients. In this section we will explain why a small pivot may cause fatal numerical errors and discuss how small is too small. Numerical error analysis is a very difficult subject in mathematics, so we will simplify things as much as possible.

SPICE OPUS represents real numbers in the 64 bit double precision IEEE floating point format. This involves a 52 bit significand with an 11 bit exponent. Consequently, real numbers have a 10^{-16} relative round-off error. Our coefficients in **G** are based on model parameters with relative errors anywhere from 0.5 to, say, 0.001. Thus, the 10^{-16} round-off error is negligible.

We further know that each arithmetic operation will produce a round-off error in the 10^{-16} order of magnitude. So even if we had the worst case of error accumulation it would take something like 10^{14} operations to endanger the numerical quality of the final result. With $\frac{n^3}{3}$ operations per LU decomposition this would take a circuit of over 60,000 nodes with a nonsparse coefficient matrix. So the round-off errors remain at a safe distance at the far end of the significand.

However, there is a situation where the round-off errors can completely flood the entire significand. The root of this evil is the subtraction of two floating point numbers which are very close in value to each other. Suppose that we only had a five digit significand. Then the simple subtraction $5.2356? \cdot 10^3 - 5.2353? \cdot 10^3$ would yield something like $3.????? \cdot 10^{-1}$. A general rule says that if we see the exponent of a sum drop several magnitudes below the largest exponent of any summand, then the round-off error advances for just as many magnitudes in the significand.

To illustrate this, consider the example matrix from (6.39) with the following specific numeric values:

$$
\begin{bmatrix}
g_{1,1} & g_{1,2} & g_{1,3} & g_{1,4} \\
g_{2,1} & g_{2,2} & g_{2,3} & g_{2,4} \\
g_{3,1} & g_{3,2} & g_{3,3} & g_{3,4} \\
g_{4,1} & g_{4,2} & g_{4,3} & g_{4,4}
\end{bmatrix}
=
\begin{bmatrix}
3.0 \cdot 10^{-1} & 8.0 \cdot 10^0 & 5.0 \cdot 10^0 & 1.0 \cdot 10^0 \\
5.0 \cdot 10^0 & 9.0 \cdot 10^0 & 0.0 \cdot 10^0 & 0.0 \cdot 10^0 \\
8.0 \cdot 10^0 & 0.0 \cdot 10^0 & 3.0 \cdot 10^0 & 0.0 \cdot 10^0 \\
7.0 \cdot 10^0 & 0.0 \cdot 10^0 & 0.0 \cdot 10^0 & 1.0 \cdot 10^0
\end{bmatrix} .
$$

To see the round-off error propagation most clearly, we use only two digit significands in LU decomposition,

$$
\begin{bmatrix}
u_{1,1} & u_{1,2} & u_{1,3} & u_{1,4} \\
l_{2,1} & u_{2,2} & u_{2,3} & u_{2,4} \\
l_{3,1} & l_{3,2} & u_{3,3} & u_{3,4} \\
l_{4,1} & l_{4,2} & l_{4,3} & u_{4,4}
\end{bmatrix}
=
\begin{bmatrix}
3.0 \cdot 10^{-1} & 8.0 \cdot 10^0 & 5.0 \cdot 10^0 & 1.0 \cdot 10^0 \\
1.7 \cdot 10^1 & -1.3 \cdot 10^2 & -8.5 \cdot 10^1 & -1.7 \cdot 10^1 \\
2.7 \cdot 10^1 & 1.7 \cdot 10^0 & 1.3 \cdot 10^1 & 1.9 \cdot 10^0 \\
2.3 \cdot 10^1 & 1.4 \cdot 10^0 & 3.1 \cdot 10^{-1} & 1.2 \cdot 10^0
\end{bmatrix} .
$$

So far everything looks just fine. Most elements have actually increased during the decomposition by one magnitude. Next we need a forward and a back substitution,

$$
\begin{bmatrix} y_1 \\ y_2 \\ y_3 \\ y_4 \end{bmatrix}
=
\begin{bmatrix} 2.0 \cdot 10^0 \\ -3.4 \cdot 10^1 \\ 3.8 \cdot 10^0 \\ 4.2 \cdot 10^{-1} \end{bmatrix} ,
\qquad
\begin{bmatrix} x_1 \\ x_2 \\ x_3 \\ x_4 \end{bmatrix}
=
\begin{bmatrix} -7.3 \cdot 10^{-2} \\ 5.9 \cdot 10^{-2} \\ 2.4 \cdot 10^{-1} \\ 3.5 \cdot 10^{-1} \end{bmatrix}
\neq
\begin{bmatrix} -8.1707 \cdot 10^{-2} \\ 4.5393 \cdot 10^{-2} \\ 2.1788 \cdot 10^{-1} \\ 5.7195 \cdot 10^{-1} \end{bmatrix} .
$$

As we can see, x turns out to be far off the correct solution on the right-hand side. Observing the computation of y_4 in (6.37), we can see that its exponent is two magnitudes lower than the largest summand involved $l_{4,1} y_1$. So we have roughly lost two digits out of two in the significand. This means that y_4 is completely wrong. Consequently, the subsequent back substitution is doomed, but not only because it depends on a wrongly calculated y_4. It includes more exponent reductions by addition.

The fate of the substitutions has been decided right at the beginning of the decomposition. The first pivot $g_{1,1}$ is actually the smallest of all nonzero elements in $\mathbf{G}$. Many elements of $\mathbf{L}$ are divided by $g_{1,1}$ in (6.34), which makes them large. These large values then propagate to $\mathbf{U}$ as well (6.35). This is why most elements have increased by one order of magnitude during the decomposition. In itself this is not a problem. However, what goes up must come down, as the final results are independent of the pivot sections. And there are only two ways for coefficients to come down, either by multiplication (division) or by addition (subtraction). Multiplication would be numerically safe. Unfortunately, this is not the case for all coefficients.

So the idea is to avoid large coefficients during the decomposition by keeping the pivots as large as possible. Let us see what happens if we make the permutations (6.40) by running the algorithm from the previous section,

$$
\begin{bmatrix}
g_{2,2} & g_{2,3} & g_{2,4} & g_{2,1} \\
g_{3,2} & g_{3,3} & g_{3,4} & g_{3,1} \\
g_{4,2} & g_{4,3} & g_{4,4} & g_{4,1} \\
g_{1,2} & g_{1,3} & g_{1,4} & g_{1,1}
\end{bmatrix}
=
\begin{bmatrix}
9.0 \cdot 10^0 & 0.0 \cdot 10^0 & 0.0 \cdot 10^0 & 5.0 \cdot 10^0 \\
0.0 \cdot 10^0 & 3.0 \cdot 10^0 & 0.0 \cdot 10^0 & 8.0 \cdot 10^0 \\
0.0 \cdot 10^0 & 0.0 \cdot 10^0 & 1.0 \cdot 10^0 & 7.0 \cdot 10^0 \\
8.0 \cdot 10^0 & 5.0 \cdot 10^0 & 1.0 \cdot 10^0 & 3.0 \cdot 10^{-1}
\end{bmatrix}.
$$

Now the decomposition yields

$$
\begin{bmatrix}
u_{2,2} & u_{2,3} & u_{2,4} & u_{2,1} \\
l_{3,2} & u_{3,3} & u_{3,4} & u_{3,1} \\
l_{4,2} & l_{4,3} & u_{4,4} & u_{4,1} \\
l_{1,2} & l_{1,3} & l_{1,4} & u_{1,1}
\end{bmatrix}
=
\begin{bmatrix}
9.0 \cdot 10^0 & 0.0 \cdot 10^0 & 0.0 \cdot 10^0 & 5.0 \cdot 10^0 \\
0.0 \cdot 10^0 & 3.0 \cdot 10^0 & 0.0 \cdot 10^0 & 8.0 \cdot 10^0 \\
0.0 \cdot 10^0 & 0.0 \cdot 10^0 & 1.0 \cdot 10^0 & 7.0 \cdot 10^0 \\
8.9 \cdot 10^{-1} & 1.7 \cdot 10^0 & 1.0 \cdot 10^0 & -2.5 \cdot 10^1
\end{bmatrix}.
$$

As expected, we have no fill-in at all and only the last row is transformed. Also, we notice that there is no general increase of coefficient values. Next, both substitutions are run to obtain

$$
\begin{bmatrix}
y_2 \\ y_3 \\ y_4 \\ y_1
\end{bmatrix}
=
\begin{bmatrix}
0.0 \cdot 10^0 \\ 0.0 \cdot 10^0 \\ 0.0 \cdot 10^0 \\ 2.0 \cdot 10^0
\end{bmatrix}
\quad
\begin{bmatrix}
x_2 \\ x_3 \\ x_4 \\ x_1
\end{bmatrix}
=
\begin{bmatrix}
4.4 \cdot 10^{-2} \\ 2.1 \cdot 10^{-1} \\ 5.6 \cdot 10^{-1} \\ -8.0 \cdot 10^{-2}
\end{bmatrix}
\approx
\begin{bmatrix}
4.5393 \cdot 10^{-2} \\ 2.1788 \cdot 10^{-1} \\ 5.7195 \cdot 10^{-1} \\ -8.1707 \cdot 10^{-2}
\end{bmatrix}.
$$

This time the results are as accurate as they can be considering that we have used a two digit significand only.

For simplicity, the involved example has been extremely downsized, which makes it unrealistic. However, the principle of round-off error propagation is realistic. The strategy of keeping pivots as large as possible by permutation actually works well in the majority of cases. However, it is not a magic wand – there are special cases where large pivots can do more harm than good.

There are two simulator parameters in SPICE OPUS to control the pivot selection criteria: an absolute tolerance `pivtol` and a relative one `pivrel` with respective default values 10^{-12} and 10^{-3}. The pivot candidates in (6.36) must meet both conditions

$$
\begin{aligned}
&|g_{i,i}| > \texttt{pivtol} \\
&|g_{i,i}| > \texttt{pivrel} \cdot |g_{j,i}| \quad \forall j = i+1, i+2, \ldots, n.
\end{aligned}
\tag{6.41}
$$

Note that the relative tolerance is checked only for elements below the pivot, because only **L** components are divided by the pivot (6.34). Although the default pivot simulator parameters can be changed, even advanced users are strongly discouraged to alter these settings, as unpredictable numeric behavior may occur.

6.3 Introducing Nonlinear Devices

So far we have seen how to set up a linear equation set from a circuit netlist and how to efficiently solve the equations. We have allowed only linear devices, which is a tough assumption in circuit design.

Now we are ready to take the next step by introducing nonlinear devices. In general, we cannot express branch currents and branch voltages explicitly anymore. We now have to use the general expression

$$\mathbf{R}(\mathbf{v}_b, \mathbf{i}_b) = \mathbf{0} \tag{6.42}$$

instead of (6.5). $\mathbf{R}(\mathbf{v}_b, \mathbf{i}_b)$ is a column vector of functions of branch voltages $(\mathbf{v}_b)$ and branch currents $(\mathbf{i}_b)$ describing the nonlinearity contained in each branch,

$$R_1(v_{b_1}, v_{b_2} \ldots v_{b_b}, i_{b_1}, i_{b_2} \ldots i_{b_b}) = 0$$
$$R_2(v_{b_1}, v_{b_2} \ldots v_{b_b}, i_{b_1}, i_{b_2} \ldots i_{b_b}) = 0$$
$$\vdots$$
$$R_b(v_{b_1}, v_{b_2} \ldots v_{b_b}, i_{b_1}, i_{b_2} \ldots i_{b_b}) = 0.$$

Of course, real circuits will not include devices with functional dependences of all circuit variables, so the above equations are the most general case. Also, under normal circumstances not all branches include nonlinear devices; hence, many of the above functions will actually be simple linear expressions as in (6.5).

Nevertheless, our complete general nonlinear equation set reads

$$\begin{bmatrix} \mathbf{A}^\mathsf{T}\mathbf{v}_n - \mathbf{v}_b \\ \mathbf{A}\mathbf{i}_b \\ \mathbf{R}(\mathbf{v}_b, \mathbf{i}_b) \end{bmatrix} = \mathbf{0} \tag{6.43}$$

instead of the more explicit form in (6.6). Although we still have a linear part, which describes the circuit topology, we cannot attempt to solve this equation set analytically. In special cases one can find direct solutions, but in general we have to resort to iterative techniques.

6.3.1 Fixed Point Iteration

One of the most fundamental principles for solving nonlinear equations with computers is fixed point iteration. The algorithm is suitable for solving equations of the form

$$x = f(x), \tag{6.44}$$

not unlike our problem in (6.43). Equation (6.44) is solved by computing the sequence

$$x^{(k+1)} = f(x^{(k)}), \text{ where } k = 0, 1, 2, \ldots, m. \tag{6.45}$$

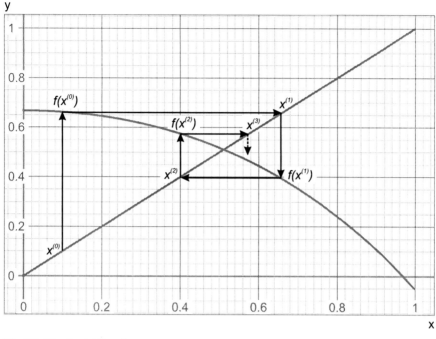

Fig. 6.3 Fixed point iteration convergence

  Obviously, the iteration series needs a starting point $x^{(0)}$ and some convergence criterion to determine the necessary number of iterations m. We assume that a value for m exists such that $|x^{(m)} - x^{(m-1)}| < \epsilon$, for an arbitrary ϵ. The last iteration $x^{(m)}$ is then accepted as the approximation of the solution $x^{(*)}$ to (6.44).

  If we are to use fixed point iteration to solve our nonlinear equation set (6.43), which has the form $g(x) = 0$ instead of (6.44), we need to transform it slightly. The idea is to add x to both sides of the equation to obtain

$$x = x + g(x). \tag{6.46}$$

  Let us look at an example of fixed point iteration. Assume the simple function $g(x) = \frac{5}{3} - e^x$. In this case we know the solution of (6.46) as $x^{(*)} = \ln(\frac{5}{3})$. Further assume that the starting point $x^{(0)} = 0.1$ and the convergence criterion is $\epsilon = 10^{-4}$.

  In Fig. 6.3 you can see a straight line representing the left-hand side of (6.46) and the curve depicting the nonlinear right-hand side. The iteration sequence starting at 0.1 needs 21 steps for convergence,

$$x^{(0)} = 0.1000000$$
$$x^{(1)} = 0.6614957$$
$$x^{(2)} = 0.3904740$$
$$x^{(3)} = 0.5794596$$

$$x^{(4)} = 0.4610527$$
$$\vdots$$
$$x^{(20)} = 0.5107529$$
$$x^{(21)} = 0.5108741.$$

The accepted approximation $x^{(21)} = 0.5108741$ is not far from the solution $x^{(*)} = \ln(\frac{5}{3}) = 0.5108256$. Although this looks promising, there are some problems we should consider.

First, there is the possibility that (6.46) does not have a solution. This means that there is no intersection between the two curves in Fig. 6.3. In this case any iteration is irrelevant. Second, there could be more than one solution, that is, more than one intersection of the curve and the straight line. However, knowing that our equation set is describing a correct electrical circuit with a stable solution, we can usually assume exactly one solution to the equation set. An exception would be a circuit with multiple operating points, like a Schmitt trigger or a flip-flop, where several stable and unstable solutions exist. In these cases fixed point iteration would find only one of the stable solutions, depending on the starting point $x^{(0)}$. So it is the responsibility of the engineer to ensure a correct circuit and manually supply a starting point (see Sect. 6.3.5) if one is needed.

A more serious problem is the question of convergence itself. It is well known that fixed point iteration works only if the absolute derivative of the curve at the intersection is less than unity,

$$\left| f'(x^{(*)}) \right| < 1 \text{ where } f'(x) = \frac{\partial f(x)}{\partial x}. \tag{6.47}$$

Moreover, the speed of convergence depends on the absolute derivative at the intersection. The closer it gets to zero, the faster the convergence. In our example we have the derivative $f'(x) = 1 + g'(x) = 1 - e^x$, which at the intersection is $f'(x^{(*)}) = -\frac{2}{3}$. This ensures convergence, as seen in Fig. 6.3, even if the iteration sequence is a little slow.

Let us now consider the function $g(x) = 5 - e^x$. The solution of (6.46) now is $x^{(*)} = \ln(5)$. We still have the same expression for the derivative, but the gradient at the solution now is $f'(x^{(*)}) = -4$, which is well outside the convergence interval $[-1 \ldots 1]$. Even if we start very close to the solution $x^{(0)} = 1.55$, our fixed point iteration now fails to find the solution, as seen from Fig. 6.4.

We have a clear divergence as a consequence of the steep intersection angle. In electronic circuits we expect a multitude of highly nonlinear devices, so we clearly cannot hope for the basic fixed point algorithm to converge. We need an improved algorithm.

6.3.2 Newton–Raphson Algorithm

Let us return to transformation (6.46) for a moment. We had to add x to both sides of the equation in order to get the right equation form for fixed point iteration. Keep in

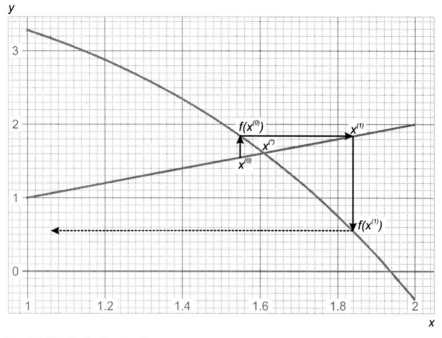

Fig. 6.4 Fixed point iteration divergence

mind that the original equation set (6.43) reads $g(x) = 0$. So before adding x we can multiply both sides by any arbitrary function $k(x)$ without altering the solution. The only condition is that $k(x)$ be nonzero at the solution $x^{(*)}$. In general, our equation now is

$$x = x + k(x)g(x). \tag{6.48}$$

So we have some freedom of choice regarding $k(x)$. We can ensure convergence (6.47) and ideally even speed up the iteration sequence. The best thing would be to achieve a zero gradient at the point of intersection.

Take the right-hand side of (6.48), $f(x) = x + k(x)g(x)$, and formulate the first derivative $f'(x) = 1 + k'(x)g(x) + k(x)g'(x)$. We want $f'(x)$ to be zero at $x = x^{(*)}$. Knowing that $g(x^{(*)}) = 0$, we get $f'(x^{(*)}) = 1 + k(x^{(*)})g'(x^{(*)}) = 0$, from which it follows that $k(x) = -g'(x)^{-1}$. Equation (6.48) now becomes

$$x = x - \frac{g(x)}{g'(x)}. \tag{6.49}$$

With a properly defined circuit there is no danger that $g'(x)$ could be zero at $x = x^{(*)}$, as this would mean multiple adjacent solutions.

This modified equation ensures that fixed point iterations always converge; moreover, the convergence should be very fast. Applying fixed point iteration to this transformation is known as the Newton–Raphson algorithm.

Before continuing, let us see how Newton–Raphson iterations will solve our problem from the previous section, $g(x) = 5 - e^x$. After the transformation we have to solve $x = x - 1 + 5e^{-x}$. Assuming the starting point $x^{(0)} = 1.1$ and the usual convergence criterion $\epsilon = 10^{-4}$, we only need a sequence of four steps,

$$
\begin{aligned}
x^{(0)} &= 1.100000 \\
x^{(1)} &= 1.764355 \\
x^{(2)} &= 1.620841 \\
x^{(3)} &= 1.609503 \\
x^{(4)} &= 1.609438.
\end{aligned}
$$

As expected, we can see the zero gradient intersection in Fig. 6.5 guaranteeing convergence and speeding up things nicely.

But there is always a price to pay. Instead of solving the simple equation $g(x) = 0$ we need to compute the complex expression (6.49) in each iteration. The expression involves the inverse of the first derivative of $g(x)$. So we are looking at a considerable additional computational effort, but we gain shorter iteration loops compensating the extra effort.

Most importantly, we now have a guarantee for convergence. Or do we? We have ensured a zero gradient, but this holds only in the close neighborhood of the solution. In other words, as long as the starting point $x^{(0)}$ is close enough to the solution $x^{(*)}$, the convergence really is guaranteed.

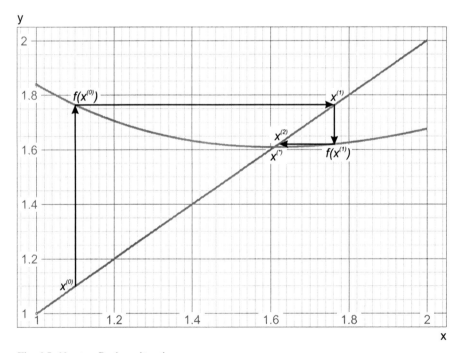

Fig. 6.5 Newton–Raphson iteration convergence

However, the right-hand side of (6.49) might not be well behaved outside the immediate neighborhood of the solution. In these cases Newton–Raphson iterations still can experience convergence problems. In real circuit simulations this is not a very likely scenario but is by no means unheard of. Circuits with strong feedback loops, for instance, tend to diverge, so it is very important to have methods which bring the iterations close enough to the region where the Newton–Raphson algorithm is convergent. These methods are discussed in Sect. 6.3.3.

Finally, let us discuss a special situation where the multiplier $k(x)$ introduces additional solutions to $k(x)g(x) = 0$. We have selected $k(x) = -g'(x)^{-1}$. So we are concerned by the fact that at some x the derivative becomes infinite. This would introduce a new and unrealistic solution. However, an infinite gradient could only be caused by a nonlinearity in (6.43) having zero conductance at some voltage. This is something that we can control on the circuit level, so we do not really have to worry about it here.

Now we are finally ready to apply Newton–Raphson iteration (6.49) to our equation set (6.43). So far we have discussed iterative algorithms for one-dimensional problems. We need to extend them to spaces of arbitrary dimensionality. Instead of a single real number, we now consider x an n-dimensional column vector of variables $x_1, x_2 \ldots x_n$ and $g(x)$ a respective column vector of n nonlinear functions $g_1(x_1, x_2 \ldots x_n), g_2(x_1, x_2 \ldots x_n) \ldots g_n(x_1, x_2 \ldots x_n)$. Our first derivative of $g'(x)$ becomes a square matrix of all partial derivatives, also known as the Jacobi matrix $\mathbf{J}(x)$,

$$\mathbf{g}'(\mathbf{x}) = \mathbf{J}(\mathbf{x}) = \begin{bmatrix} \frac{\partial g_1(\mathbf{x})}{\partial x_1} & \frac{\partial g_1(\mathbf{x})}{\partial x_2} & \cdots & \frac{\partial g_1(\mathbf{x})}{\partial x_n} \\ \frac{\partial g_2(\mathbf{x})}{\partial x_1} & \frac{\partial g_2(\mathbf{x})}{\partial x_2} & \cdots & \frac{\partial g_2(\mathbf{x})}{\partial x_n} \\ \vdots & \vdots & & \vdots \\ \frac{\partial g_n(\mathbf{x})}{\partial x_1} & \frac{\partial g_n(\mathbf{x})}{\partial x_2} & \cdots & \frac{\partial g_n(\mathbf{x})}{\partial x_n} \end{bmatrix}. \tag{6.50}$$

Let us substitute the general expressions for the Jacobi matrix with our circuit tableau notation, where

$$\mathbf{x} = \begin{bmatrix} \mathbf{v}_n \\ \mathbf{v}_b \\ \mathbf{i}_b \end{bmatrix} \text{ and } \mathbf{g}(\mathbf{x}) = \begin{bmatrix} \mathbf{A}^T \mathbf{v}_n - \mathbf{v}_b \\ \mathbf{A}\mathbf{i}_b \\ \mathbf{R}(\mathbf{v}_b, \mathbf{i}_b) \end{bmatrix}. \tag{6.51}$$

In general, the Jacobi matrix appears in the following form:

$$\mathbf{g}'(\mathbf{x}) = \begin{bmatrix} \frac{\partial(\mathbf{A}^T \mathbf{v}_n - \mathbf{v}_b)}{\partial \mathbf{v}_n} & \frac{\partial(\mathbf{A}^T \mathbf{v}_n - \mathbf{v}_b)}{\partial \mathbf{v}_b} & \frac{\partial(\mathbf{A}^T \mathbf{v}_n - \mathbf{v}_b)}{\partial \mathbf{i}_b} \\ \frac{\partial(\mathbf{A}^T \mathbf{i}_b)}{\partial \mathbf{v}_n} & \frac{\partial(\mathbf{A}^T \mathbf{i}_b)}{\partial \mathbf{v}_b} & \frac{\partial(\mathbf{A}^T \mathbf{i}_b)}{\partial \mathbf{i}_b} \\ \frac{\partial \mathbf{R}(\mathbf{v}_b, \mathbf{i}_b)}{\partial \mathbf{v}_n} & \frac{\partial \mathbf{R}(\mathbf{v}_b, \mathbf{i}_b)}{\partial \mathbf{v}_b} & \frac{\partial \mathbf{R}(\mathbf{v}_b, \mathbf{i}_b)}{\partial \mathbf{i}_b} \end{bmatrix}. \tag{6.52}$$

We can partially simplify this expression, because the topological part of the circuit tableau is linear. We get

$$g'(\mathbf{x}) = \begin{bmatrix} \mathbf{A}^T & -\mathbf{I} & \mathbf{0} \\ \mathbf{0} & \mathbf{0} & \mathbf{A}^T \\ \mathbf{0} & \dfrac{\partial \mathbf{R}(\mathbf{v}_b,\mathbf{i}_b)}{\partial \mathbf{v}_b} & \dfrac{\partial \mathbf{R}(\mathbf{v}_b,\mathbf{i}_b)}{\partial \mathbf{i}_b} \end{bmatrix}. \tag{6.53}$$

This expression is very similar to a linear equation set as in (6.6). Actually, we are looking at the linear equivalent of our nonlinear circuit, where all nonlinearities are replaced by gradients at a specific operating point $\mathbf{v}_b$, $\mathbf{i}_b$.

Unfortunately, we will have to compute the inverse of the Jacobi matrix if we want to use the Newton–Raphson iteration form (6.49). The only alternative is to solve a linear equation set instead. Let us first apply the iterative algorithm to (6.49),

$$\mathbf{x}^{(k+1)} = \mathbf{x}^{(k)} - [g'(\mathbf{x}^{(k)})]^{-1} g(\mathbf{x}^{(k)}). \tag{6.54}$$

Let us multiply both sides of this equation by $g'(\mathbf{x})$. We have to be careful because $g'(\mathbf{x})$ is a square matrix, so we must premultiply by $g'(\mathbf{x})$,

$$g'(\mathbf{x}^{(k)})\mathbf{x}^{(k+1)} = g'(\mathbf{x}^{(k)})\mathbf{x}^{(k)} - g(\mathbf{x}^{(k)}). \tag{6.55}$$

The result looks a bit confusing, because the left-hand side now includes the variable vector from the current iteration $\mathbf{x}^{(k)}$ as well as the new one $\mathbf{x}^{(k+1)}$. Keep in mind that the latter is to be computed from (6.55). We are actually looking at a linear equation set of the type $\mathbf{Gx} = \mathbf{c}$. The entire right-hand side is known from iteration k and so is the Jacobi matrix on the left-hand side. From (6.55), (6.51), and (6.53) we have

$$\begin{bmatrix} \mathbf{A}^T & -\mathbf{I} & \mathbf{0} \\ \mathbf{0} & \mathbf{0} & \mathbf{A} \\ \mathbf{0} & \frac{\partial \mathbf{R}}{\partial \mathbf{v}_b} & \frac{\partial \mathbf{R}}{\partial \mathbf{i}_b} \end{bmatrix}^{(k)} \begin{bmatrix} \mathbf{v}_n \\ \mathbf{v}_b \\ \mathbf{i}_b \end{bmatrix}^{(k+1)} = \begin{bmatrix} \mathbf{A}^T & -\mathbf{I} & \mathbf{0} \\ \mathbf{0} & \mathbf{0} & \mathbf{A} \\ \mathbf{0} & \frac{\partial \mathbf{R}}{\partial \mathbf{v}_b} & \frac{\partial \mathbf{R}}{\partial \mathbf{i}_b} \end{bmatrix}^{(k)} \begin{bmatrix} \mathbf{v}_n \\ \mathbf{v}_b \\ \mathbf{i}_b \end{bmatrix}^{(k)} - \begin{bmatrix} \mathbf{A}^T \mathbf{v}_n - \mathbf{v}_b \\ \mathbf{A}\mathbf{i}_b \\ \mathbf{R} \end{bmatrix}^{(k)}. \tag{6.56}$$

For simplicity, we have dropped the explicit functional dependence of $\mathbf{R}$ on $\mathbf{v}_b$, $\mathbf{i}_b$. For the same reason, the iteration superscripts are symbolically placed outside the matrices rather than at each variable instance. We can further resolve part of the right-hand side, so that (6.56) simplifies to

$$\begin{bmatrix} \mathbf{A}^T & -\mathbf{I} & \mathbf{0} \\ \mathbf{0} & \mathbf{0} & \mathbf{A} \\ \mathbf{0} & \frac{\partial \mathbf{R}}{\partial \mathbf{v}_b} & \frac{\partial \mathbf{R}}{\partial \mathbf{i}_b} \end{bmatrix}^{(k)} \begin{bmatrix} \mathbf{v}_n \\ \mathbf{v}_b \\ \mathbf{i}_b \end{bmatrix}^{(k+1)} = \begin{bmatrix} \mathbf{0} \\ \mathbf{0} \\ \frac{\partial \mathbf{R}}{\partial \mathbf{v}_b}\mathbf{v}_b + \frac{\partial \mathbf{R}}{\partial \mathbf{i}_b}\mathbf{i}_b - \mathbf{R} \end{bmatrix}^{(k)}. \tag{6.57}$$

The algorithm of (6.57) resembles our well-known linear equation set in (6.6). The topological part is identical, but the branch equations are obviously a linearization at the point of the previous iteration. This can be illustrated separately as

$$\frac{\partial \mathbf{R}}{\partial \mathbf{v}_b}\mathbf{v}_b^{(k+1)} + \frac{\partial \mathbf{R}}{\partial \mathbf{i}_b}\mathbf{i}_b^{(k+1)} = \frac{\partial \mathbf{R}}{\partial \mathbf{v}_b}\mathbf{v}_b + \frac{\partial \mathbf{R}}{\partial \mathbf{i}_b}\mathbf{i}_b - \mathbf{R}. \tag{6.58}$$

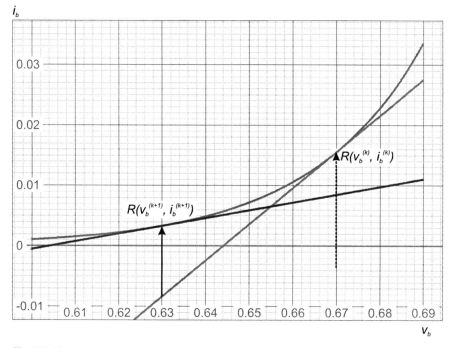

Fig. 6.6 Newton–Raphson equivalent device

All expressions above originate from the k-th iteration except the variable vectors $\mathbf{v}_b$ and $\mathbf{i}_b$ marked with $(k+1)$. The expression clearly describes a $2b$ dimensional plane touching the branch curves at the point where $\mathbf{v}_b^{(k+1)} = \mathbf{v}_b^{(k)}$ and $\mathbf{i}_b^{(k+1)} = \mathbf{i}_b^{(k)}$. The plane is tilted in the same direction as the curve at the point of inflection. In other words, we have a tangential plane at $\mathbf{R}(\mathbf{v}_b^{(k)}, \mathbf{i}_b^{(k)})$.

To illustrate this point, consider a simple diode with the branch relation $R(v_b, i_b) = I_s(e^{\frac{v_b}{v_t}} - 1) - i_b = 0$ where $I_s = 10^{-14}$ A and $v_t = 26$ mV. In Fig. 6.6 this nonlinear relation is depicted by a curve. Substituting R in (6.58) with the diode expression, we get a straight line $\frac{I_s}{v_t} e^{\frac{v_b}{v_t}} \cdot v_b^{(k+1)} - i_b^{(k+1)} = I_s e^{\frac{v_b}{v_t}} (\frac{v_b}{v_t} - 1) + I_s$ touching $R(v_b, i_b) = 0$ at $v_b^{(k+1)} = v_b$.

Suppose the branch voltage in iteration k is $v_b = 0.67$V. We get a line representing the diode as a linear element for the duration of iteration $k + 1$. After solving the linear equation set (6.57), we get a solution somewhere on that line. Suppose we have obtained $v_b = 0.63$V. The next iteration is going to be based on a linearization at that point.

The iterative process will continue in this manner until at some point the iteration falls very close to the previous one for all nonlinear branches. In this case the approximation of the nonlinearity by a straight line is justified, and the result of the linear equation set is considered to be sufficiently close to the solution of the underlying nonlinear equation set.

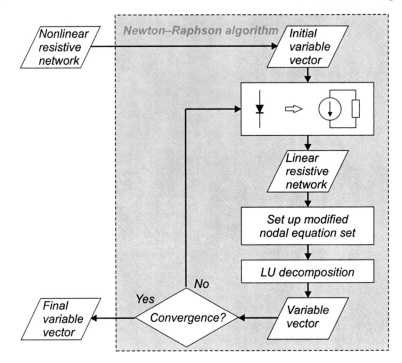

Fig. 6.7 Newton–Raphson algorithm

In this section we have seen that the Newton–Raphson algorithm very efficiently solves nonlinear circuit equations. Moreover, the algorithm can be applied at the device level, as illustrated in Fig. 6.7.

Before any circuit equations are formulated, all branches containing nonlinear devices are transformed according to (6.58); that is, the nonlinear relations are replaced with respective tangents. In terms of devices, the tangent is realized by the parallel combination of an independent current source and a conductance. After replacing all nonlinearities, we obtain a linear resistive network, for which an equation set is formulated, as described in Sect. 6.1, and solved exactly, as explained in Sect. 6.2.

Although we have not yet discussed dynamic devices, we can now consider the operating point and the DC sweep analysis in SPICE OPUS. In both cases we are interested in steady state circuit solutions. By definition, after all transients die out, the voltages across inductors and the currents through capacitors become equal to zero. Consequently, branches containing capacitances are replaced with independent zero current sources while inductances are replaced with independent zero voltage sources. In this way, we get a nonlinear resistive network which can be solved by the Newton–Raphson algorithm, as depicted in Fig. 6.8.

This is the program flow on SPICE OPUS when running the operating point analysis as described in Sect. 4.1. The DC sweep from Sect. 4.2 is also very similar.

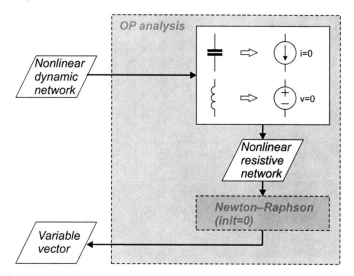

Fig. 6.8 Operating point analysis

It consists of a series of operating point calculations while one or two sources are varied. The individual steps in the sweep are theoretically independent operating point analyses, but SPICE OPUS takes advantage of the fact that adjacent steps in the sweep should produce adjacent operating point solutions. So the solution of each sweep step is wisely used as the initial variable vector in Fig. 6.8 for the next Newton–Raphson iteration. In other words, the initial iteration $k = 0$ in (6.58) is the result of the previous run. In this way the number of Newton–Raphson iterations is greatly reduced. Formally, the program flow of a DC sweep analysis is shown in Fig. 6.9.

So far we have succeeded in solving nonlinear circuits by iteratively solving a linearized transformation until the variable vectors of two consecutive iterations differ by less than a predefined value.

A number of open questions still remain, like how to determine the initial Newton–Raphson iteration, how to detect convergence or the lack thereof, and what to do in cases of divergence.

6.3.3 Convergence Detection

We have already mentioned that the Newton–Raphson algorithm is stopped if two consecutive iterations in (6.57) are close enough. In fact, SPICE OPUS employs a more advanced convergence detection mechanism.

There are three simulator parameters involved in the convergence detection criteria with the default values $\texttt{abstol} = 10^{-12}$, $\texttt{reltol} = 10^{-3}$, and $\texttt{vntol} = 10^{-6}$. SPICE OPUS actually checks not only the difference between iteration k and $k + 1$ but also the one before, $k - 1$. The latest iteration pair must satisfy

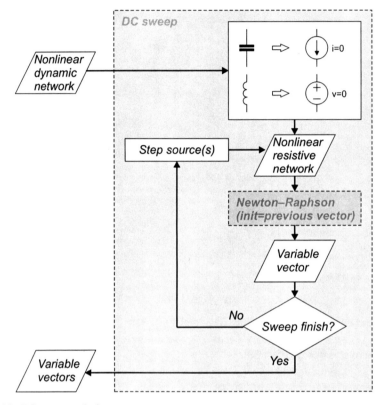

Fig. 6.9 DC sweep analysis

$$|v_{n_i}^{(k+1)} - v_{n_i}^{(k)}| \leq \texttt{reltol} \cdot \max(|v_{n_i}^{(k+1)}|, |v_{n_i}^{(k)}|) + \texttt{vntol}$$
$$|i_{b1_i}^{(k+1)} - i_{b1_i}^{(k)}| \leq \texttt{reltol} \cdot \max(|i_{b1_i}^{(k+1)}|, |i_{b1_i}^{(k)}|) + \texttt{abstol}. \qquad (6.59)$$

All nodal voltage differences must be small enough in a relative as well as an absolute sense. Moreover, all current defined branches (6.28) must fulfill the same criteria. The same applies for the previous iteration pair,

$$|v_{n_i}^{(k)} - v_{n_i}^{(k-1)}| \leq \texttt{reltol} \cdot \max(|v_{n_i}^{(k)}|, |v_{n_i}^{(k-1)}|) + \texttt{vntol}$$
$$|i_{b1_i}^{(k)} - i_{b1_i}^{(k-1)}| \leq \texttt{reltol} \cdot \max(|i_{b1_i}^{(k)}|, |i_{b1_i}^{(k-1)}|) + \texttt{abstol}. \qquad (6.60)$$

After all the above criteria are met, there still is one more check left. The three points form a triangle. The angle at the second point must be smaller than 90°.

$$|v_{n_i}^{(k+1)} - v_{n_i}^{(k-1)}| \leq \sqrt{|v_{n_i}^{(k)} - v_{n_i}^{(k-1)}|^2 + |v_{n_i}^{(k+1)} - v_{n_i}^{(k)}|^2}$$
$$|i_{b1_i}^{(k+1)} - i_{b1_i}^{(k-1)}| \leq \sqrt{|i_{b1_i}^{(k)} - i_{b1_i}^{(k-1)}|^2 + |i_{b1_i}^{(k+1)} - i_{b1_i}^{(k)}|^2}. \qquad (6.61)$$

This complex convergence detection scheme is necessary, because sometimes two adjacent iteration points can be very close to each other only to drift apart again. By defining the simulator parameter `noconviter` one can switch off the double convergence check, but without a very special reason the user should not change any of the simulator parameters that control convergence.

Basically, there are two things that can go wrong in connection with convergence. Although we have chosen the very stable and robust Newton–Raphson iteration algorithm in Sect. 6.3.2, there still is no guarantee of convergence. Large numbers of transistors with their exponential characteristics, interconnected in a certain way, can cause havoc in any iteration algorithm. The iterations either oscillate instead of zooming in on the solution or they even diverge. In both cases the maximum iteration limit is exceeded `itl1` = 100 with still no convergence detected.

The other problem arises when at some iteration in Fig. 6.7 the LU decomposition fails. This can happen for a number of reasons, e.g., diverging iterations can reach absurd numerical values. Whatever the reason, the alarm always goes off in the pivoting algorithm when at some step it cannot find a single coefficient that would satisfy (6.41).

But this is not the end of the Newton–Raphson iteration algorithm in SPICE OPUS; it is rather the beginning of a whole series of tricks. The convergence helpers described in the following section are among the most important quality marks of any circuit simulator.

6.3.4 Automatic Convergence Helpers

SPICE OPUS does not give up easily. There are six tricks built into SPICE OPUS, and all of them are activated automatically. Normally the user does not have to intervene, and the average user does not even have to know about them.

All six automatic convergence helpers rely on two assumptions. First, at least one solution of the circuit must exist. Obviously, no convergence helper can find a nonexisting solution. A circuit with no solution is not unheard of. Very often an erroneous netlist causes singularities; for instance, two parallel voltage sources, or an ideal transformer generating a floating subcircuit. In these cases all convergence helpers will fail.

The second assumption is that the proximity of the solution is convergent for the Newton–Raphson algorithm. In other words, if we can find an initial vector $k = 0$ for (6.57) which is close enough to the final solution, the iteration series will converge.

Normally, Newton–Raphson iterations start from a zero vector. If convergence problems arise, convergence helper algorithms are used to find a suitable starting point.

The simulator tries the convergence helper algorithms in the order in which numbers 1, 2, and 3 are assigned to simulator parameters `gminpriority`, `srcspriority`, and `srclpriority`. The three simulator parameters specify the order in which the GMIN stepping (conductance stepping), source value stepping, and source value lifting convergence helpers are used.

The source value lifting convergence helper can be completely disabled by specifying the `nosrclift` simulator parameter. This is sometimes necessary for circuits that oscillate in the time-domain analysis.

The `noopiter` simulator parameter disables the initial Newton–Raphson iteration, and the simulator goes straight to solving the circuit with convergence helpers. This parameter is useful for circuits that are particularly hard to solve.

The `opdebug` parameter makes the simulator print messages on what exactly is happening with the circuit as various convergence helpers are tried.

The simulator tries to fine-tune convergence helpers so that subsequent analyses find the operating point much faster. This fine-tuning can be disabled by setting the `noautoconv` simulator parameter.

6.3.4.1 Junction Voltage Limitation

This is a very crude measure that prevents Newton–Raphson iterations from running amok. Especially in the beginning of an iteration series, the exponential nature of p–n junctions can cause extreme numerical values going right to the limit of floating point numbers. Sometimes the limit is even violated, causing exceptions in the mathematical package.

In order to prevent these rare occasions, SPICE OPUS sets an upper limit on voltages across a p–n junction regardless of their polarity. By changing the simulator parameter `voltagelimit`, the default limit can even be overridden.

So after each new iteration in (6.57), the branch voltages $\mathbf{v}_b^{(k+1)}$ containing p–n junctions are checked. Any value above or below `voltagelimit` is reset to the respective limit,

$$
v_{b_i}^{(k+1)} = \begin{cases} v_{b_i}^{(k+1)}; & \forall \, |v_{b_i}^{(k+1)}| \leq \texttt{voltagelimit} \\ -\texttt{voltagelimit}; & \forall \, v_{b_i}^{(k+1)} < -\texttt{voltagelimit} \\ \texttt{voltagelimit}; & \forall \, v_{n_i}^{(k+1)} > \texttt{voltagelimit}. \end{cases} \tag{6.62}
$$

Junction voltage limitation not only prevents numerical exceptions – in most cases it also speeds up convergence by limiting the iteration space. Normally the default value of `voltagelimit` does not need to be changed.

6.3.4.2 Damping Newton–Raphson Steps

As we have mentioned, one of the things that can prevent convergence is an oscillating iteration sequence. Instead of quickly zooming in on the solution, the iterations start oscillating around it. A major problem is the detection of oscillations as they are not necessarily periodic. Also note that we are looking for oscillating iterations in (6.57), where each vector has $2b + n$ components. In the worst scenario, we need to detect a stochastic limit cycle in a $2b + n$ dimensional space. This is not a trivial problem, and we do not want to spend more computational effort on the detection of oscillations than on solving the problem itself.

Let us put aside the question of detection and look at the solution first. An oscillation can also be seen as an infinite series of iterates in the proximity of the solution. Intuitively, one would limit the iteration steps. This approach is termed the damped Newton–Raphson algorithm, whatever limiting strategy is employed.

In SPICE OPUS the simulator parameter `sollim` (default value is 10) regulates the damping of iteration steps. The damping algorithm is based on the same voltage and current tolerance thresholds as the convergence detection in (6.59). We first define the thresholds for iteration $k + 1$ (note that i is the index of a component in the solution vector $\mathbf{v}_n, \mathbf{i}_{b1}$),

$$v_{s_i}^{(k+1)} = \texttt{reltol} \cdot \max(|v_{n_i}^{(k+1)}|, |v_{n_i}^{(k)}|) + \texttt{vntol}$$
$$i_{s_i}^{(k+1)} = \texttt{reltol} \cdot \max(|i_{b1_i}^{(k+1)}|, |i_{b1_i}^{(k)}|) + \texttt{abstol}. \tag{6.63}$$

With damping turned on, the Newton–Raphson algorithm still runs exactly as defined in (6.57). Moreover, the same convergence detection is used as in (6.59). However, before feeding the variable vector back to the netlist, all variables which have stepped over their respective threshold (6.63) are limited. The Newton–Raphson flow now has one more transformation in the loop, as can be seen in Fig. 6.10.

Actually, the steps are limited to a fraction determined by `sollim` of the respective threshold. The damping of voltages is expressed in the following equation:

$$v_{n_i}^{(k+1)} = \begin{cases} v_{n_i}^{(k+1)}; & \forall \, |v_{n_i}^{(k+1)} - v_{n_i}^{(k)}| \leq v_{s_i}^{(k+1)} \\ v_{n_i}^{(k)} - \frac{v_{s_i}^{(k+1)}}{\texttt{sollim}}; & \forall \, v_{n_i}^{(k+1)} - v_{n_i}^{(k)} < -v_{s_i}^{(k+1)} \\ v_{n_i}^{(k)} + \frac{v_{s_i}^{(k+1)}}{\texttt{sollim}}; & \forall \, v_{n_i}^{(k+1)} - v_{n_i}^{(k)} > v_{s_i}^{(k+1)}. \end{cases} \tag{6.64}$$

There are three possibilities. If the variable step was already within its convergence tolerance, nothing happens. If the new nodal voltage $v_{n_i}^{(k+1)}$ is lower than its predecessor $v_{n_i}^{(k)}$ by more than the threshold, then this step is reduced to a fraction of the threshold. An excessively long positive voltage step is limited accordingly.

Similarly, all branch currents that occur in the set of modified nodal equations are damped,

$$i_{b1_i}^{(k+1)} = \begin{cases} i_{b1_i}^{(k+1)}; & \forall \, |i_{b1_i}^{(k+1)} - i_{b1_i}^{(k)}| \leq i_{s_i}^{(k+1)} \\ i_{b1_i}^{(k)} - \frac{i_{s_i}^{(k+1)}}{\texttt{sollim}}; & \forall \, i_{b1_i}^{(k+1)} - i_{b1_i}^{(k)} < -i_{s_i}^{(k+1)} \\ i_{b1_i}^{(k)} + \frac{i_{s_i}^{(k+1)}}{\texttt{sollim}}; & \forall \, i_{b1_i}^{(k+1)} - i_{b1_i}^{(k)} > i_{s_i}^{(k+1)}. \end{cases} \tag{6.65}$$

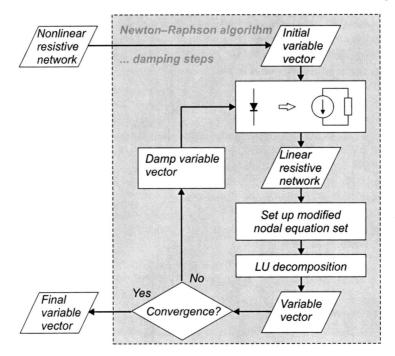

Fig. 6.10 Newton–Raphson iterations with damped steps

With a default damping fraction of 10, it is clear that the damped steps will be extremely tiny. This ensures robust convergence at the expense of many more iterations. The normal iteration limit itl1 with its default value of 100 would thus be exceeded immediately, so SPICE OPUS has a multiplier simulator parameter sollimiter with the default value 10. Therefore, whenever damping is turned on, the iteration limit actually becomes itl1 · sollimiter.

The damping factor sollim, which is actually defining a fraction of the threshold, would normally be greater than one. However, SPICE OPUS also accepts values smaller than one, in which case the steps are limited to more than the tolerance threshold. In this case the damped step is additionally limited by the original step. The idea is to reduce the original Newton–Raphson steps rather than enlarging them.

As we shall see in the following sections, the damped Newton–Raphson algorithm is used when the GMIN stepping and source value stepping algorithms fail to produce a solution. This removes the need for a sophisticated oscillation detection algorithm and works well in practice.

6.3.4.3 GMIN Stepping

The idea behind this convergence helper is strikingly simple and efficient. SPICE OPUS adds a conductance between each node and ground. In this way convergence is greatly improved, at the expense of not solving the original circuit, but rather a transformed version. Intuitively, it is clear that any circuit which causes problems for the Newton–Raphson algorithm can be tamed by adding large conductances between the nodes.

Formally this can be verified. As discussed in Sect. 6.2.1, an arbitrary admittance y_k between nodes l and m contributes to four coefficients in the sparse equation set (6.29), namely to the diagonal $g_{l,l} = +y_k$, $g_{m,m} = +y_k$ and the anti-diagonal $g_{l,m} = -y_k$, $g_{m,l} = -y_k$. In our case one terminal is always the ground node, $m = 0$. Because the reference node is reduced from the incidence matrix, only one contribution is left on the diagonal, $g_{l,l} = +y_k$.

By adding a huge conductance y_k to each node, we get a diagonal with large terms in the equation set. However, a large and linear diagonal guarantees a solution. We have sure convergence because the linearity outweighs all nonlinear problems and the strong diagonal ensures perfect pivots for (6.41).

With large enough conductances we always get a solution. However, this is the solution of the transformed circuit, whereas we want the solution of the original circuit. The trick is to gradually fade out the additional conductances (and that is where the name GMIN stepping comes from). In each step the conductances are reduced and the Newton–Raphson algorithm is run from the previous solution. Eventually, the conductances are stepped down to zero, rendering the original circuit. This procedure, called GMIN stepping, is very efficient although also very computationally intensive. Note that each step involves one complete Newton–Raphson run.

So far we have conveyed just the general idea. Now let us be more specific. There are three simulator parameters with respective default values controlling this process in SPICE OPUS: the minimal conductance for the AC and TRAN analysis, gmin $= 10^{-12}$, the minimal conductance for DC analysis, gmindc $= 10^{-12}$, and the maximal number of GMIN steps before giving up, gminsteps $= 10$.

The simplest way would be to start with an extremely large conductance g and then step it down gradually. However, it is uneconomical to start down-stepping from a needlessly high value. Therefore, SPICE OPUS starts with the default minimal conductance g_{min} and steps it up in huge three decade increments until convergence is reached. In the second part the conductance is gradually stepped down, always starting the Newton–Raphson iteration from the previous solution. If everything works out, the conductance g eventually falls below g_{min} and is finally completely removed. The three steps are illustrated in Fig. 6.11.

The up-stepping part is simple. SPICE OPUS starts with $g = g_{min}$ and multiples it by 1,000 for each unsuccessful Newton–Raphson run, as seen in Fig. 6.12.

The overall step counter n is initialized at the beginning and is limiting the total number of Newton–Raphson runs to the predefined value $n_{max} =$ gminsteps. Whenever this number is reached, the entire GMIN stepping procedure is considered to have failed.

The up-stepping is fast and should take only a few runs. However, singular circuits are bound to fail already during the up-stepping part.

Given a successful up-stepping, we have our first solution. From this point on we need to step-down the conductance. This process is fragile because the individual steps always need to be close enough not to break convergence. The algorithm is depicted in Fig. 6.13.

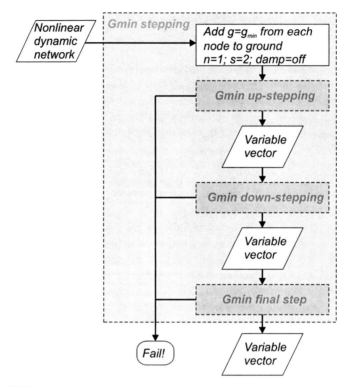

Fig. 6.11 GMIN stepping algorithm

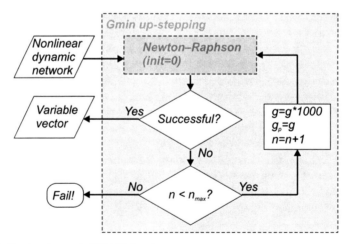

Fig. 6.12 Up-stepping part of GMIN stepping

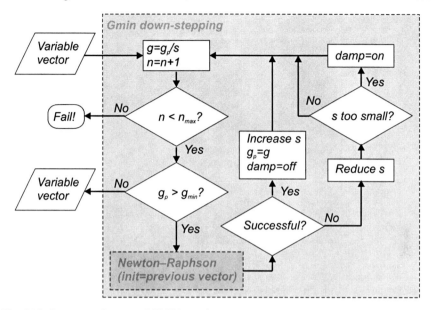

Fig. 6.13 Down-stepping part of GMIN stepping

Each successful step increases the stepping fraction s, speeding up the down-stepping. However, a too large step is likely to break the convergence chain, so the algorithm has to revert to the previous conductance $g = g_p$ and accordingly reduce s. This adaptive step sizing is computationally very efficient, but can lead to extremely small step sizes, in which case Newton–Raphson damping is turned on automatically.

The fact that at some point a very small step in g is causing a loss of convergence does not necessarily mean that the Newton–Raphson algorithm is experiencing oscillating iterations! Putting the Newton–Raphson algorithm in damping mode is based on pure heuristic reasoning. It just turns out that this measure very often works.

In the case of a successful down-stepping without exceeding the maximal number of Newton–Raphson iterations, we have convergence for some $g < g_{min}$. With the default value of $g_{min} = 10^{-12}$ this is almost our original circuit. Only a very small resistive network is still superimposed on each node. So the final part of the algorithm is to completely remove g from the network and do one final Newton–Raphson series starting from the last known solution, as shown in Fig. 6.14.

One could argue that this part is not necessary as the difference between having $g = 10^{-12}$ and $g = 0$ is irrelevant. With normal circuits this is the case, but some singularities in the circuits can prevent convergence in this very last part. For instance, suppose that there is no DC path from some node to ground. By removing the smallest of g, we are facing a dangling node with a theoretically undeterminable voltage.

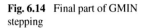

Fig. 6.14 Final part of GMIN stepping

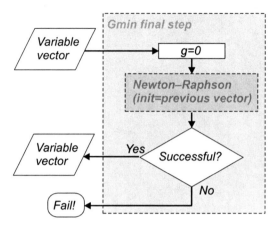

Looking back at the entire GMIN stepping procedure in Fig. 6.11, it is obvious that the default maximal Newton–Raphson count of `gminsteps` $= 10$ is rather low. In fact, the default is only meant to do a quick trial of GMIN stepping, producing successful results only for relatively simple circuits. In difficult cases GMIN stepping will run out of steps and will report this to the user. In this situation one should first review the circuit for systematic singularities before increasing `gminsteps` because the resulting GMIN stepping runs may take a long time to complete.

6.3.4.4 Source Value Stepping

Instead of looking at convergence problems from a strictly mathematical point of view, let us examine the electrical engineer's perspective. Under normal circumstances any circuit which is powered down will have an obvious trivial solution. With zero voltage and zero current convergence is no issue. So the trick is to start a Newton–Raphson iteration with all independent DC sources reduced to a fraction of their original value. A successful run is then used to start the next Newton–Raphson iteration with DC sources turned up one notch. The source value stepping is repeated in this manner until full power is achieved.

We can easily verify our thinking in a formal way. With all independent sources set to zero, the entire right-hand side of the circuit tableau (6.6) becomes zero. Thus we have a trivial solution. The same applies for nodal equations (6.17) and any modification thereof (6.28). As soon as we have a zero excitation vector $\mathbf{e}$, the circuit is effectively powered down.

The specifics of the source value stepping algorithm are presented in Fig. 6.15. A reduction step $0 \leq s \leq 1$ is introduced and all independent DC sources are multiplied by s. With each unsuccessful Newton–Raphson iteration, s is reduced until convergence is achieved. Then s is gradually increased until it becomes 1.

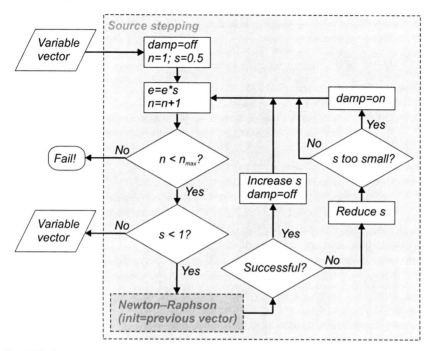

Fig. 6.15 Source value stepping algorithm

Each successful run delivers the starting point for the next run. There is also an iteration counter n limiting the computation effort to $n_{\max}$ rounds. This upper limit is controlled by simulator parameter `srcsteps` with the default value 10.

In fact, the algorithm is very similar to the down-stepping part of GMIN stepping shown in Fig. 6.13. It is also based on the speculation that a too small step s indicates oscillating iterations, so subsequent Newton–Raphson iterations are run in damping mode.

6.3.4.5 Source Value Lifting and *cmin* Value Stepping

Normally all attempts to reach convergence are based on the assumption of a resistive network. To this end all dynamic devices are removed from the circuit; that is, all capacitances and inductances are replaced by independent zero current sources and zero voltages, respectively. This should work with all circuits which will eventually reach steady state. Some circuits, however, have several operating point solutions, e.g., a flip-flop. In these cases there are several convergence points for the Newton–Raphson iteration. In reality, it is the dynamic devices which determine the DC behavior of such circuits. Therefore, removing dynamic devices prior to a Newton–Raphson run is not always a good idea.

Another way of obtaining a good initial guess for the Newton–Raphson algorithm is by running a time-domain analysis of the circuit. More details on the time-domain analysis are given in Sect. 6.5. The time-domain analysis of the circuit takes into account dynamic devices like capacitors and inductors.

The time-domain analysis is started by assuming the dynamic devices have no stored energy (e.g., the initial voltages of capacitors and the initial currents of inductors are all zero). The values of independent sources are gradually ramped up from zero to their respective values used in the operating point analysis. The final steady state solution (after the initial transient phenomena die off) is a good initial guess for the Newton–Raphson algorithm.

The `srclriseiter` simulator parameter specifies the number of time points in which the independent sources are ramped up to their respective values used in the operating point analysis. The `srclrisetime` simulator parameter specifies the time in which the independent sources are ramped up. It takes precedence over the `srclriseiter` simulator parameter.

The time step of the analysis is bound from below by the `srclminstep` simulator parameter. The `srclmaxtime` simulator parameter specifies the maximal time up to which the time-domain analysis is performed in an attempt to find a good initial guess for the Newton–Raphson algorithm. If all of the node voltages and device currents stabilize within their respective tolerances and remain stabilized for `srclconviter` time steps, the time-domain analysis is interrupted and the final time point is used as the initial guess. `srclmaxiter` specifies the maximal number of time points simulated in the time-domain analysis before it is interrupted.

If the analysis is interrupted due to the number of time points reaching `srclmaxiter` or the time reaching `srclmaxtime`, the final time point is used as the initial guess for the Newton–Raphson algorithm.

In some circuits very fast changes of some node voltages occur during time-domain analysis, which can cause problems in the analysis. The problems can sometimes be alleviated by temporarily connecting small capacitors (`cmin`) between every node and the ground. This is the main idea behind *cmin* value stepping.

If *cmin* value stepping is enabled (simulator parameter `cminsteps` is set to a value greater than 0), source value lifting is repeated with a modified circuit where a capacitor (with capacitance specified by the `cmin` simulator parameter) is connected between every node and the ground. If this source value lifting fails to produce a good initial guess for the Newton–Raphson algorithm, the value of the capacitors is increased by `cmin` and the process is repeated. The maximal number of repetitions is specified by the `cminsteps` simulator parameter. If the `noinitsrcl` parameter is set, the initial source value lifting without capacitors on every node is skipped and the simulator starts immediately with the first cmin step.

6.3.4.6 Shunting

The last resort before giving up the attempt to find the operating point is to permanently connect resistors from every node to ground. If the resistances are small enough, any circuit converges. On the other hand, if the resistances are too small, they significantly change the operating point of the circuit. The value of the shunt resistors can be specified with the `rshunt` simulator parameter.

A similar simulator parameter is available for the time-domain analysis (`cshunt`). The parameter permanently connects capacitors to all nodes.

6.3.5 Specifying the Initial Point for the Newton–Raphson Algorithm

The initial point for the Newton–Raphson algorithm is set to 0 by default. This can be changed with the use of the `.nodeset` netlist directive or `nodeset` NUTMEG command (see Sects. 3.10 and 5.5.6).

The initial trial point (V) of the Newton–Raphson algorithm is forced by connecting a current source $g_i V$ in parallel with conductance g_i to the respective node, where i denotes the index of the Newton–Raphson iteration. g_0 is large (10^{10} A/V) and is gradually reduced to 0 as i reaches `nssteps`,

$$g_i = \begin{cases} g_0^{1-\frac{2i}{N}} & 0 \leq i < \texttt{nssteps} \\ 0 & i \geq \texttt{nssteps} \end{cases}. \tag{6.66}$$

The value of g_0 is specified by the `nsfactor` simulator parameter.

6.4 Dynamic Devices and Frequency-Domain Analysis

There is one more step we need to make towards simulating real circuits. We have to find a way to include dynamic devices. We will not worry about nonlinear energy storing devices as they can be represented by a combination of a nonlinear controlled source and either a linear capacitor or inductor. So we are really talking about introducing simple linear capacitance and inductances into branch relations (6.5).

There are two ways to do this, either in the frequency or the time domain, each having its respective advantages and disadvantages.

6.4.1 Small Signal Response (AC) Analysis

Small signal response analysis (also referred to as frequency-domain analysis, AC analysis, or AC sweep), is based on the Fourier transformation [7] of the circuit's equations. Unfortunately, we have to assume either linear circuits or small signals. As we are primarily interested in nonlinear circuits, we will first discuss the notion of small signals.

Observe any nonlinear analog circuit in the steady state. From previous sections we know how to obtain an operating point solution. Now imagine that some independent source between two nodes starts oscillating with an arbitrary frequency but with an extremely small magnitude. Unless the circuit has some discontinuity in the immediate proximity of the operating point, this tiny oscillation will propagate linearly through the entire circuit. In other words, the circuit can be considered linear in the neighborhood of the operating point, provided that the neighborhood is small enough. With this assumption small signal response analysis needs to calculate only the ratios of magnitudes and the phase delays to completely describe any sinusoidal signal.

This is done with complex arithmetic, where all resistive devices are represented by admittances equal to the gradients in the operating point. These gradients are already computed as part of each Newton–Raphson iteration, as shown in Fig. 6.6. Capacitances and inductances become imaginary functions of the frequency $y_c = j\omega C$ and $y_l = \frac{1}{j\omega L}$, respectively. Consequently, independent voltage sources become short circuits, and independent current sources are replaced by open circuits. With these replacements we are facing a simple linear resistive network, but not like that described by (6.29). The only differences are the parametric frequency dependence and the complex arithmetic. The latter is no problem; the procedure is exactly equivalent to that described in Sect. 6.2. The frequency dependence is usually solved by sweeping ω over a relevant interval.

The program flow of the small signal analysis in SPICE OPUS from Sect. 4.4 is summarized in Fig. 6.16.

New users often have misunderstandings regarding the AC analysis. We clarify some of them here.

- The linearized resistive network in Fig. 6.16 needs at least one AC excitation. So unless we have defined at least one AC source, the result will be zero, because the right-hand side of (6.29) is all zero.
- The AC excitation magnitude and phase basically are arbitrary, because we have a linear system where the superposition theorem applies. Therefore, it makes sense to chose unit magnitude and zero phase. In this way the results can be directly interpreted as ratios. Do not be disturbed by SPICE OPUS reporting 1 MV at the output of an operational amplifier. Remember to interpret this as the amplification gain of a unit excitation.
- You can specify more than just one AC excitation source, in which case all excitation sources are swept with the same frequency. Thus, any output will be the product of the combined excitation. This questions the purpose of the entire analysis, but in some special cases one might make use of this possibility.
- Unknown x's in (6.29) should be considered in polar form where the absolute values signify the magnitude and the angle represents the relative phase shift. This puts some natural limits on the phase response calculation because the complex number space only allows angles up to 2π radians. On the other hand, especially with modern high gain operational amplifiers, we experience phase shifts well outside the $\pm 180°$ range. The small signal analysis will still deliver

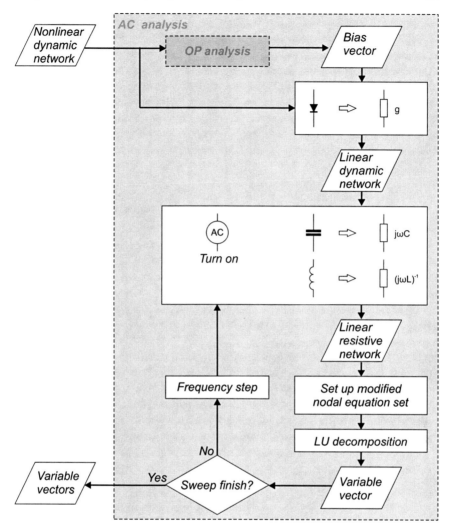

Fig. 6.16 Small signal response analysis

correct results; however, the phase will be wrapped. For instance, $\varphi = 14°$ should actually be interpreted as $\varphi = 14° + k \cdot 360°$, where k can be any integer including zero. As circuit designers we can guess the value of k for each specific case. Nevertheless, there is the special NUTMEG command unwrap() which attempts to make an intelligent guess of k.

- Small signal analysis results depend heavily on the operating point of the circuit. Always double-check the resulting operating point before focusing on the AC analysis data.

One could say that small signal response analysis reveals the differential properties of a circuit in the neighborhood of a specific operating point. With a different bias (resulting in a different operating point), the gradients of the characteristics of nonlinear resistive devices change, and the AC analysis might yield completely different results. Despite the fact that the small signal analysis results in much information on the differential behavior of the circuit, it does not provide us with a picture of the circuit's nonlinear behavior. Nevertheless, the analysis is fast and almost never results in convergence problems. Convergence problems are usually the result of the initial operating point analysis that precedes AC analysis.

6.4.2 Poles and Zeros of the Small Signal Transfer Function (PZ)

Pole-zero analysis is basically an extension of the AC sweep discussed in the previous section. Pole-zero analysis is useful for finding all poles and zeros of a circuit's transfer function in the complex s-plane. Designers sometimes use this information to determine properties of control circuits.

The circuit is viewed as a two-port small signal transfer function, specified by the syntax as explained in Sect. 4.4. The user defines two input nodes n_{ip}, n_{im} and two respective output nodes n_{op}, n_{om}. SPICE OPUS adds a unit AC current source across the input while turning off all other AC excitations. Next, the set of small signal equations is formulated just as described in Sect. 6.4.1. The equations are based on the complex frequency $s = (\sigma + j\omega)$ instead of just the imaginary frequency $j\omega$. Now we have two parameters (σ and ω) defining the s-plane. In control theory the s-plane is also called the Laplace domain. The connection between the time domain and the s-plane is the well-known Laplace transformation [47].

After solving the equation set (6.29) for an arbitrary s, two transfer functions $f(s)$ are observed, the amplification ratio $a(s)$ or the transimpedance $z(s)$,

$$a(s) = \left| \frac{v_{n_{op}} - v_{n_{om}}}{v_{n_{ip}} - v_{n_{im}}} \right|, \quad z(s) = \left| \frac{v_{n_{op}} - v_{n_{om}}}{1\text{A}} \right|. \tag{6.67}$$

We are not interested in the transfer function values themselves, but rather wish to find all respective poles and zeros, that is, those values s_{p_i} and s_{z_i} where the transfer functions become infinite and zero, respectively. In principle, SPICE OPUS needs to do a complex frequency sweep in order to find all poles and zeros. However, this would be computationally rather intensive and quite unnecessary – there are much more efficient ways to search for extrema of a transfer function. In fact, SPICE OPUS utilizes a suboptimal heuristic search algorithm to find the neighborhood of an extremum and then switches to the Müller method [32] to zoom in on the solutions.

The overall pole-zero analysis flow is depicted in Fig. 6.17. In comparison to the one-dimensional sweep of the AC analysis in Fig. 6.16, the two-dimensional parameter space is now searched for extrema.

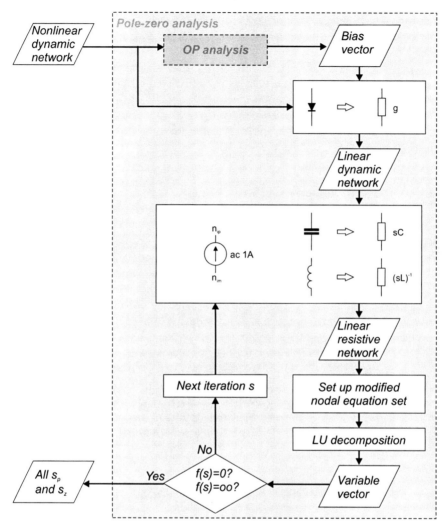

Fig. 6.17 Pole-zero analysis

6.4.3 *Small Signal Noise Analysis (NOISE)*

Small signal noise analysis is another small signal analysis derivative (also in the frequency domain), which is very important and is being heavily used in modern chip design. The goal is to find the total noise spectrum produced by a network of individual electronic devices. The noise level in electronic circuits puts a major constraint on the dynamic range, so small signal noise analysis is crucial to modern chip design. It is based on the knowledge of each individual noise source. In theory there are three noise sources: thermal noise from every resistance as well as shot noise and flicker noise from every semiconductor device. Noise sources can depend on the frequency (e.g., flicker noise) and on the operating point (shot noise, flicker noise).

A noise signal $v(t)$ can be represented by its power spectrum density $S_{vv}^+(f)$ (see Sect. 1.9). If the signal is a node voltage, the corresponding unit for the signal is V and the unit for the power spectrum density is V^2/Hz. If the signal is filtered by an ideal bandpass filter with frequency range $f_1 \leq f \leq f_2$, a new noise signal $v'(t)$ is obtained. By integrating the power spectrum density $S_{vv}^+(f)$ from f_1 to f_2, the mean square of $v'(t)$ is obtained,

$$\lim_{\tau \to \infty} \frac{1}{2\tau} \int_{-\tau}^{\tau} |v'(t)|^2 dt = \int_{f_1}^{f_2} S_{vv}^+(f) df. \tag{6.68}$$

Every resistor and every semiconductor device is a source of noise (see Sect. 1.10). Because the noise is a small signal, it is assumed that the circuit is linear within the noise magnitude around the circuit's operating point. Therefore, the circuit can be linearized and an ordinary small signal response analysis applied to obtain the noise characteristics of the linearized circuit.

The noise introduced by a device is represented by one or more noise current sources in the AC small signal model of the device. The phase of such noise current sources is 0, whereas their magnitude generally depends on the frequency and the circuit's operating point. Figure 6.18 depicts the noise model of a resistor. The dashed rectangle in Fig. 2.19 surrounds the noise model of a bipolar transistor.

Generally, there are n_{noise} noise current sources in the circuit. Let $S_{nn,i}^+(f)$ denote the power spectrum density of i-th noise source. Because individual noise sources are uncorrelated, the power spectrum density of the output noise $S_{out,out}^+(f)$ is the sum of power spectrum density contributions from individual noise sources (Sect. 1.9),

$$S_{out,out}(f) = \sum_{i=1}^{n_{noise}} |A_i(f)|^2 S_{nn,i}^+(f), \tag{6.69}$$

where $A_i(f)$ is the transfer function from the i-th noise source to the output at frequency f.

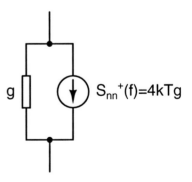

Fig. 6.18 Resistor noise model. k is the Boltzmann constant, T is the absolute temperature, $g = 1/R$ is the conductance of the resistor, and $S_{nn}^+(f)$ is the current noise power spectrum density (A^2/Hz) representing the thermal noise of the resistor

g $S_{nn}^+(f)=4kTg$

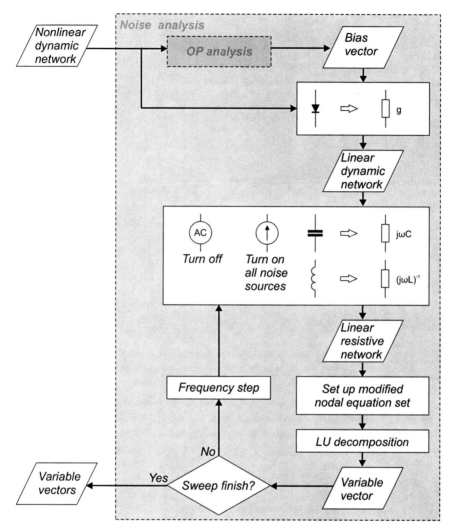

Fig. 6.19 Noise analysis

The equivalent input noise is defined via a noiseless circuit (one in which the circuit elements generate no noise). It is the noise signal that must enter a noiseless circuit at its input in order to produce the same output noise power spectrum density as one would get with a noisy circuit and the input signal equal to zero.

The equivalent input noise power spectrum density is obtained by dividing the output noise power spectrum density by the squared magnitude of the circuit's transfer function from the input to the output. This transfer function can be obtained with an ordinary small signal response analysis.

The block diagram of the noise analysis in SPICE OPUS is depicted in Fig. 6.19.

6.4.4 Frequency-Domain Model of an Inductor in SPICE OPUS

Although the preceding sections treated an inductor as an element with admittance equal to $Y_L = \frac{1}{j\omega L}$, there is a reason against using such a model. Suppose we have coupled inductors in the circuit. A pair of such inductors (see Fig. 6.20) is described by a pair of equations

$$V_1 = j\omega L_1 I_1 + j\omega M_{12} I_2$$
$$V_2 = j\omega M_{12} I_1 + j\omega L_2 I_2,$$

where $M_{12} = k_{12}\sqrt{L_1 L_2}$ and $0 < k_{12} < 1$ is the coupling factor. Because we are describing the circuit in the frequency domain, all currents and voltages are complex numbers that depend on the frequency. Therefore, they are denoted by capital letters. Also note that the voltage V_i is the branch voltage of inductor L_i.

In the case of n inductances this generalizes to

$$V_i = j\omega L_i I_i + \sum_{l \neq i} j\omega M_{il} I_l, \qquad (6.70)$$

where $M_{il} = k_{il}\sqrt{L_i L_l}$.

Such a system of coupled inductors is not trivial to describe, particularly if the procedure for assembling the circuit equations must be kept simple even if there are multiple cross-coupled inductors present.

The approach used in SPICE OPUS introduces an additional node per inductor. This node is called Li#branch and its potential represents the current I_i flowing through the inductor L_i. Now this may look strange at first, but such an approach has the advantage of making the procedure for assembling the circuit equations simple. If we rearrange (6.70) we get

$$V_i - j\omega L_i I_i - \sum_{l \neq i} j\omega M_{il} I_l = 0. \qquad (6.71)$$

Equation (6.71) represents the current balance for node V_{xi}. With this in mind, we arrive at the model depicted in Fig. 6.21. The depicted model is used in SPICE

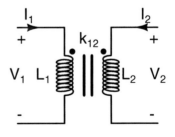

Fig. 6.20 A pair of coupled inductors

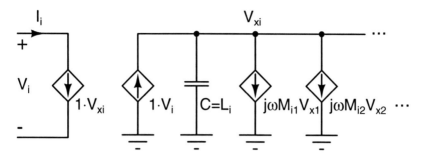

Fig. 6.21 SPICE OPUS model of a coupled inductor

OPUS small signal analyses. The first term, the second term, and the sum in (6.71) represent the $1 \cdot V_i$ controlled current source, the capacitor with $C = L_i$, and the chain of controlled current sources to the capacitor's right, respectively.

6.5 Transient (Time-Domain) Analysis (TRAN)

Time-domain analysis is the most computationally intensive approach, but also the most realistic in the sense that real circuits actually perform in the time domain. Besides the dynamic excitation from independent sources it is the reactances that influence the transient behavior of any circuit. In other words, we need to find a way to supplement branch equations (6.5) with equations for capacitors and inductors.

6.5.1 Backward Euler Integration

The time-domain model of a capacitor (inductor) is given by the following equation.

$$i_c(t) = C \frac{dv_c(t)}{dt}, \quad v_l(t) = L \frac{di_l(t)}{dt}. \tag{6.72}$$

A direct inclusion would turn coefficient matrices Y and Z into operators (e.g., terms of the form $\frac{d}{dt}$ would appear as matrix elements) and the linear equation set (6.6) into a set of ordinary differential equations. In order to avoid this, we use the trick of partitioning time into small steps. The basic idea can be considered a type of finite element approach to time-domain analysis.

The simplest way is to assume constant currents through capacitors and constant voltages across inductors. First, let us examine the consequences on capacitors.

We are observing the time interval $[t_k, t_{k+1}]$. Inside this time interval we assume a constant capacitor current,

$$i(t) = i(t_{k+1}). \tag{6.73}$$

By substituting this expression for the current in the integral form of (6.72) and solving the simple integral, we get

$$v(t) = \frac{1}{C} \int i(t)dt = \frac{1}{C} \int i(t_{k+1})dt = \frac{i(t_{k+1})}{C}t. \qquad (6.74)$$

As expected, the constant current through the capacitance causes a linear voltage ramp across it. However, we are not as interested in the transient inside the time slice as in the variable increments of the time step, that is the difference between variable values at times t_k and t_{k+1},

$$v(t_{k+1}) - v(t_k) = \frac{i(t_{k+1})}{C}(t_{k+1} - t_k). \qquad (6.75)$$

The time step is usually referred to as $h = t_{k+1} - t_k$. With this substitution we get the following admittance form:

$$i(t_{k+1}) = \frac{C}{h}v(t_{k+1}) - \frac{C}{h}v(t_k). \qquad (6.76)$$

This actually represents an impedance $g_c = \frac{C}{h}$ in parallel with a constant current source $i_c = -\frac{C}{h}v(t_k)$. So by knowing the circuit variables at t_k, we can compute the variables in the next time step t_{k+1} by solving a resistive network where each capacitor has been replaced by a respective g_c and i_c (Fig. 6.22).

Naturally, the procedure for impedances is exactly dual. We assume a constant voltage across inductances,

$$v(t) = v(t_{k+1}). \qquad (6.77)$$

Next we solve the basic integral

$$i(t) = \frac{1}{L} \int v(t)dt = \frac{1}{L} \int v(t_{k+1})dt = \frac{v(t_{k+1})}{L}t \qquad (6.78)$$

and formulate the current step

$$i(t_{k+1}) - i(t_k) = \frac{v(t_{k+1})}{L}(t_{k+1} - t_k). \qquad (6.79)$$

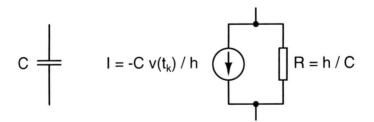

Fig. 6.22 The model of a capacitor in time-domain analysis

At this point the duality ends, because we need an admittance expression in both cases, so the final time-step equation for the inductors is

$$i(t_{k+1}) = \frac{h}{L}v(t_{k+1}) + i(t_k). \tag{6.80}$$

In each time step, inductors are thus replaced by the impedance $z_l = \frac{L}{H}$ in parallel with the independent current source $i_l = i(t_k)$.

In summary, we are able to compute a transient response of a dynamic nonlinear circuit by successively solving resistive nonlinear networks for each individual time step. Actually, we have just reinvented the well-known backward Euler integration algorithm [10]. The program flow is illustrated in Fig. 6.23.

Each step $k+1$ is computed based on the previous one, k, under the assumption of constant currents through capacitances and constant voltages across inductances. But what about the first step? We need some initialization for the integration algorithm. SPICE OPUS always preforms an operating point analysis prior to any integration steps. The results of the operating point analysis are then used as the initial time step $k = 0$, as depicted in Fig. 6.23.

Before we go into any more details, let us review the basic backward Euler integration on a very simple example. Consider the circuit in Fig. 6.24 with a 1 K resistor, a 10 μF capacitor, and a 5 mA step excitation at $t = 0$.

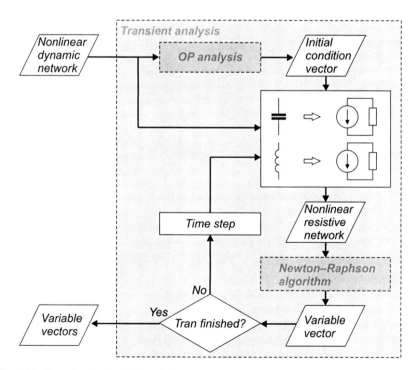

Fig. 6.23 Time-domain transient analysis

Fig. 6.24 Example circuit demonstrating integration algorithms

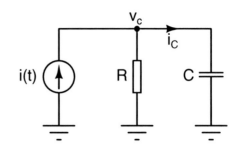

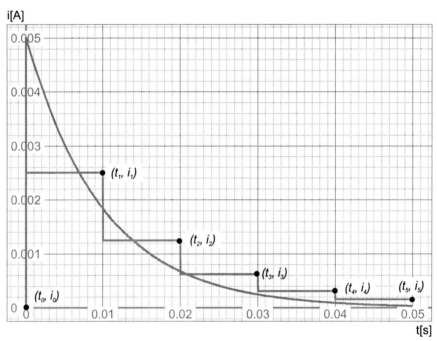

Fig. 6.25 Backward Euler integration current through capacitor

The case is linear and simple enough so that we can guess the transient. We expect a 5 mA surge in the capacitor current $i_c(t)$ dying exponentially while the voltage across the capacitor $v_c(t)$ will rise exponentially toward 5 V.

The initial operating point is trivial as the excitation at $t = 0$ is zero. The selected time step is intentionally large, $h = 10$ ms, which is actually too large for the circuit time constant. In this way we exaggerate the inexact behavior of the integration algorithm. The observed time interval is 50 ms, so we have five integrations steps besides the initial operating point.

The resulting current through the capacitor and the calculated voltage across it are displayed in Figs. 6.25 and 6.26, respectively. The integration steps are clearly marked. Each figure also shows the exact exponential solution for comparison. In

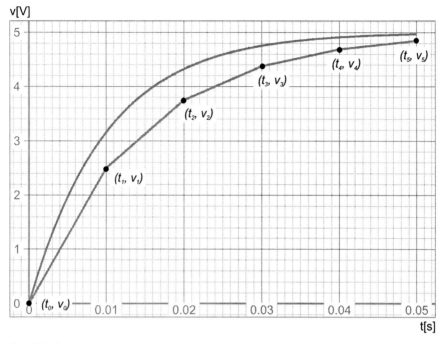

Fig. 6.26 Backward Euler integration voltage across capacitor

both figures we have explicitly represented the backward Euler approximations between the data points, that is, the rectangular current transient and its trapezoidal
voltage integration.

Let us observe the current step function in Fig. 6.25. Because of the huge time
step, it is way off the correct solution. The absolutely worst part is the beginning,
because the assumption that nothing much is happening inside the time slice is most
seriously violated at the beginning. This was to be expected as we have a discontinuous excitation at $t = 0$. Unfortunately, this is a typical scenario in real circuits.
So ideally, instead of a constant integration time step h we would rather have an
intelligently adapting time step.

There is another interesting observation to be made in Fig. 6.25. By considering
the fact that each integration step is numerically based on the results of the previous
calculations (6.76) and (6.80), one would think that the calculation errors will have
a cumulative nature. In other words, calculations based on erroneous data cannot
be expected to produce results with improved accuracy. However, by looking at the
current steps we observe the surprising fact that toward the end of the transient the
integration algorithm comes much closer to the correct solution than it started out.
This is by no means a rule, but neither is it very exceptional. Section 6.5.9 will
discuss these effects in more detail.

Turning our attention to the voltage transient in Fig. 6.26, we find one more typical effect. A rough glance at the characteristics reveals that the voltage transient

is more accurate than its current counterpart. If we were observing an inductor instead of the capacitor, we would find the reverse situation. The reason for this is that SPICE controls the numerical error by keeping an eye on the circuit variables that are subject to numerical integration (voltages for capacitors and currents for inductors). Other variables may exhibit larger numerical errors.

With a smaller time step h, the presented backward Euler integration could produce useful results with real circuits.

In SPICE OPUS backward Euler integration is selected by setting the method simulator parameter to `trap` and the `maxord` simulator parameter to 1. Besides backward Euler integration, SPICE OPUS also provides more advanced algorithms.

6.5.2 Trapezoidal Integration

For trapezoidal integration, instead of assuming constant currents through capacitors and constant voltages across inductors with a time slice, we assume respective linear functions. Instead of having to integrate rectangles, we must now integrate a piecewise linear function with each segment having a trapezoidal form.

We will examine the situation on capacitors first. The expression will be a bit more complicated this time, so we will simplify the notation. We drop the explicit functional dependence on time and use time-step indices directly on current and voltage symbols. So instead of $i(t), i(t_k), v(t_k)$ we will use i, i_k, v_k, respectively.

Consequently, a constant current (6.73) is now replaced by the linear ramp

$$i = \frac{(i_{k+1} - i_k)}{(t_{k+1} - t_k)}(t - t_k) - i_k. \tag{6.81}$$

We obtain the corresponding voltage transient by integration, resulting in a parabolic characteristic,

$$v = \frac{1}{C} \int i\, dt = \frac{1}{C}\left[\frac{(i_{k+1} - i_k)}{(t_{k+1} - t_k)}(\frac{t^2}{2} - t_k t) + i_k t\right]. \tag{6.82}$$

The expression is somewhat complicated, but as we are only interested in the voltage increment, we can calculate the difference $v_{k+1} - v_k$ between time steps t_{k+1} and t_k using (6.82),

$$v_{k+1} - v_k = \frac{1}{2C}(t_{k+1} - t_k)(i_{k+1} + i_k). \tag{6.83}$$

We now resolve the expression to admittance form and substitute h for the time step

$$i_{k+1} = \frac{2C}{h}v_{k+1} - (\frac{2C}{h}v_k + i_k). \tag{6.84}$$

Again we arrive at an admittance $g_c = \frac{2C}{h}$ in parallel with an independent current source $i_c = -\frac{2C}{h}v(t_k) + i(t_k)$. Comparing (6.84) to (6.76), we can see that this time the current source depends not only on the voltage from the previous step but also involves the current. This is very important. As the trapezoidal integration algorithm [10] proceeds through the time steps, it takes more information from the signal history than backward Euler integration.

The same can be observed with inductors. First we assume a linear voltage ramp across the inductor,

$$v = \frac{(v_{k+1} - v_k)}{(t_{k+1} - t_k)}(t - t_k) - v_k. \tag{6.85}$$

After integration we get

$$i = \frac{1}{L}\int v dt = \frac{1}{L}\left[\frac{(v_{k+1} - v_k)}{(t_{k+1} - t_k)}(\frac{t^2}{2} - t_k t) + v_k t\right]. \tag{6.86}$$

The resulting current step now is

$$i_{k+1} - i_k = \frac{1}{2L}(t_{k+1} - t_k)(v_{k+1} + v_k), \tag{6.87}$$

and the admittance form with the h substitution finally yields

$$i_{k+1} = \frac{h}{2L}v_{k+1} + (\frac{h}{2L}v_k + i_k). \tag{6.88}$$

Inductors are thus replaced by the impedance $g_l = \frac{h}{2L}$ in parallel with the independent current source $i_l = \frac{h}{2L}v_k + i(t_k)$. In comparison with (6.80) we again notice the additional dependence on $v(t_k)$.

In conclusion, we expect the assumption of piecewise linear derivatives of reactances to give us a better approximation of real transients than the simplistic step model in backward Euler integration.

Let us repeat the five integration steps for the example in Fig. 6.24. For comparison, we have plotted the trapeziodal current steps against the correct current transient as well as the results from backward Euler integration in Fig. 6.27.

Interestingly, the first step is just as badly off the correct current surge as in backward Euler integration. Actually it is even worse, because the average current in the first time slice is even less than that in the backward Euler step. The remaining steps, on the other hand, catch up with the correct signal nicely.

This becomes even more evident by comparing the voltage responses in Fig. 6.28. Not only is the trapezoidal algorithm with its parabolic segments smoother than the backward Euler integration curve, it also approximates the correct voltage trace much better.

The discontinuous nature of the circuit excitation does not match well with trapezoidal integration. The more an integration algorithm is drawing information from previous time slices, the worse it reacts to discontinuous phenomena. Unfortunately, in electrical engineering discontinuous signals are quite frequent.

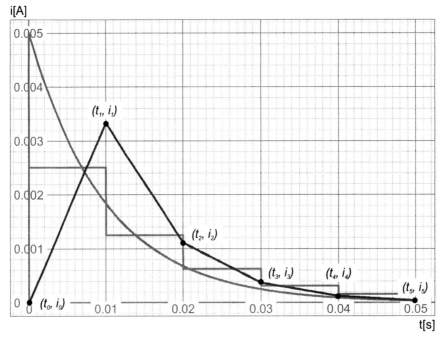

Fig. 6.27 Trapezoidal integration current through capacitor

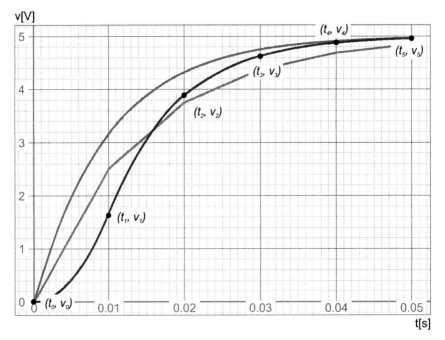

Fig. 6.28 Trapezoidal integration voltage across capacitor

SPICE OPUS uses trapezoidal integration if the `method` simulator parameter is set to `trap` and the `maxord` simulator parameter is set to 2.

By default, SPICE OPUS utilizes trapezoidal integration. In everyday use this integration method has proved itself as the most efficient. Backward Euler integration is used only when the simulator encounters a discontinuity or when the transient analysis is started. However, for special cases the user can manually switch to backward Euler integration.

The trapezoidal integration algorithm is prone to numerical oscillations. If oscillations are detected, SPICE OPUS switches to an integration algorithm that is a hybrid of backward Euler and trapezoidal integration.

The general formula for trapezoidal and backward Euler integration of a capacitor used in SPICE OPUS is

$$i(t_{k+1}) = \frac{C}{(1 - \mathtt{xmu})h}(v(t_{k+1}) - v(t_k)) - \frac{\mathtt{xmu}}{1 - \mathtt{xmu}}i(t_k). \qquad (6.89)$$

`xmu` $= 0$ and `xmu` $= 0.5$ result in backward Euler integration and trapezoidal integration, respectively. A value between 0 and 0.5 results in a mixture of both algorithms.

Oscillations are detected by monitoring the capacitor charge. Let q and i denote the charge and the current of a capacitor. Failure to satisfy the following relation results in an assumption that the capacitor is the cause of numerical oscillations (large capacitor current with only a small change of capacitor charge):

$$|i(t_k)| \leq \mathtt{trapratio} \cdot |q(t_{k+1}) - q(t_k)|/h.$$

The value of `trapratio` must be above 1 and is 10 by default.

`xmu` is normally set to 0.5 (trapezoidal integration). If numerical oscillations are detected at any capacitor in the circuit, `xmu` is reduced to `xmu` $\cdot$ `xmumult` and the time step is repeated. This causes the integration algorithm to behave more like backward Euler integration. `xmu` is kept at the reduced value while numerical oscillations are being detected. When a time step is performed without detecting numerical oscillations, `xmu` is restored to its original value.

`xmumult` must be below 1 and is 0.8 by default. Automatic switching to a more Euler-like algorithm is turned off if `xmumult` is set to 1. On the other hand, a value $0 < \mathtt{xmu} < 0.5$ permanently selects an Euler-like algorithm that is used instead of pure trapezoidal integration.

6.5.3 General Linear Multistep Numerical Integration

This section deals with the numerical integration of nonlinear capacitors, the way it is implemented in all Berkeley-based SPICE simulators. Similar results can be obtained for inductors, if we use flux and current instead of charge and voltage.

To simplify the notation, we denote $\frac{dx(t)}{dt}$ and $\frac{dy(x)}{dx}$ by $\dot{x}(t)$ and $y'(t)$, respectively. Suppose we have a first order ordinary differential equation (first order ODE)

$$\dot{x}(t) = f(x(t), t), \tag{6.90}$$

and we try to find the approximate solution comprising a finite number of pairs (t_k, x_k) approximating the response $x(t)$ on a finite time interval.

Let h_k denote the time step ($h_k = t_{k+1} - t_k$). A general linear multistep numerical integration algorithm [10] for solving (6.90) can be expressed as

$$x_{k+1} = \sum_{i=0}^{p} a_i x_{k-1} + h_k \sum_{i=-1}^{r} b_i f(x_{k-i}, t_{k-i}). \tag{6.91}$$

If $b_{-1} \neq 0$ the algorithm is implicit, otherwise it is explicit.

All terms on the right-hand side of (6.91) depend on past values of x, except for the one with b_{-1} that depends on the still unknown x_{k+1}. Because x_{k+1} cannot be explicitly expressed from $f(x_{k+1}, t_{k+1})$, (6.91) represents an implicit numerical integration algorithm.

If we set $p = 0$ and $r = n-2$, the Adams–Moulton integration algorithm [10] of order n is obtained. The Adams–Moulton algorithm with $n = 1$ and $n = 2$ is actually the backward Euler and the trapezoidal integration algorithm, respectively. On the other hand, if we set $p = n-1$ and $r = -1$ ($b_0 = b_1 = \cdots = 0$), Gear's integration algorithm [10] (in the literature often referred to as the backward differentiation formula (BDF)) of order n is obtained. Gear's algorithm for $n = 1$ is actually the backward Euler algorithm.

Equation (6.91) is capable of predicting the correct value of x_{k+1} if $x(t)$ is a polynomial. The order of the algorithm is actually the maximal order of the polynomial for which the algorithm exactly predicts x_{k+1}.

The coefficients a_i and b_i are computed using the requirement that an algorithm of order n must be exact if the circuit's response $x(t)$ is a polynomial of order not exceeding n. As there are $n + 1$ degrees of freedom in choosing the coefficients of the polynomial, the algorithm has exactly $n + 1$ unknown coefficients. Initially, $x(t)$ is assumed to be constant ($x(t) = \alpha$). By inserting it into (6.91), the following equation is obtained:

$$\sum_{i=0}^{p} a_i = 1. \tag{6.92}$$

By assuming $x(t)$ to be $x(t) = \alpha t^j$, $j = 1, 2, \ldots, n$ and inserting it into (6.91), n additional equations are obtained,

$$\sum_{i=0}^{p} t_{k-i}^{j} a_i + \sum_{i=-1}^{r} j h_k t_{k-i}^{j-1} b_i = t_{k+1}^{j} \quad j = 1, 2, \ldots, n. \tag{6.93}$$

Table 6.3 Coefficients of implicit linear multistep integration algorithms for a constant time step ($h_k = h_{k-1} = h_{k-2} = \cdots$). Only nonzero coefficients are listed

Algorithm	n	$a_0, a_1, \ldots$	$b_{-1}, b_0, \ldots$
Adams–Moulton	1 (back. Euler)	1	1
	2 (trapezoidal)	1	$\frac{1}{2}, \frac{1}{2}$
Gear	1 (back. Euler)	1	1
	2	$\frac{4}{3}, -\frac{1}{3}$	$\frac{2}{3}$
	3	$\frac{18}{11}, -\frac{9}{11}, \frac{2}{11}$	$\frac{6}{11}$
	4	$\frac{48}{25}, -\frac{36}{25}, \frac{16}{25}, -\frac{3}{25}$	$\frac{12}{25}$
	5	$\frac{300}{137}, -\frac{300}{137}, \frac{200}{137}, -\frac{75}{137}, \frac{12}{137}$	$\frac{60}{137}$
	6	$\frac{360}{147}, -\frac{450}{137}, \frac{400}{147}, -\frac{225}{147}, \frac{72}{147}, -\frac{10}{147}$	$\frac{60}{147}$

The terms t_{k-i}^{j} in (6.93) can be very small and can result in a floating point over- or underflow. Therefore, the equation is usually divided by t_{k+1}^{j}. With no loss of generality, an additional assumption of $t_k = 0$ can significantly simplify the system; as a consequence, $t_{k+1} = h_k$ follows. Note that 0^0 does not really occur. Terms with 0^0 originate from the first derivative of t at $t = 0$ and are equal to 1.

Equations (6.92) and (6.93) are used to obtain the coefficients of a general linear multistep integration algorithm. In the course of simulation these coefficients must be recalculated at every time step if the time step is not constant. Table 6.3 lists the algorithm coefficients for a constant time step.

Equation (6.91) can also be viewed as a formula for predicting the derivative of $x(t)$ at t_{k+1}. It actually holds perfectly if $x(t)$ is a polynomial of order not exceeding n. To express $\dot{x}(t_{k+1})$ (6.90) must be recalled,

$$\dot{x}(t_{k+1}) = f(x_{k+1}, t_{k+1}) = \frac{1}{h_k b_{-1}} x_{k+1} - \sum_{i=0}^{p} \frac{a_i}{h_k b_{-1}} x_{k-i} - \sum_{i=0}^{r} \frac{b_i}{b_{-1}} \dot{x}_{k-i}. \quad (6.94)$$

Suppose a nonlinear capacitor is voltage defined so that the charge is a function of the capacitor voltage ($q = q(v)$). By assuming $x = q$ and $i(t) = \dot{q}(t)$, we can rewrite (6.94) as

$$i_{k+1} = \frac{1}{h_k b_{-1}} q(v_{k+1}) - \sum_{i=0}^{p} \frac{a_i}{h_k b_{-1}} q(v_{k-i}) - \sum_{i=0}^{r} \frac{b_i}{b_{-1}} i_{k-i}. \quad (6.95)$$

This is an implicit nonlinear equation for the capacitor voltage at t_{k+1}. In the spirit of the Newton–Raphson iteration, it can be linearized as

$$i_{k+1}^{j+1} = G_{k+1}^{j} v_{k+1}^{j+1} + I_{k+1}^{j} \quad (6.96)$$

$$G_{k+1}^{j} = \frac{1}{h_k b_{-1}} q'(v_{k+1}^{j}) \quad (6.97)$$

$$I_{k+1}^{j} = \frac{1}{h_k b_{-1}} q(v_{k+1}^{j}) - \sum_{i=0}^{p} \frac{a_i}{h_k b_{-1}} q(v_{k-i}) - \sum_{i=0}^{r} \frac{b_i}{b_{-1}} i_{k-i}$$

$$- \frac{1}{h_k b_{-1}} q'(v_{k+1}^{j}) v_{k+1}^{j}, \tag{6.98}$$

where j is the index of the previous Newton–Raphson iteration. v_{k+1}^{j+1} and i_{k+1}^{j+1} are the capacitor voltage and current that are the subject of iteration $j + 1$. Voltages and currents with no superscript come from past time points for which the Newton–Raphson algorithm converged successfully. G_{k+1}^{j} and I_{k+1}^{j} are calculated using the value of v_{k+1}^{j} from the previous iteration (j) of the Newton–Raphson algorithm. $q'(v) = \frac{dq(v)}{dt} = c(v)$ is the differential capacitance.

As we can see, the simulator needs to store $\max(p, r) + 1$ past values of capacitor charge and current. The capacitor is modeled as a conductance in parallel with an independent current source. The value of these two elements is determined by (6.97) and (6.98).

Recall Sects. 6.5.1 and 6.5.2. We assumed a piecewise constant current in the former section and a piecewise linear current in the latter section. This corresponds to a first and a second order polynomial time dependence of charge, and confirms that the order of the backward Euler and trapezoidal algorithm is 1 and 2, respectively.

SPICE OPUS supports the Adams–Moulton algorithm up to the second order and the Gear's algorithm up to the sixth order.

6.5.4 Local Truncation Error (LTE)

Ideally (assuming it is smooth enough), the circuit's response can be perfectly represented by means of Taylor's series [53],

$$x(t) = \sum_{j=0}^{\infty} \frac{x^{(j)}(a)}{j!} (t - \tau)^{j}, \tag{6.99}$$

where τ is some point in time and $x^{(j)}(t)$ is the j-th derivative of $x(t)$ with respect to time.

A truncated series (a polynomial $p_n(t)$ of order n) is used in the derivation of the linear multistep integration method coefficients. This truncation means that there is a difference between the polynomial and the exact circuit's response. A linear multistep integration method is capable of predicting the correct circuit's response at t_{k+1} if the past values of the response are exact and $x(t)$ is a polynomial of order not exceeding n. For any other response there is an inherent error in the method's prediction, even if the past values are known exactly. This error is called the local truncation error (LTE) [10].

To derive an estimate of the error, the derivative of the response with respect to time must be expressed,

$$\dot{x}(t) = \sum_{j=1}^{\infty} \frac{x^{(j)}(a)}{j!} j(t - \tau)^{j-1}. \tag{6.100}$$

By inserting (6.90), (6.99), and (6.100) in the right-hand side of 6.91 and subtracting (6.99) the LTE can be expressed as a series,

$$\epsilon_{k+1} = x_{k+1} - x(t_{k+1}) = \sum_{j=0}^{\infty} C_j x^{(j)}(t_k) h_k^j \tag{6.101}$$

$$C_0 = -1 + \sum_{i=0}^{\infty} a_i \tag{6.102}$$

$$C_j =$$
$$\frac{1}{j!} \left(-1 + \sum_{i=0}^{p} a_i \left(\frac{t_{k-i} - t_k}{h_k} \right)^j + j \sum_{i=-1}^{r} b_i \left(\frac{t_{k-i} - t_k}{h_k} \right)^{j-1} \right). \tag{6.103}$$

Here $\tau = t_k$, which makes sense if $x(t_{k+1})$ is being expressed as a series where terms of the form h_k^j occur. The occurrence of terms with 0^0 originates from the first derivative of t at $t = 0$. The correct interpretation of such terms is to treat them as 1.

Coefficients C_j are also referred to as error coefficients. The requirement that the LTE of a linear multistep integration algorithm of order n must be zero if the circuit's response is a polynomial of order not exceeding n can be expressed in terms of error coefficients as

$$C_j = 0 \quad j = 0, 1, \ldots, n. \tag{6.104}$$

By setting $t_k = 0$ in (6.104), (6.92) and (6.93) are obtained. This confirms that a linear multistep integration algorithm of order n has no LTE terms of order lower than $n + 1$.

After rearranging (6.101) the following expression is obtained:

$$\epsilon_{k+1} = x_{k+1} - x(t_{k+1}) = C_{n+1} x^{(n+1)}(t_k) h_k^{n+1} + h_k^{n+2} O(h_k), \tag{6.105}$$

where $O(t)$ is some function. If h_k is sufficiently small, the LTE can be approximated as

$$\epsilon_{k+1} = x_{k+1} - x(t_{k+1}) \approx C_{n+1} x^{(n+1)}(t_k) h_k^{n+1}. \tag{6.106}$$

Table 6.4 lists the coefficients C_n for the integration algorithms in SPICE OPUS when the time step is constant.

Table 6.4 Error coefficients for Adams–Moulton and Gear's algorithms (constant time step is assumed)

Algorithm	n	C_{n+1}
Adams–Moulton	1 (back. Euler)	$\frac{1}{2}$
	2 (trapezoidal)	$\frac{1}{12}$
Gear	1 (back. Euler)	$\frac{1}{2}$
	2	$\frac{2}{9}$
	3	$\frac{3}{22}$
	4	$\frac{12}{125}$
	5	$\frac{10}{137}$
	6	$\frac{60}{1029}$

The $n+1$-th derivative of the charge ($q(t)$) can be estimated by means of interpolation. The Newton interpolation polynomial of n-th order interpolating a set of $n+1$ data points $(x_0, y(x_0)), (x_1, y(x_1)), \ldots, (x_n, y(x_n))$ is defined as

$$N_n(x) = \sum_{i=0}^{n} a_i \eta_i(x),$$

where a_i are the coefficients and $\eta_i(x)$ are the Newton basis polynomials. The latter are expressed as

$$\eta_i(x) = \prod_{j=0}^{i-1} (x - x_j).$$

So the first three basis polynomials are $\eta_0(x) = 1$, $\eta_1(x) = x - x_0$, and $\eta_2(x) = (x - x_0)(x - x_1)$. The coefficients a_i can be expressed as divided differences in a recursive manner,

$$a_i = y[x_0, x_1, \ldots, x_i],$$

where the divided difference $y[x_0, x_1, \ldots, x_i]$ is defined as

$$y[x_i] = y(x_i)$$
$$y[x_0, x_1, \ldots, x_i] = \frac{y[x_1, x_2, \ldots, x_i] - y[x_0, x_1, \ldots, x_{i-1}]}{x_i - x_0}.$$

Take, for instance, the divided difference $y[x_0, x_1, x_2]$. In order to evaluate it, we must first obtain $y[x_0, x_1] = \frac{y(x_1) - y(x_0)}{x_1 - x_0}$ and $y[x_1, x_2] = \frac{y(x_2) - y(x_1)}{x_2 - x_1}$, upon which the final result follows,

$$y[x_0, x_1, x_2] = \frac{\frac{y(x_2) - y(x_1)}{x_2 - x_1} - \frac{y(x_1) - y(x_0)}{x_1 - x_0}}{x_2 - x_0}.$$

The n-th derivative of $N_n(x)$ is simply

$$N_n^{(n)}(x) = n! \, y[x_0, x_1, \ldots, x_n].$$

With this in mind, the LTE (6.106) of the capacitor charge can be estimated as

$$\epsilon_k \approx (n+1)! C_{n+1} h_k^{n+1} q[t_{k-n}, t_{k-n+1}, \ldots, t_{k+1}]. \tag{6.107}$$

6.5.5 Inductors in SPICE OPUS Time-Domain Analysis

A set of multiple coupled inductors (see also Sect. 6.4.4) is described by the following time-domain equation:

$$v_i = L_i \frac{di_i}{dt} + \sum_{l \neq i} M_{il} \frac{di_l}{dt}. \tag{6.108}$$

Equation (6.108) is the time-domain version of (6.70). Here all currents and voltages are denoted by lowercase letters (u and i) and are represented by real numbers (time domain). Note that v_i denotes the branch voltage of inductor L_i.

The previous section on time-domain analysis considered inductors as elements that can be modeled in a way dual to capacitors. In SPICE OPUS a different approach is used that models an inductor with a capacitor and an internal node representing the inductor's current (v_{xi}, see also Sect. 6.4.4). This greatly simplifies the construction of circuit equations, especially in the case of coupled inductors. By rearranging (6.108) we get

$$v_i - L_i \frac{di_i}{dt} - \sum_{l \neq i} M_{il} \frac{di_l}{dt} = 0, \tag{6.109}$$

which is the time-domain version of (6.71) representing the current balance at node v_{xi}. The coupling to other inductors ($M_{il} \frac{dv_{xl}}{dt}$, $l \neq i$) is represented by controlled current sources. The corresponding circuit is depicted by Fig. 6.29.

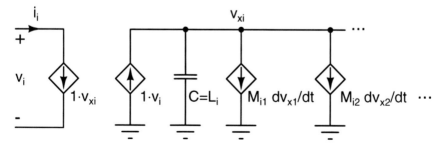

Fig. 6.29 SPICE OPUS model of a coupled inductor in time domain

The integration algorithm is applied to node voltage v_{xi} for every inductor L_i. The controlled current sources $M_{il}\frac{dv_{xl}}{dt}$ are actually capacitors with capacitances M_{il}. The voltages integrated by these capacitors are voltages at nodes v_{xl}. The current flowing through these capacitors is, however, pulled from node v_{xi}. Note that the integration algorithm does not have to be applied to voltages v_{xl} as it is already performed by capacitors $C = L_l$ connected to internal nodes of inductors L_l, $l \neq i$.

6.5.6 Initial Conditions

The simulator starts the transient analysis with initial conditions for the capacitors. One initial condition is required per every capacitor,

$$q(v(0)) = q_0. \tag{6.110}$$

The initial capacitor charge depends on the initial capacitor voltage. This voltage represents the initial condition and can be obtained in two ways. If the uic parameter (use initial condition parameter) is not given to the tran analysis command, the initial conditions are obtained by solving for the operating point of the circuit where all capacitors are removed (initial operating point analysis). Note that inductors are modeled with capacitors in SPICE (Sect. 6.5.5). The results of the initial operating point analysis are used for calculating the initial capacitor voltages and represent the circuit's solution at $t = 0$ (first time point).

If the uic parameter is given, the initial operating point analysis is omitted. The initial condition (capacitor voltage) is obtained from the ic parameter of every capacitor. If the ic parameter is not given, the initial capacitor voltage is calculated from the difference between the initial values of the node voltages. These initial node voltage values are 0 by default, unless they are specified with the .ic netlist directive or the ic NUTMEG command. The initial node voltage values represent the circuit's solution at $t = 0$.

Unfortunately, the initial node voltage values sometimes do not represent a physically consistent circuit's solution (remember that they can be specified by the user). To obtain a physically consistent initial circuit solution that takes into account the initial conditions specified for the capacitors, the icstep simulator parameter can be used. If it is set to a nonzero value, the circuit is first solved at icstep · t_1. The obtained solution represents the circuit's solution at $t = 0$. If icstep is set to a small value, the contribution of the numerical integration is negligible and the circuit is solved almost as in the operating point analysis (with the addition that the initial conditions are taken into account). Typically, icstep is set to a value around 10^{-9}.

Initially (at $t = 0$) and after every break point (see Sect. 6.5.9), the integration order is set to 1. This way the linear multistep integration algorithm requires only a single past charge value (provided by the initial condition). Later the integration order is increased by the simulator up to maxord.

6.5.7 Predictor-Corrector Integration

Whenever the simulator tries to solve the circuit at a new time point t_{k+1}, the initial point for the Newton–Raphson algorithm is chosen to be the circuit's solution at t_k. This does not always guarantee a low number of Newton–Raphson iterations because the solution at the last time point (t_k) may be too far from the one at t_{k+1}.

In order to further reduce the number of Newton–Raphson iterations per time point, the initial point for the iterative process can be predicted from the solutions of the circuit at past time points. This approach is called predictor-corrector numerical integration. The Newton–Raphson algorithm still converges to the same solution as it would do without the use of a predictor, except that the convergence is much faster in most cases.

The predictor can be a linear multistep integration algorithm, or even plain polynomial extrapolation. Using an implicit numerical integration algorithm for the predictor makes no sense as such an approach would require the solution of another system of nonlinear equations. Explicit integration algorithms are used instead. By setting $p = 0$, $r = n - 1$, and $b_{-1} = 0$ in (6.91) the Adams–Bashforth explicit integration algorithm [10] of order n is obtained (6.111). By applying it to node voltages instead of capacitor charges the following predictor algorithm is obtained:

$$v_{k+1}^P = v_k + h_k \sum_{i=0}^{n-1} \beta_i \dot{v}_{k-i}. \qquad (6.111)$$

v_{k+1}^P denotes the predicted voltage value at t_{k+1}, and $\dot{v}_{k-i}$ is the derivative of the voltage with respect to time at t_{k-i}. The coefficients of various Adams–Bashforth algorithms for a constant time step are listed in Table 6.5 (note that $\beta_i = b_i$). If the time step is not constant, the coefficients β_i must be recalculated at every time step. The simulator does not compute $\dot{v}_k$. The simplest approach for calculating $\dot{v}_k$ is approximation with backward difference,

$$\dot{v}_k \approx \frac{v_k - v_{k-1}}{t_k - t_{k-1}} = \frac{v_k - v_{k-1}}{h_{k-1}}. \qquad (6.112)$$

Table 6.5 Coefficients of Adams–Bashforth algorithms for a constant time step ($h_k = h_{k-1} = h_{k-2} = \cdots$). Only nonzero coefficients are listed

n	a_0	$b_0, b_1, \ldots$				
1 (forward Euler)	1	1				
2	1	$\frac{3}{2}, -\frac{1}{2}$				
3	1	$\frac{23}{12}, -\frac{16}{12}, \frac{5}{12}$				
4	1	$\frac{55}{24}, -\frac{59}{24}, \frac{37}{24}, -\frac{9}{24}$				
5	1	$\frac{1901}{720}, -\frac{2774}{720}, \frac{2616}{720}, -\frac{1274}{720}, \frac{251}{720}$				
6	1	$\frac{4277}{1440}, -\frac{7923}{1440}, \frac{9982}{1440}, -\frac{7298}{1440}, \frac{2877}{1440}, -\frac{475}{1440}$				

Table 6.6 Polynomial extrapolation coefficients for a constant time step $(h_k = h_{k-1} = h_{k-2} = \cdots)$. Only nonzero coefficients are listed

n	$\gamma_0, \gamma_1, \ldots$
1 (linear)	2, −1
2 (quadratic)	3, −3, 1
3 (cubic)	4, −6, 4, −1
4	5, −10, 10, −5, 1
5	6, −15, 20, −15, 6, −1
6	7, −21, 35, −35, 21, −7, 1

The Adams–Bashforth predictor of order n is used in conjunction with the n-th order Adams–Moulton corrector (integrator). For the n-th order Gear corrector (integration), polynomial extrapolation with a polynomial of order n is used as predictor (6.113). Equation (6.91) turns into polynomial extrapolation if $p = n$, $b_i = 0$ ($r = -2$), and $a_i = \gamma_i$. Note that in this case (6.91) no longer represents an integration algorithm as it does not include terms with $f(x, t)$. The polynomial extrapolation predictor can be expressed as

$$v_{k+1}^P = \sum_{i=0}^{n} \gamma_i x_{k-i}. \tag{6.113}$$

The coefficients for polynomial extrapolation for a constant time step are listed in Table 6.6. If the time step is not constant, the coefficients γ_i must be recalculated at every time step.

The use of a predictor-corrector integration with a predictor of order n implies that the node voltages for $n + 1$ past time steps must be stored. A special case is the predictor for the first time point t_1. Even with the first order integration algorithm ($n = 1$) a first order predictor requires two past time points. Therefore, the predictor is not used for solving the circuit at t_1. The initial point for the Newton–Raphson algorithm is specified by the initial node voltage values (.ic netlist directive).

The coefficients of a predictor (β_i or γ_i) can be calculated using (6.92) and (6.93) where $a_i == \beta_i$ and $b_i = \gamma_i$. SPICE OPUS uses predictor-corrector integration by default. The simulator parameter nopredictor disables the use of a predictor in numerical integration.

The use of predictor-corrector integration simplifies the estimation of the local truncation error. Let C_j^P denote the error coefficients of the predictor. Then the following holds:

$$x_{k+1}^P = x(t_{k+1}) + \epsilon_{k+1}^P \tag{6.114}$$

$$x_{k+1} = x(t_{k+1}) + \epsilon_{k+1}. \tag{6.115}$$

Subtraction of the predicted value from the corrected value results in

$$x_{k+1} - x_{k+1}^P = \epsilon_{k+1} - \epsilon_{k+1}^P \approx (C_{n+1} - C_{n+1}^P) x^{(n+1)}(t_k) h_k^{n+1}. \tag{6.116}$$

Table 6.7 Error coefficients for predictor algorithms used in SPICE OPUS (constant time step is assumed)

Algorithm	n	C_{n+1}^P
Adams–Bashforth	1 (forward Euler)	$-\frac{1}{2}$
	2	$-\frac{5}{12}$
Polynomial extrapolation	All $n > 0$	-1

After taking into account (6.106), the following expression for the corrector LTE is obtained:

$$\epsilon_{k+1} = x_{k+1} - x(t_{k+1}) \approx \frac{C_{n+1}}{C_{n+1} - C_{n+1}^P} \cdot (x_{k+1} - x_{k+1}^P). \qquad (6.117)$$

The error coefficients for the predictor algorithms used in SPICE OPUS are listed in Table 6.7 (constant time step is assumed). Just like all other coefficients for (6.91), these must also be recalculated at every time step if the time step is not constant.

6.5.8 Choosing the Integration Algorithm

In SPICE OPUS the integration algorithm can be selected with the simulator parameter `method`. The default value is `trap` (Adams–Moulton algorithm). To choose the Gear's algorithm, set `method` to `gear`.

The order of the algorithm is dynamically controlled by the simulator. The upper bound on the integration order can be specified with the `maxord` simulator parameter. If Gear integration is used, its value can be between 1 and 6. For Adams–Moulton integration `maxord` can be 1 or 2. The simulator sets the integration order to 1 at the beginning of the simulation ($t = 0$) and after every break point (see Sect. 6.5.9).

Of the second order algorithms trapezoidal integration (assuming a fixed time step) is more accurate (smaller C_{n+1}) and allows longer time steps resulting in a shorter simulation run time. Stable differential equations remain stable and unstable ones remain unstable with the trapezoidal algorithm. On stiff circuits (circuits with time constants much smaller than the time step) it exhibits numerical oscillations (point-to-point ringing) which can be fixed by reducing the time step (decreasing `trtol`) or by using the `xmu`, `xmumult`, and `trapratio` simulator parameters.

All stable and some unstable differential equations are stable with the backward Euler and second order Gear integration. Both algorithms exhibit artificial numerical damping. This results in the damping of the simulated response being greater than it is in reality. The effect is more pronounced with the backward Euler algorithm.

In general, it is not recommended to use higher order integration algorithms, because they perform badly in connection with discontinuous excitation, their polynomial dependence on the recent signal history being the major drawback.

6.5.9 Integration Time-Step Control

A very important aspect of time-domain simulation is the choice of time step. The first and most obvious choice is probably a uniform time step that is small enough to sufficiently sample any dynamic behavior that the circuit may exhibit.

Such a choice can lead to a large waste of computing resources. Take, for instance, an inverter circuit that switches from a low to a high output in 10 ns. To sample 10 points during this transition the time step must be at most 1 ns. Suppose that the input of the inverter is constant for 10 ms, upon which it transitions from the initial high level to a low level. If we want to capture the transition, we must simulate at least until 10 ms + 10 ns with a 1 ns step. This results in no less than 10,000,011 time points, although most of the time nothing interesting happens in the circuit.

Let t_{k+1} denote the time point at which the simulator tries to solve the circuit. $h_k = t_{k+1} - t_k$ is the current time step.

Sometimes the time step must adapt so that the circuit is evaluated at a particular time point. This is achieved through break points.

6.5.9.1 Break Points

Break points are time points at which the circuit must be evaluated no matter how large or small the time step may be. Break points are requested by devices (typically independent current and voltage sources) and organized in a sorted list.

A typical example of a break point is the point at which the first derivative is discontinuous for a piecewise linear or pulse source.

Let $t_{b1} < t_{b2} < \cdots$ denote the break points. The simulator ensures that outdated break points are discarded so that $t_k \leq t_{b1}$ always holds.

If a break point is hit ($t_k < t_{b1} \leq t_{k+1}$), the time step h_k is reduced so that $t_{k+1} = t_k + h_k = t_{b1}$. In the next time step (after the circuit is solved at t_{b1}) the integration order is reduced to 1. In some sense the time-step control is restarted after a break point is hit.

If two break points differ by less than $10h_{min}$, they are considered to be one single break point. See the next subsection for the definition of h_{min}.

Two break points are always present. They are placed at $t = 0$ and $t = \texttt{stop}$, where stop is the final time specified with the tran command.

6.5.9.2 Initial, Maximal, and Minimal Time Step

Let step and maxstep denote the initial time step and the upper time-step limit specified with the tran command.

The initial time step is calculated as

$$h_0 = \texttt{fs} \cdot \min(\texttt{step}, \texttt{maxstep}, t_{b2}, \texttt{fs}/(2 f_{max})),$$

where f_{max} specifies the highest frequency of a sinusoidal or frequency modulated independent source in the circuit. If there are no such sources in the circuit, the last term in min() is omitted. The same goes for maxstep if it is not given with the tran command. t_{b2} represents the next break point ($t_{b2} > 0$). fs is a simulator parameter (0.25 by default).

The maximal time step h_{max} is calculated from the following equation:

$$h_{max} = \min(\text{rmax} \cdot \text{step}, \text{maxstep}, \text{fs} \cdot (t_{b2} - t_{b1}), \text{fs}/(2 f_{max})).$$

Again, the last term in min is omitted if no sinusoidal or frequency modulated independent sources are present. The term maxstep is left out if it is not given with the tran command. t_{b2} and t_{b1} represent the next and the last break point, where $t_{b1} \le t_k < t_{b2}$. rmax is a simulator parameter (5 by default). h_{max} is recalculated at every time step. If the time step exceeds h_{max}, it is set to h_{max}.

The minimal time step is calculated only at the beginning of the transient analysis,

$$h_{min} = \text{rmin} \cdot \text{tstep}.$$

rmin is a simulator parameter (10^{-9} by default). The minimal time step affects the handling of break points. If two break points are less than $10 h_{min}$ apart, they are considered to be one single break point.

The time step is initially not allowed to go below h_{min}. If this happens, the time step is set to h_{min}. If it happens again, the time step is allowed to go below h_{min}, but temporarily (until the time step rises above h_{min}) the allowed number of transient Newton–Raphson iterations increases to $10 \cdot \text{imax}$, and the iteration count time-step control algorithm is used.

6.5.9.3 Accepting Time Points and Controlling the Order of the Integration Algorithm

Even if the Newton–Raphson algorithm converges at t_{k+1}, the simulator does not always proceed to the next time point t_{k+2}. Sometimes the time step between t_k and t_{k+1} is too long and the integration algorithm produces a considerable error. The error is usually not detected directly by calculation. Often the dynamics of the circuit and the number of required Newton–Raphson iterations are monitored as indirect indicators for the growing integration error.

In such cases the time point is said to be rejected. A new (shorter) time step is used and the circuit is solved again at a new point t'_{k+1}, where $t_k < t'_{k+1} < t_{k+1}$. Whenever a time point is rejected, the integration order is set to 1.

If the simulator moves on to the next time point t_{k+2}, the time point t_{k+1} is said to be accepted. The integration order is not decreased. Often it is increased, but never above maxord.

6.5.9.4 Rejecting Time Points Due to Slow Convergence of the Newton–Raphson Algorithm

In transient analysis the number of Newton–Raphson iterations per trial time point is limited to a much lower value than in operating point analysis (8 by default). This value is set by the imax (itl4) simulator parameter. The reason for this is the fact that the initial trial point for the Newton–Raphson algorithm (solution at the last accepted time point t_k) is close to the expected solution at t_{k+1}. If the nonlinear solver requires more then imax iterations, the time step (h_k) is assumed to be too long. Time point t_{k+1} is rejected and the time step is reduced by a factor defined by simulator parameter ft (0.25 by default).

This cannot go on indefinitely. If the time step becomes too small compared to the time of the last accepted time point t_k (below double floating point precision), the simulator gives up and aborts the simulation.

If the Newton–Raphson algorithm converges, the solution may still be inaccurate due to the time step being too long. On the other hand, successful convergence of the Newton–Raphson algorithm is an indicator for a possible time step increase. Various algorithms for time-step control can be employed at this point. SPICE OPUS offers three such algorithms: iteration count, dVdt, and LTE.

6.5.9.5 Iteration Count Algorithm

Let n_{NR} denote the number of iterations that are required for the Newton–Raphson algorithm to converge at time point t_{k+1}. This number equals 2 if the circuit is linear. If the circuit is not linear, the number of iterations grows with the circuit's nonlinearity.

If the time step ($h_k = t_{k+1} - t_k$) is long, the solution of the circuit moves further away from the solution at t_k and it is more likely that nonlinearity is encountered along this path. Therefore, the number n_{NR} is more likely to be large. Short time steps (with the assumption of continuous device characteristics) imply linear behavior along the path from t_k to t_{k+1} and result in a low n_{NR}.

If $n_{NR} \leq$ imax, the solution at t_{k+1} is accepted and the simulator moves on to the next time point. The integration order is increased. If $n_{NR} \leq$ imin also holds (where imin < imax), the time step is also increased ($h_{k+1} = 2h_k$). imin is a simulator parameter and its value is 3 by default.

6.5.9.6 The dVdt Algorithm

The dVdt algorithm monitors the rate of slope change of node voltages. The algorithm ignores current nodes (branch currents of voltage sources and inductors). If the rate of slope change is too high, the time step is reduced. The slope at the previous and the slope at the current accepted time point are defined as

$$g_{k-1} = \frac{y_k - y_{k-1}}{h_{k-1}}$$

$$g_k = \frac{y_{k+1} - y_k}{h_k},$$

where y denotes a node voltage.

If t_k is a break point, the simulator searches for the node for which

$$\Delta_y = |y_{k+1} - y_k - g_{k-1}h_k| \tag{6.118}$$

is maximal. Using the values of `absvar`, `relvar`, y_{k+1}, and $y_k + g_{k-1}h_k$, the corresponding voltage norm for that node is calculated. If Δ_y is of similar magnitude or greater than the voltage norm, the solution at t_{k+1} is rejected and the time step is reduced by a factor not smaller than `ft`. Otherwise, the solution is accepted and $h_{k+1} = h_k$.

If t_k is not a break point, the maximal value of

$$\Delta_g = \frac{|g_k - g_{k-1}|}{\max(|g_k|, |g_{k-1}|)}$$

across voltage nodes is calculated.

If Δ_g is above `slopetol`, the time point t_{k+1} is rejected and the time step is reduced to `ft` $\cdot h_k$. If Δ_g is below `slopetol`, the solution at t_{k+1} is accepted and the time step is kept unchanged or even increased (if Δ_g is well below `slopetol`).

The dVdt algorithm does not allow two consecutive time-step reductions. To avoid such situations, the solution at t_{k+1} is automatically accepted if the time step was reduced in the last invocation of the dVdt algorithm. So even if the time point t_{k+1} would have been rejected due to $\Delta_g > $ `slopetol`, it is still accepted and the time step remains unchanged.

If the number of Newton–Raphson iterations that were needed to solve the circuit at t_{k+1} is not above `imin`, the time step is further increased by a factor of 2.

The dVdt algorithm keeps the second derivatives of node voltages bounded. Assuming linear capacitances, this implies a bound on the second derivatives of capacitor charges. By looking at (6.106) one can quickly see that the dVdt algorithm can keep the LTE bounded only for first order algorithms. Therefore, it is not recommended for algorithms of order $n > 1$.

6.5.9.7 The LTE Algorithm

The LTE algorithm first estimates the new time step that will keep the LTE within a user-defined bound. This is the default time-step control algorithm in SPICE OPUS. Two algorithms that deliver the estimate are available.

The old algorithm is used if the `newtrunc` simulator parameter is not set (default). For every capacitor (note that inductors are treated as capacitors connected to internal current nodes in SPICE OPUS) the LTE bound is first calculated,

$$LTE_{cur} = h_k(\texttt{lteabstol} + \texttt{ltereltol} \cdot \max(|i_{k+1}|, |i_k|))$$
$$LTE_{chg} = \texttt{relq} \cdot \max(|q_{k+1}|, |q_k|, \texttt{chgtol})$$
$$LTE_{bound} = \max(LTE_{cur}, LTE_{chg}). \tag{6.119}$$

Next the divided difference $q[t_{k-n}t_{k-n+1}, \ldots, t_{k+1}]$ is evaluated. The upper bound on the time step that will keep the LTE below $\texttt{trtol} \cdot LTE_{bound}$ is calculated using (6.106),

$$h_{bound} = \left(\frac{\texttt{trtol} \cdot LTE_{bound}}{\max(\texttt{lteabstol} \cdot h_k, (n+1)!|C_{n+1}q[t_{k-n}t_{k-n+1}, \ldots, t_{k+1}]|)} \right)^{\frac{1}{n+1}} \tag{6.120}$$

where n and C_{n+1} denote the order and the error coefficient of the integration algorithm, respectively.

$\texttt{lteabstol}$ prevents the time step from increasing beyond any limit when the truncation error goes to zero. $\texttt{trtol}$ is the truncation error underestimation factor. Larger values of $\texttt{trtol}$ result in longer time steps. $\texttt{trtol} = 1$ results in the simulator using the true value of the LTE (no underestimation).

The new time step candidate h'_k is then calculated as

$$h'_k = \min(2h_k, h_{bound,min}), \tag{6.121}$$

where $h_{bound,min}$ is the smallest value of h_{bound} across all capacitors in the circuit.

If the $\texttt{newtrunc}$ simulator parameter is set, a different approach is used for estimating the LTE. Node voltages are used for LTE estimation instead of capacitor charges. This is justifiable if the difference between the predicted and the corrected values is small (which generally is true). Then the predictor-corrector difference of capacitor charge can be approximated as

$$q_{k+1} - q_{k+1}^P \approx \frac{dq(v_{k+1}^P)}{dv}(v_{k+1} - v_{k+1}^P) = c(v_{k+1}^P)(v_{k+1} - v_{k+1}^P). \tag{6.122}$$

As the relation between the charge difference and the voltage difference is approximately linear, charge (x) can be replaced by voltage (v) in (6.117). Furthermore, as the capacitor voltage is the difference between the two terminal node voltages, the LTE bound can be enforced on node voltages.

For this alternative approach to work, predictor-corrector integration must be enabled (the $\texttt{nopredictor}$ simulator parameter must be turned off). This happens automatically when $\texttt{newtrunc}$ is enabled. The LTE is estimated using (6.117). The bound on the LTE is obtained from the node voltage,

$$LTE_{bound} = \texttt{lteabstol} + \texttt{ltereltol} \cdot \max(|v_{k+1}|, |v_{k+1}^P|), \tag{6.123}$$

where v_{k+1}^P is the predicted node voltage and v_{k+1} is the value obtained from the Newton–Raphson algorithm.

By taking into account the fact that the LTE grows with the $n + 1$-th power of the time step, the upper bound on the time step h_{bound} that will keep the LTE below `trtol` $\cdot LTE_{bound}$ can be estimated as

$$
h_{bound} = h_k \left| \frac{\texttt{trtol} \cdot LTE_{bound}}{\epsilon_{k+1}} \right|^{\frac{1}{n+1}}
$$

$$
\approx h_k \left| \frac{C_{n+1} - C_{n+1}^P}{C_{n+1}} \frac{\texttt{trtol} \cdot LTE_{bound}}{v_{k+1} - v_{k+1}^P} \right|^{\frac{1}{n+1}}, \tag{6.124}
$$

where C_{n+1}^P is the error coefficient of the predictor. Larger values of `trtol` result in longer time steps.

In SPICE OPUS only nodes where the difference $v_{k+1} - v_{k+1}^P$ is nonzero are used for calculating the new time step. Let $h_{bound,min}$ denote the smallest h_{bound} across all such nodes. The new time step candidate is obtained as

$$
h_k' = \min(2h_k, h_{bound,min}). \tag{6.125}
$$

From this point on the time-step control algorithm is independent of the `newtrunc` setting. If $h_k' < 0.9h_k$ the time point t_{k+1} is rejected. If a rapid drop in the time step is detected ($h_k < $ `ft` $\cdot h_{k-1}$) and the integration order is above 1, the integration order is set to 1, the time step is set to h_{max}, and the integration order is not allowed to increase until $t \geq t_k + 100h_{k-1}$. This helps prevent oscillations when higher order integration methods are used. If, however, $h_k \geq $ `ft` $\cdot h_{k-1}$ or the integration order is 1, the time step is set to h_k'.

On the other hand, if h_k' is above $0.9h_k$ the order of integration is increased by 1, but not above `maxord`. The truncation error is reestimated for the increased order of integration and the new value of h_k' is calculated. If the new value of h_k' is below $1.05h_k$, the integration order is reset to 1 and h_k' is recalculated. At this point the time point is accepted regardless of the integration order and the time step is set to h_k'.

Chapter 7
Examples

Although circuit simulators are basically analytical tools, they are ultimately used in the process of circuit synthesis. Circuit synthesis is a complex procedure. One could describe this procedure as a synergy between human inspiration and extensive analytical legwork. SPICE OPUS is the perfect tool for the latter. The simulation examples in this chapter have been selected to emphasize the unique scripting capabilities of SPICE OPUS employed to speed up the tedious legwork – enabling circuit designers to focus on the creative side of their work.

In Sect. 7.1 the power of using parametrized subcircuits is demonstrated by showing how simple and efficient circuit topology reuse can be. Next, in Sect. 7.2, topology switching via the netclass syntax is explained. At the same time, this example shows that any physical (nonelectrical) unit can be seamlessly incorporated in the analysis. Extracting specific (standard) circuit properties is often referred to as taking measurements. This can be efficiently automated, as seen in Sect. 7.3 where some other integrated circuit design specifics are also demonstrated. Section 7.4 goes one step further in this direction, demonstrating techniques for simple handling of design corners. Two important analyses in modern integrated circuit design are addressed in Sects. 7.5 and 7.6: the assessment of device mismatch influence and the prediction of the production yield. So far the examples have dealt with typical analog circuit problems. However, SPICE OPUS can also be used to greatly improve the design of digital and mixed mode building blocks. In Sects. 7.7–7.12 a series of digital circuit phenomena is presented hierarchically, ranging from simple logic gates to complex phase-locked loops.

7.1 Parametrized Attenuator and Transmission Lines

This example demonstrates the use of parametrized subcircuits as well as lossless and lossy transmission lines.

An attenuator is a symmetric circuit with two ports. To satisfy the no-reflection condition, the input impedance of an attenuator at every port must be equal to Z_0 if the remaining port is terminated with an impedance Z_0. Z_0 is also referred to as the characteristic impedance of the attenuator. If the characteristic impedance

T. Tuma and Á. Bűrmen, *Circuit Simulation with SPICE OPUS*, Modeling and Simulation in Science, Engineering and Technology, DOI 10.1007/978-0-8176-4867-1_7, © Birkhäuser Boston, a part of Springer Science+Business Media, LLC 2009

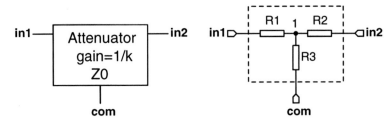

Fig. 7.1 Attenuator symbol and circuit

of the attenuator matches the characteristic impedances of the transmission lines connected to the attenuator, the no-reflection condition is satisfied.

If a voltage source with output impedance Z_0 is connected to port 1 with the port 2 terminated by Z_0, the ratio of the voltage at port 1 to the voltage at port 2 equals the attenuation factor (k). An attenuator can be implemented as a purely resistive circuit (Fig. 7.1).

Let k denote the attenuation factor. The no-reflection condition then results in the following values for the resistors:

$$R_1 = R_2 = Z_0 \cdot \frac{k-1}{k+1}$$
$$R_3 = Z_0 \cdot \frac{2k}{k^2 - 1}.$$

An attenuation of 6 dB is equivalent to $k = 10^{6/20} \approx 2$. Now suppose you want to build an attenuator with $k = 2$ and $Z_0 = 50$. Then $R_1 = R_2 = 16.67\,\Omega$ and $R_3 = 66.67\,\Omega$. The subcircuit of such an attenuator is described by the following netlist:

```
.subckt atn6 in1 in2 com
r1 (in1 1) r=16.67
r2 (in2 1) r=16.67
r3 (1 com) r=66.67
.ends
```

To use the 6 dB attenuator subcircuit by naming it x1 and connecting it between nodes 10, 20, and 0, the following line must be added to the netlist:

```
x1 (10 20 0) atn6
```

Unfortunately, such an approach requires a new subcircuit definition if one wants to use a different attenuator in the circuit (e.g., one with 3 dB attenuation). To overcome this it is practical to define a parametrized attenuator subcircuit. Such a subcircuit takes two parameters: characteristic impedance and attenuation factor. The values of the three resistors are then calculated from the two input parameters.

```
.subckt atn in1 in2 com param: z0=50 k
.param r1=z0*(k-1)/(k+1)
.param r3=z0*2*k/(k^2-1)
r1 (in1 1) r={r1}
```

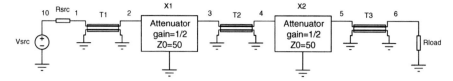

Fig. 7.2 A series of transmission lines and attenuators

```
r2 (in2 1) r={r1}
r3 (com 1) r={r3}
.ends
```

Parameter z0 represents the characteristic impedance. As its default value is specified at subcircuit definition, it does not have to be specified when the subcircuit is instantiated in the netlist (its value defaults to 50). The attenuation factor (k) has no default value and therefore must be specified at invocation. To put a 6 dB attenuator named x1 in the circuit between nodes 10, 20, and 0 the following line must be added to the netlist:

```
x1 (10 20 0) atn param: z0=50 k=2
```

Attenuators are often used in combination with transmission lines. The circuit in Fig. 7.2 comprises three ideal transmission lines with a propagation delay of 10 μs. The characteristic impedance of transmission lines and attenuators is 50 Ω. Both attenuators have 6 dB attenuation. The input impedance of the voltage source and the load impedance are both 50 Ω.

The netlist for the circuit in Fig. 7.2 is as follows:

```
Ideal transmission line and attenuator

.subckt atn in1 in2 com param: z0=50 k
.param r1=z0*(k-1)/(k+1)
.param r3=z0*2*k/(k^2-1)
r1 (in1 1) r={r1}
r2 (in2 1) r={r1}
r3 (com 1) r={r3}
.ends

vsrc (10 0) dc=0 pulse=(0 3.6 1u 1n 1n 5u)
rsrc (10 1) r=50
t1 (1 0 2 0) z0=50 td=10u
x1 (2 3 0) atn param: k=2
t2 (3 0 4 0) z0=50 td=10u
x2 (4 5 0) atn param: k=2
t3 (5 0 6 0) z0=50 td=10u
rload (6 0) r=50

.end
```

The input source supplies a 5 μs wide and 3.6 V high pulse at 1 μs with rise time and fall time equal to 1 ns. The pulse magnitude is halved as it enters the first transmission line (voltage divider between `rsrc` and input impedance of `t1`).

The circuit must first be loaded with the `source` command. A transient analysis reveals what happens in the circuit.

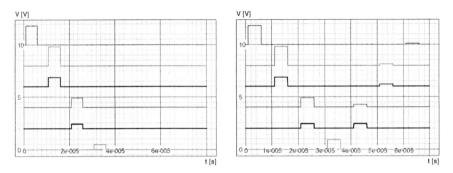

Fig. 7.3 Response of circuit in Fig. 7.2 for two values of load resistance (50 Ω *left* and 1 MΩ *right*)

```
tran 0.01u 80u
plot v(1)+10 v(2)+8 v(3)+6 v(4)+4 v(5)+2 v(6) xlabel 't [s]'
```

After the analysis is finished, the node voltages are plotted in the same graph. To make the graph more readable, individual node voltages are shifted upwards by multiples of 2. The resulting graph is shown in Fig. 7.3, left.

The topmost trace is the voltage at the input of the first transmission line (t1). The next trace is the voltage at the output of t1. It is identical to the voltage at the input of t1. This voltage represents the input for attenuator x1. The third trace represents the voltage at the output of x1. It is halved with respect to the input voltage of the attenuator. This behavior is then repeated at t2 and x2. Finally, the third delay is added by transmission line t3 resulting in the bottom trace representing the voltage at the load resistance. Because the load resistance equals the characteristic impedance of transmission line t3, the wave is not reflected back into t3. Its whole energy is dissipated by rload. All characteristic impedances, the load, and the source resistance are identical, so no reflected waves traveling back toward the source are observed.

On the other hand, if the load resistance is changed to 1 MΩ, the load is effectively removed, terminating the output port of transmission line t3 with an open circuit. As an open circuit cannot dissipate energy, the incoming wave is reflected back into transmission line t3 with the same polarity with which it arrived at the output of t3. The reflected wave propagates back toward the signal source, where it is completely dissipated by the input resistance of the signal source. Note that if the input resistance of vsrc was not equal to the characteristic impedance of t1, some of the reflected wave would be reflected again and travel toward t3. Eventually, the ping-pong would die off as the attenuators take their toll every time the wave passes them by. The behavior of the circuit is depicted in Fig. 7.3, right. The graph can be reproduced with the following NUTMEG commands:

```
let @rload[r]=1meg
tran 0.01u 70u
plot v(1)+10 v(2)+8 v(3)+6 v(4)+4 v(5)+2 v(6) xlabel 't [s]'
```

Besides ideal transmission lines SPICE also supports lossy transmission lines. A simple circuit with a lossy transmission line is depicted in Fig. 7.4.

Fig. 7.4 Testing a lossy
transmission line

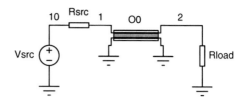

The transmission line has the following characteristics: inductance of 50 nH/m,
capacitance of 20 pF/m, and resistance of 2 Ω/m. This yields a propagation delay of
1 ns/m and a characteristic impedance of 50 Ω/m. The transmission line in Fig. 7.4
is 100 m long which results in a 100 ns delay. The input impedance of the voltage
source and the load resistance both equal 50 Ω.

The circuit is described by the following netlist:

```
Lossy transmission line

vsrc (10 0) dc=0 acmag=1 pulse=(0 3.6 1n 1p 1p 50n)
rin (10 1) r=50

o0 (1 0 2 0) lossy
.model lossy ltra (r=2 l=50n c=20p len=100)

rload (2 0) r=50

.end
```

Note that the `acmag` parameter of `vsrc` is set to 1, providing a small signal
excitation for the AC analysis. The circuit must first be loaded using the `source`
command. To obtain the magnitude of the transfer function, the following commands must be issued. The magnitude is plotted against a logarithmic frequency
scale (Fig. 7.5, left).

```
ac dec 100 1 1g
let h=v(2)/v(1)
plot db(h)
```

The fundamental difference when compared to the response of an ideal transmission line (which must have a gain of 0 dB for all frequencies) is immediately
obvious. The lossy transmission line with a length of 100 m attenuates the signal by
14 dB (in other words, has 0.14 dB/m attenuation). The attenuation is even higher
for frequencies above 1 MHz.

When plotting the phase, a plot against a linear scale is more revealing. Therefore, a different `ac` analysis must be performed (one with a linear sweep).

```
ac lin 1000 1 10meg
let h=v(2)/v(1)
plot unwrap(phase(h))
```

The phase is depicted in Fig. 7.5 (right). The phase should have a linear dependence on the frequency, and the figure demonstrates this with a marker line
positioned beside the phase response.

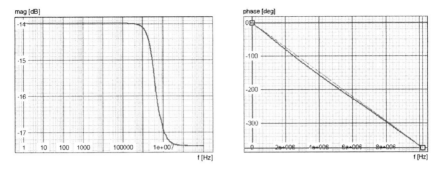

Fig. 7.5 The transfer function of the lossy transmission line in Fig. 7.4. The magnitude is plotted against a logarithmic frequency scale (*left*) and the phase is plotted against a linear scale (*right*)

Fig. 7.6 Response of a lossy transmission line to an input pulse for different values of resistance per unit length. The *top curve* represents the input voltage (increased by 2 V) and the *bottom curve* represents the output voltage at the ports of the transmission line

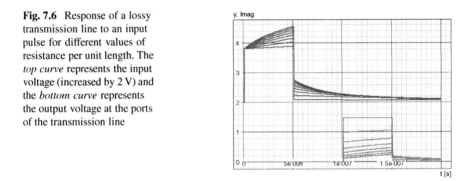

As the signal passes through a lossy transmission line, its shape is distorted. The following script plots a family of responses to the input pulse for different values of the resistance per unit length (from $0.2\,\Omega/\text{m}$ to $3.0\,\Omega/\text{m}$ in $0.4\,\Omega/\text{m}$ steps).

```
* Create an empty plot window.
plot create lossy

* Create a group of vectors for storing the current resistance per unit length.
setplot new
nameplot ctl

* Start loop, r = 0.2 .. 3.0
let ctl.r=0.2
while ctl.r le 3
  * Change the lossy transmission line's parameter.
  let @@lossy[r]=ctl.r
  * Perform a transient analysis.
  tran 0.1n 200n
  * Append the input and output voltage to the plot window.
  plot append lossy v(1)+2 v(2) xlabel 't [s]'
  * Increase resistance per unit length.
  let ctl.r=ctl.r+0.4
end
```

The result of the script is depicted in Fig. 7.6.

7.2 Modeling a Nonlinear Transformer

Suppose you want to model the effects of nonlinearity in a transformer core. A transformer is schematically depicted in Fig. 7.7. The cross section of the core is denoted by A. L denotes the mean length of the magnetic force lines inside the core (dashed line in Fig. 7.7). N_1 and N_2 denote the number of windings in the primary and secondary coils. R_1 and R_2 are the resistances of the primary and secondary windings.

The currents in the primary ($i_1(t)$) and the secondary winding ($i_2(t)$) are sources of magnetization resulting in a magnetic field strength in the core given by

$$H(t) = g(i_1, i_2) = \frac{N_1 i_1(t) + N_2 i_2(t)}{L}. \tag{7.1}$$

The corresponding magnetic flux density depends on the material and is generally related to H in a nonlinear manner,

$$B(t) = f(H(t)). \tag{7.2}$$

The change in magnetic flux density induces a voltage in the primary (e_1) and secondary (e_2) windings,

$$e_1(t) = \frac{d}{dt}(N_1 A B(t)) \tag{7.3}$$

$$e_2(t) = \frac{d}{dt}(N_2 A B(t)). \tag{7.4}$$

The function $f(H)$ depends on the type of material the core is made of. In this example the following characteristic will be used:

$$f(H) = \arctan(H/40) + 4\pi \cdot 10^{-7} \cdot 110 H. \tag{7.5}$$

The units of $f(H)$ are Vs/m^2 and the units for H are A/m. Note that this function does not take into account hysteresis.

The above equations can be described by a model circuit with two electric inputs and an internal magnetic circuit, as depicted by Fig. 7.8.

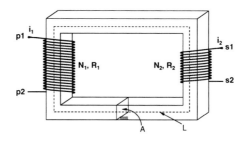

Fig. 7.7 A transformer with two windings

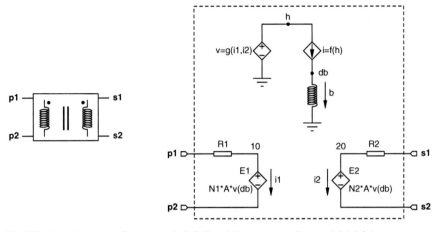

Fig. 7.8 A nonlinear transformer symbol (*left*) and the corresponding model (*right*)

The magnetic part of the model circuit comprises a linear current-controlled voltage source providing excitation to the h node representing the magnetic field strength. This source is controlled by the currents of the primary and the secondary windings and is therefore implemented as a nonlinear SPICE source (B source). Its function is denoted by $g(i_1, i_2)$.

The voltage at the h node is transformed into a current representing the magnetic flux density by another nonlinear B source. The current of this source is determined by the $f(H)$ function. This current is pushed through an inductor with inductance equal to 1H. As a result the voltage at node db represents the derivative of B with respect to time ($\frac{dB}{dt}$).

This derivative controls the controlled voltage sources E1 and E2 that represent the induced voltage in the primary and secondary windings.

The model circuit can be described by the following parametrized subcircuit:

```
.subckt xform p1 p2 s1 s2 param: n1 n2 r1 r2 l a
bh (h 0) v=({n1}*i(e1)+{n2}*i(e2))/{l}
bl (h db) i=atan(v(h)/40)+110*v(h)*4e-7*3.14159265
ll (db 0) l=1

e1 (10 p2 db 0) gain={n1*a}
e2 (20 s2 db 0) gain={n2*a}
r1 (p1 10) r={r1}
r2 (s1 20) r={r2}
.ends
```

Note how the subcircuit parameters are enclosed in curly braces when used in an expression of a B source (e.g., in bh).

First let us plot the B–H curve of the transformer's core. For this purpose the model is connected as depicted by Fig. 7.9 (left). The corresponding part of the netlist is

```
.netclass test bh
x1 (1 0 2 0) xform param: n1=4470 n2=570 r1=300 r2=5 l=0.2 a=2e-4
i1 (0 1) dc=0
.endn
```

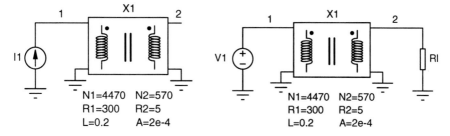

Fig. 7.9 Measuring the B–H curve of transformer's core (*left*) and response to sinusoidal voltage at the primary coil (*right*)

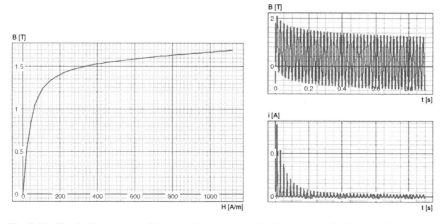

Fig. 7.10 The B–H response of the transformer's core (*left*), the magnetic flux density transient (*top right*), and the current transient (*bottom right*) for unloaded secondary. Note that the secondary current is 0

To run the simulation and plot the result, the corresponding netclass must be activated first.

```
netclass select test::bh
netclass rebuild
```

Next a dc analysis is performed to obtain the core's response.

```
dc i1 0 0.05 0.001
plot create bh i(l1:x1) vs v(h:x1) xlabel 'H [A/m]' ylabel 'B [T]'
```

The magnetic flux density is represented by the current flowing through the inductance l1 within subcircuit instance x1. The magnetic field strength is represented by voltage at node h within x1. The result is depicted in Fig. 7.10 (left).

Now let us look at the response of the transformer when a sinusoidal voltage is applied at the primary. The following netlist fragment describes the circuit that is used for obtaining the response:

```
.netclass test sine
x1 (1 0 2 0) xform param: n1=4470 n2=570 r1=300 r2=5 l=0.2 a=2e-4
v1 (1 0) dc=0 sin=(0 311 50)
rl (2 0) r=1meg
.endn
```

Let us take a look at the response when the secondary is open-circuited. Initially the correct netclass is selected and the load resistance is set to a high value (100 MΩ). This is followed by a simulation and plotting of two graphs (*B* and transformer currents). The resulting graphs are depicted in Fig. 7.10 (right).

```
netclass select test::sine
netclass rebuild

let @rl[r]=100meg

tran 0.1m 900m 0 1m uic
plot create oc:b i(l1:x1) xlabel 't [s]' ylabel 'B [T]'
plot create oc:i i(e1:x1) i(e2:x1) xlabel 't [s]' ylabel 'i [A]'
```

The initial high values of *B* and primary current are the cause of the jolt that is often heard when a transformer is connected to the main power. The steady state response of the unloaded transformer can be obtained by performing a transient analysis until all initial transient effects die out and then recording only the last few periods of the signal.

```
tran 0.1m 5000m 4950m 1m uic
plot create ocs:b i(l1:x1) xlabel 't [s]' ylabel 'B [T]'
plot create ocs:i i(e1:x1) i(e2:x1) xlabel 't [s]' ylabel 'i [A]'
```

The result is depicted in Fig. 7.11 (top left and right). The magnetic flux density is sinusoidal as it must induce a sinusoidal voltage in the transformer's primary winding that counteracts the main power (Fig. 7.11, top left). But because the B–H characteristic of the core is nonlinear, the magnetizing current flowing into the primary is highly nonlinear (Fig. 7.11, top right).

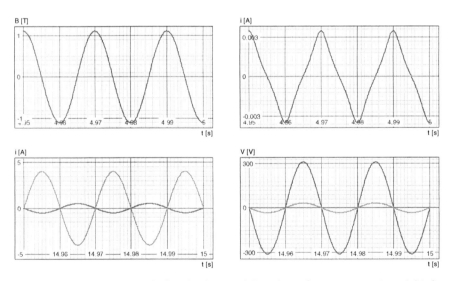

Fig. 7.11 Steady state magnetic flux density (*top left*) and transformer currents (*top right*) for unloaded secondary. Transformer currents for shorted secondary (*bottom left*) and transformer voltages for secondary loaded with a 50 Ω resistance (*bottom right*)

Now suppose you want to observe the response of the transformer with a shorted secondary. The only thing that must be changed is the load resistance. This time it is set to 1 mΩ. The following lines produce a plot of the steady state currents in the transformer:

```
let @rl[r]=1m
```

```
tran 0.1m 15000m 14950m 1m uic
plot create scs:i i(e1:x1) i(e2:x1) xlabel 't [s]' ylabel 'i [A]'
```

The result is depicted in Fig. 7.11 (bottom left). Note how the secondary current (higher amplitude) and the primary current (lower amplitude) have opposite phase. This way the magnetization of the core is kept below saturation.

To obtain the response of the transformer with a 50 Ω load, the following commands must be entered:

```
let @rl[r]=50
```

```
tran 0.1m 15000m 14950m 1m uic
plot create lds:v v(1) v(2) xlabel 't [s]' ylabel 'V [V]'
```

The result (primary and secondary voltage) is plotted in Fig. 7.11 (bottom right).

Usually a transformer is used in conjunction with a rectifier to construct power supplies. Figure 7.12 depicts a transformer in conjunction with a bridge rectifier. The following is the netlist fragment corresponding to the circuit:

```
.netclass test rect
x1 (1 0 2 0) xform param: n1=4470 n2=570 r1=300 r2=5 l=0.2 a=2e-4
v1 (1 0) dc=0 sin=(0 311 50)

d1 (2 p) dn4001
d2 (0 p) dn4001
d3 (n 2) dn4001
d4 (n 0) dn4001

.model dn4001 d (is=1.4n rs=0.04 n=1.7 tt=5u cjo=55p vj=0.34 m=0.38 bv=75)

cl (p n) c=1000u
rl (p n) r=50
.endn
```

In order to simulate the circuit, one must first switch to the correct netclass. Then a transient analysis is performed, and the transformer currents are plotted along with the load voltage and magnetic flux density (Fig. 7.13 top, also bottom left).

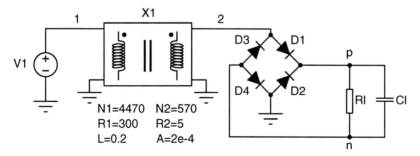

Fig. 7.12 A bridge rectifier in conjunction with a transformer

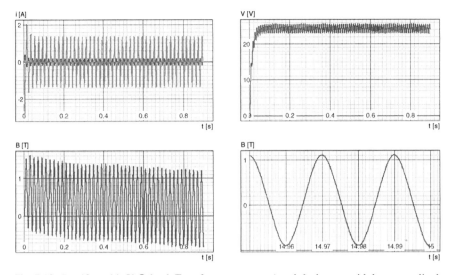

Fig. 7.13 Rectifier with 50 Ω load. Transformer currents (*top left*, the one with lower amplitude is the primary current), load voltage (*top right*), magnetic flux density transient (*bottom left*), and steady state (*bottom right*)

```
netclass select test::rect
netclass rebuild

tran 0.1m 900m 0 1m uic
plot create rec:i i(e1:x1) i(e2:x1) xlabel 't [s]' ylabel 'i [A]'
plot create rec:v v(p,n) xlabel 't [s]' ylabel 'V [V]'
plot create rec:b i(l1:x1) xlabel 't [s]' ylabel 'B [T]'
```

From the transformer current and magnetic flux density it can be seen that probably a mechanical jolt will occur as the transformer is connected to the main power. The load voltage is gradually rising until it reaches its steady state. The load capacitance is responsible for smoothing the voltage.

To plot the steady state response, a different prolonged transient analysis must be performed.

```
tran 0.1m 15000m 14950m 1m uic
plot create recs:b i(l1:x1) xlabel 't [s]' ylabel 'B [T]'
plot create recs:i i(e1:x1) i(e2:x1) xlabel 't [s]' ylabel 'i [A]'
plot create recs:v v(p,n) xlabel 't [s]' ylabel 'V [V]'
```

The steady state magnetic flux density is depicted in Fig. 7.13 (bottom right). The steady state transformer currents and load voltage for a 50 Ω load resistance are shown in Fig. 7.14 (top). Particularly interesting are the transformer currents as they exhibit pulse behavior. The transformer is supplying current to the rectifier only in those parts of the sinusoidal cycle when the load capacitance is being charged.

To see what happens if the load resistance is increased to 500 Ω, the following commands must be entered:

```
let @rl[r]=500

tran 0.1m 15000m 14950m 1m uic
```

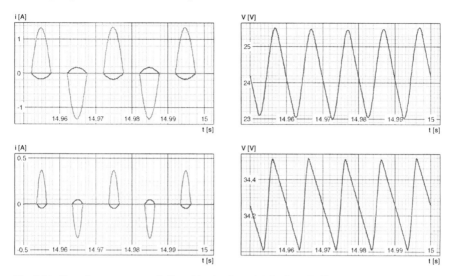

Fig. 7.14 Transformer currents (*left*) and load voltage (*right*) for a 50 Ω load (*top*) and 500 Ω load (*bottom*). The current with lower amplitude is the primary current

```
plot create recs1:i i(e1:x1) i(e2:x1) xlabel 't [s]' ylabel 'i [A]' yl -0.5 0.5
plot create recs1:v v(p,n) xlabel 't [s]' ylabel 'V [V]'
```

Figure 7.14 (bottom) depicts the transformer currents and load voltage for a 500 Ω load. When compared to the response at a 50 Ω load, one can see that the current pulses are narrower and lower at the higher load resistance. The load ripple voltage decreases and the load voltage increases with increasing load resistance.

7.3 Analyzing a CMOS Differential Amplifier

The CMOS differential amplifier [2] which is the subject of this section is depicted in Fig. 7.15. The basic building blocks of the amplifier are MOS transistors. Every MOS transistor is described with its channel width (W) and channel length (L). Besides these two parameters there are several others available. They are used for parasitics evaluation (AD, AS, PD, PS, NRD, and NRS). Due to design layout rules, these parameters can be expressed with W and L as follows:

$$AD = AS = W \cdot 0.18 \,\mu m \tag{7.6}$$

$$PD = PS = 2 \cdot (W + 0.18 \,\mu m) \tag{7.7}$$

$$NRD = NRS = 0.18 \,\mu m / W \tag{7.8}$$

To simplify the netlist and leave the calculation of dependent MOS parameters to the simulator, the basic MOS devices are wrapped in subcircuits that automatically evaluate AD, AS, PD, PS, NRD, and NRS.

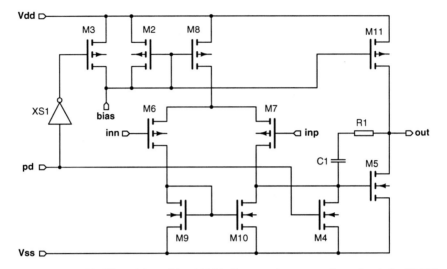

Fig. 7.15 A CMOS differential amplifier. MOS bulk connections are not drawn for clarity. NMOS
(PMOS) bulks are connected to Vss (Vdd)

```
.subckt submodn drain gate source bulk param: w l m=1
m0 (drain gate source bulk) nmosmod w={w} l={l} m={m}
+ ad={w*0.18u} as={w*0.18u}
+ pd={2*(w+0.18u)} ps={2*(w+0.18u)}
+ nrd={0.18u/w} nrs={0.18u/w}
.ends

.subckt submodp drain gate source bulk param: w l m=1
m0 (drain gate source bulk) pmosmod w={w} l={l} m={m}
+ ad={w*0.18u} as={w*0.18u}
+ pd={2*(w+0.18u)} ps={2*(w+0.18u)}
+ nrd={0.18u/w} nrs={0.18u/w}
.ends
```

To demonstrate the use of these two MOS subcircuits, let us build the definition
of a CMOS inverter. Note that every MOS transistor name is prefixed by x. The
x prefix indicates that the device is described by a subcircuit (either submodp or
submodn). The power supplies of the inverter (inputs vsp and vsn) are not displayed
in Fig. 7.15. They are implicitly assumed to be connected to the vdd and the vss
inputs of the amplifier, respectively.

```
.subckt tinv in out vdd vss
xm1 (out in vdd vdd) submodp param: w=0.5u l=0.18u
xm2 (out in vss vss) submodn param: w=0.5u l=0.18u
.ends
```

Now we can describe the amplifier subcircuit (Fig. 7.15).

```
.subckt amp inp inn out vdd vss pd bias
xm2 (bias bias vdd vdd) submodp w=7.14u l=0.44u m=1
xm8 (a2 bias vdd vdd)   submodp w=7.14u l=0.44u m=8
xm11 (out bias vdd vdd) submodp w=7.14u l=0.44u m=8

xm3 (bias w_6 vdd vdd)  submodp w=0.5um l=0.18u
xm4 (a6 pd vss vss)     submodn w=0.5um l=0.18u
```

```
xm6 (a5 inn a2 vdd)     submodp w=4.5u  l=0.51u m=4
xm7 (a6 inp a2 vdd)     submodp w=4.5u  l=0.51u m=4

xm9 (a5 a5 vss vss)     submodn w=1.53u l=1.18u m=4
xm10 (a6 a5 vss vss)    submodn w=1.53u l=1.18u m=4

xm5 (out a6 vss vss)    submodn w=13.5u l=0.77u m=4

xs1 (pd w_6 vdd vss) tinv

c1 (a6 w_7)  c=6.6p
r1 (out w_7) r=750
.ends
```

Besides the two inputs, the output, and the two power supply terminals, the amplifier has a pd input for powerdown. In normal operation this input is pulled low. Powerdown functionality is implemented by transistors M3 and M4 together with inverter XS1. The bias input is used to set the amplifier's bias current (see Fig. 7.16). Transistors M6 and M7 represent the differential pair. M9 and M10 act as an active load for the differential pair. Transistor M2 sets the reference current that is mirrored by M8 and M11 (multiplied by factor 8). Transistors M5 and M11 represent the second stage of the amplifier. C1 and R1 improve the stability of the amplifier.

The netlist of the test circuit is as follows:

```
vdd    (vdd 0)       dc=1.8
vagnd  (agnd 0)      dc=0.9
vcm    (inp agnd)    dc=0
```

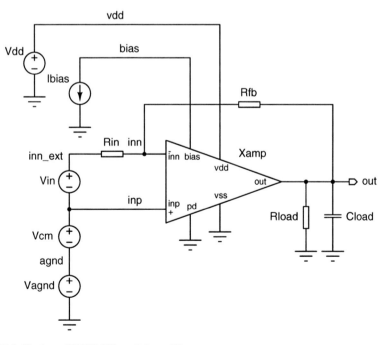

Fig. 7.16 Testing a CMOS differential amplifier

```
vin    (inn_ext inp)  dc=0 acmag=1 pulse=(0.5 -0.5 0.1u 2n 2n 2u)
rin    (inn_ext inn)  r=1meg
rfb    (inn out)      r=1meg
rload  (out 0)        r=100k
cload  (out 0)        c=1p
ibias  (bn 0)         dc=5u
xamp   (inp inn out vdd 0 0 bn) amp
```

The test circuit is shown in Fig. 7.16. 1 MΩ resistors `rin` and `rfb` constitute a negative feedback. This feedback helps put the amplifier in the active region regardless of its offset. The gain of the system (with `vin` voltage as input) is -1, but we can still measure the amplifier's DC characteristic by looking directly at its noninverting and inverting inputs instead of taking the `vin` voltage as the input signal. `vin` also serves as the source of excitation in the `ac` analysis (parameter `acmag`) and in the `tran` analysis (pulse between -0.5 and 0.5 V with 2 μs width and 2 ns rise and fall time.

The circuit is powered by `vdd` set to 1.8 V. The `vagnd` voltage source sets the analog ground level. It is set to one-half of the supply voltage (0.9 V). `vcm` sets the common mode voltage which is 0 V by default. The output load is represented by `rload` (100 kΩ) and `cload` (1 pF). The bias current is set by `ibias` to 5 μA.

Finally, to make the circuit work in SPICE, we need the PMOS and NMOS models. The model parameters are described in [9]. The basis for the listed CMOS model was obtained from the Predictive Technology Model page [41].

```
.param vth0n=0.3999
.param u0n=3.5000000E-02
.param vth0p=-0.42
.param u0p=8.0000000E-03

.model nmosmod NMOS
+ Level = 53
+ Lint = 4.e-08 Tox = 4.e-09
+ Vth0 = {vth0n} Rdsw = 250
+ lmin=1.8e-7 lmax=4u wmin=1.8e-7 wmax=1.0e-4
+ version =3.1
+ Xj= 6.0000000E-08        Nch= 5.9500000E+17
+ lln= 1.0000000           lwn= 1.0000000           wln= 0.00
+ wwn= 0.00                ll= 0.00
+ lw= 0.00                 lwl= 0.00                wint= 0.00
+ wl= 0.00                 ww= 0.00                 wwl= 0.00
+ Mobmod=  1               binunit= 2               xl=  0
+ xw=  0
+ Dwg= 0.00                Dwb= 0.00
+ K1= 0.5613000            K2= 1.0000000E-02
+ K3= 0.00                 Dvt0= 8.0000000          Dvt1= 0.7500000
+ Dvt2= 8.0000000E-03      Dvt0w= 0.00              Dvt1w= 0.00
+ Dvt2w= 0.00              Nlx= 1.6500000E-07       W0= 0.00
+ K3b= 0.00                Ngate= 5.0000000E+20
+ Vsat= 1.3800000E+05      Ua= -7.0000000E-10       Ub= 3.5000000E-18
+ Uc= -5.2500000E-11       Prwb= 0.00
+ Prwg= 0.00               Wr= 1.0000000            U0= {u0n}
+ A0= 1.1000000            Keta= 4.0000000E-02      A1= 0.00
+ A2= 1.0000000            Ags= -1.0000000E-02      B0= 0.00
+ B1= 0.00
+ Voff= -0.12350000        NFactor= 0.9000000       Cit= 0.00
+ Cdsc= 0.00               Cdscb= 0.00              Cdscd= 0.00
+ Eta0= 0.2200000          Etab= 0.00               Dsub= 0.8000000
+ Pclm= 5.0000000E-02      Pdiblc1= 1.2000000E-02   Pdiblc2= 7.5000000E-03
+ Pdiblcb= -1.3500000E-02  Drout= 1.7999999E-02     Pscbe1= 8.6600000E+08
+ Pscbe2= 1.0000000E-20    Pvag= -0.2800000         Delta= 1.0000000E-02
```

```
+ Alpha0= 0.00            Beta0= 30.0000000
+ kt1= -0.3700000         kt2= -4.0000000E-02      At= 5.5000000E+04
+ Ute= -1.4800000         Ua1= 9.5829000E-10       Ub1= -3.3473000E-19
+ Uc1= 0.00               Kt1l= 4.0000000E-09      Prt= 0.00
+ Cj= 0.00365             Mj= 0.54                 Pb= 0.982
+ Cjsw= 7.9E-10           Mjsw= 0.31               Php= 0.841
+ Cta= 0                  Ctp= 0                   Pta= 0
+ Ptp= 0                  JS=1.50E-08              JSW=2.50E-13
+ N=1.0                   Xti=3.0                  Cgdo=2.786E-10
+ Cgso=2.786E-10          Cgbo=0.0E+00             Capmod= 2
+ NQSMOD= 0               Elm= 5                   Xpart= 1
+ Cgsl= 1.6E-10           Cgdl= 1.6E-10            Ckappa= 2.886
+ Cf= 1.069e-10           Clc= 0.0000001           Cle= 0.6
+ Dlc= 4E-08              Dwc= 0                   Vfbcv= -1

.model pmosmod PMOS
+ Level = 53
+ Lint = 3.e-08 Tox = 4.2e-09
+ Vth0 = {vth0p} Rdsw = 450
+ lmin=1.8e-7 lmax=4u wmin=1.8e-7 wmax=1.0e-4
+ version =3.1
+ Xj= 7.0000000E-08       Nch= 5.9200000E+17
+ lln= 1.0000000          lwn= 1.0000000           wln= 0.00
+ wwn= 0.00               ll= 0.00
+ lw= 0.00                lwl= 0.00                wint= 0.00
+ wl= 0.00                ww= 0.00                 wwl= 0.00
+ Mobmod=  1              binunit= 2               xl= 0.00
+ xw= 0.00
+ Dwg= 0.00               Dwb= 0.00
+ ACM= 0                  ldif=0.00                hdif=0.00
+ rsh= 0                  rd= 0                    rs= 0
+ rsc= 0                  rdc= 0
+ K1= 0.5560000           K2= 0.00
+ K3= 0.00                Dvt0= 11.2000000         Dvt1= 0.7200000
+ Dvt2= -1.0000000E-02    Dvt0w= 0.00              Dvt1w= 0.00
+ Dvt2w= 0.00             Nlx= 9.5000000E-08       W0= 0.00
+ K3b= 0.00               Ngate= 5.0000000E+20
+ Vsat= 1.0500000E+05     Ua= -1.2000000E-10       Ub= 1.0000000E-18
+ Uc= -2.9999999E-11      Prwb= 0.00
+ Prwg= 0.00              Wr= 1.0000000            U0= {u0p}
+ A0= 2.1199999           Keta= 2.9999999E-02      A1= 0.00
+ A2= 0.4000000           Ags= -0.1000000          B0= 0.00
+ B1= 0.00
+ Voff= -6.40000000E-02   NFactor= 1.4000000       Cit= 0.00
+ Cdsc= 0.00              Cdscd= 0.00              Cdscb= 0.00
+ Eta0= 8.5000000         Etab= 0.00               Dsub= 2.8000000
+ Pclm= 2.0000000         Pdiblc1= 0.1200000       Pdiblc2= 8.0000000E-05
+ Pdiblcb= 0.1450000      Drout= 5.0000000E-02     Pscbe1= 1.0000000E-20
+ Pscbe2= 1.0000000E-20   Pvag= -6.0000000E-02     Delta= 1.0000000E-02
+ Alpha0= 0.00            Beta0= 30.0000000
+ kt1= -0.3700000         kt2= -4.0000000E-02      At= 5.5000000E+04
+ Ute= -1.4800000         Ua1= 9.5829000E-10       Ub1= -3.3473000E-19
+ Uc1= 0.00               Kt1l= 4.0000000E-09      Prt= 0.00
+ Cj= 0.00138             Mj= 1.05                 Pb= 1.24
+ Cjsw= 1.44E-09          Mjsw= 0.43               Php= 0.841
+ Cta= 0.00093            Ctp= 0                   Pta= 0.00153
+ Ptp= 0                  JS=1.50E-08              JSW=2.50E-13
+ N=1.0                   Xti=3.0                  Cgdo=2.786E-10
+ Cgso=2.786E-10          Cgbo=0.0E+00             Capmod= 2
+ NQSMOD= 0               Elm= 5                   Xpart= 1
+ Cgsl= 1.6E-10           Cgdl= 1.6E-10            Ckappa= 2.886
+ Cf= 1.058e-10           Clc= 0.0000001           Cle= 0.6
+ Dlc= 3E-08              Dwc= 0                   Vfbcv= -1
```

The commands that follow must be entered into the control block of the netlist or typed in the command window. First let us destroy all plots and set degrees for angle units.

```
destroy all
set units=degrees
```

We then perform a simple operating point analysis to obtain the power consumption.

```
* Operating point analysis
op
* Get supply current
let isup=-i(vdd)
* Print it
print isup
```

Next we use the dc analysis to obtain the DC transfer function. The result is shown in Fig. 7.17 (top left). Note how we plot the output voltage vs. the voltage between the noninverting and inverting input instead of plotting it against the default scale. In the latter case we would get the transfer function of an inverting amplifier with voltage gain equal to -1.

```
* DC sweep
dc vin -2 2 lin 400
* Plot it
plot v(out,agnd) vs v(inp,inn) xl -20m 20m xlabel "V(+,-) [V]" ylabel "Vout [V]"
* Measure input offset (input when output is at agnd)
let c=0
cursor c right v(out,agnd) 0
let inoffs=v(inp,inn)[%c]
* Print it
print inoffs
```

The ac analysis evaluates the frequency response of the circuit. Figure 7.17 (top, right) shows the magnitude and the phase of the AC response. Again, the input signal is the voltage between the noninverting and inverting input of the amplifier. After the magnitude and phase are plotted, the phase margin is measured and printed. Finally, the plot with the ac analysis results is named acres so that we can refer to it later.

```
* Prepare for AC analysis
let @vdd[acmag]=0
let @vin[acmag]=1
* AC analysis
ac dec 50 1 100meg
* Frequency response
let h=v(out,agnd)/v(inp,inn)
plot db(h) unwrap(phase(h)) xlabel "f [Hz]" ylabel "mag [dB], phase [deg]"
* Get maximal gain
let acgain=max(db(h))
* Get phase margin (move cursor c to gain=0dB)
let c=0
cursor c right db(h) 0
let pm=unwrap(phase(h))[%c]+180
* Get unity gain bandwidth
let ugbw=abs(frequency[%c])
* Print the results
print acgain pm ugbw
* Rename the plot
nameplot acres
```

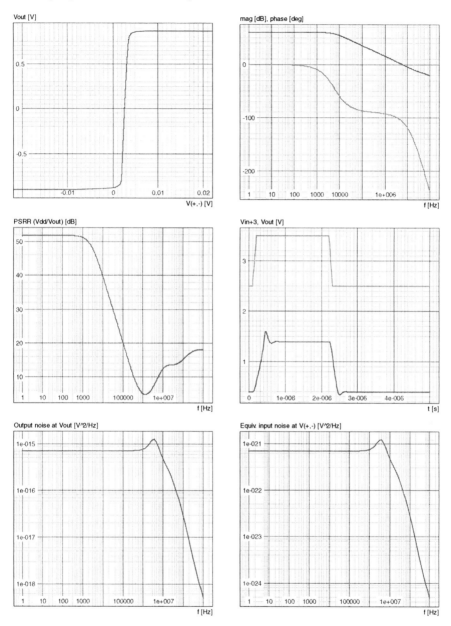

Fig. 7.17 CMOS differential amplifier simulation results. DC characteristic (*top left*), AC response (*top right*, *top curve* is the magnitude, *bottom curve* is the phase), PSRR (*middle left*), transient response (*middle right*, *top curve* is input shifted up by 3 V, *bottom curve* is output), output noise power spectrum density (*bottom left*), and equivalent input noise power spectrum density (*bottom right*)

To analyze the power supply rejection we set the input source's acmag parameter to 0 and the power supply's ac parameter to 1. The power supply rejection ratio (PSRR) is measured with feedback which sets the voltage gain to 10. An ac analysis results in the frequency response of the circuit, this time to the small signal excitation coming from the power supply. The response is shown in Fig. 7.17 (middle left).

```
* Prepare for PSRR measurement, feedback for A=10
let @vdd[acmag]=1
let @vin[acmag]=0
let @rfb[r]=10meg
* AC analysis (PSRR)
ac dec 50 1 10meg
let h=v(vdd)/v(out,agnd)
plot db(h) xlabel "f [Hz]" ylabel "PSRR (Vdd/Vout) [dB]"
* PSRR at 0Hz
let psrr0=db(h)[0]
* Print it
print psrr0
* Restore original feedback (A=1)
let @rfb[r]=1meg
```

The noise analysis produces the output noise spectrum. Unfortunately, the output noise spectrum contains the noise from the amplifier and three unwanted contributions: from rin, rfb, and rload. These contributions must be subtracted from the output noise spectrum.

The equivalent input noise contributions are not calculated by SPICE, so the equivalent input noise power spectrum density must be determined by dividing the corrected output noise power spectrum density with the ac squared ac response of the amplifier (remember that the ac analysis results were named acres). The division can be done with no additional vector manipulation because the scale of the ac analysis is the same as the scale of the noise analysis (dec 10 1 1g). The output and equivalent input spectra are shown in Fig. 7.17 (bottom left and right).

```
* Noise analysis
noise v(out,agnd) vin dec 10 1 1g 1
* Move to plot containing power spectra
setplot previous
* Correct output power spectrum density
let ns=onoise_spectrum-onoise_rfb-onoise_rin-onoise_rload
* Calculate equivalent input power spectrum density
let nsin=ns/abs(acres.h)^2
* Plot both spectra
plot ns    ylog xlabel "f [Hz]" ylabel "Output noise at Vout [V^2/Hz]"
plot nsin ylog xlabel "f [Hz]" ylabel "Equiv. input noise at V(+,-) [V^2/Hz]"
* Measure power spectrum density at 100 Hz and 100 kHz
let c=0
cursor c right frequency 100
let inoise100=nsin[%c]
cursor c right frequency 100k
let inoise100k=nsin[%c]
* Print power spectrum density (V^2/Hz)
print inoise100 inoise100k
* Move back to the plot containing integrated noise
setplot next
```

Finally, the transient analysis simulates the circuit's response to a pulse generated at vin. The result is shown in Fig. 7.17 (middle right).

```
* Transient analysis
tran 0.01u 5u
* Plot response
plot v(out) v(inp,inn_ext)+3 xlabel "t [s]" ylabel "Vin+3, Vout [V]"

* Measurements at input rising edge
* Get initial and final level (at 0.1us and 2.1us)
let c1=0
cursor c1 right time 0.1u
let out1=v(out)[%c1]
cursor c1 right time 2.1u
let out2=v(out)[%c1]
* Calculate 10% and 90% level
let out10=out1+(out2-out1)/10
let out90=out2-(out2-out1)/10
* Move cursors c1 and c2 to 10% and 90% level
let c1=0
let c2=0
cursor c1 right v(out) out10 1
cursor c2 right v(out) out90 1
* Calculate rise time (s)
*    and slew rate in V/s (slope between 10% and 90% level)
let trise=time[%c2]-time[%c1]
let slewrise=(out90-out10)/trise
* Move cursor c3 to the end of the input rising edge response
let c3=0
cursor c3 right time 2.1u
* Calculate absolute deviations from the final level
let aux=abs(v(out)-out2)
* Move back to the time when deviations rise to 101%
cursor c3 left aux 0.01*(out2-out1)
* Extract settling time
*   (between c3 and the start of rising edge at 0.1us)
let tsetrise=time[%c3]-0.1u

* Do the same measurements for the input falling edge
* Initial and final level from rising edge change roles
let c1=0
let c2=0
cursor c1 right v(out) out90 2
cursor c2 right v(out) out10 2
let tfall=time[%c2]-time[%c1]
let slewfall=(out90-out10)/tfall
let c3=0
cursor c3 right time
let aux=abs(v(out)-out1)
cursor c3 left aux 0.01*(out2-out1)
let tsetfall=time[%c3]-2.1u

* Get relative overshoot and undershoot
let over=(max(v(out))-out2)/(out2-out1)
let under=(out1-min(v(out)))/(out2-out1)

* Print the results
print trise tfall slewrise slewfall over under tsetrise tsetfall
```

If all of the above-mentioned scripts are run, we obtain the following output:

```
isup = 8.477974e-005
inoffs = 2.599784e-003
acgain = 6.018926e+001
pm = 7.238533e+001
ugbw = 6.036387e+006
psrr0 = 5.180083e+001
inoise100 = 6.855232e-022
inoise100k = 6.861164e-022
trise = 2.302304e-007
tfall = 1.637719e-007
```

```
slewrise = 3.467799e+006
slewfall = 4.875031e+006
over = 2.028948e-001
under = 5.397594e-002
tsetrise = 5.824900e-007
tsetfall = 4.849213e-007
```

7.4 Using Corner Models

The variations of the IC manufacturing process result in varying characteristics of MOS transistors. A part of these variations affects all devices on a chip in the same manner. These are called global process variations. They can be taken into account if the circuit is designed to operate correctly not only for the typical device parameter values, but also for combinations of extreme device parameter variations.

The two MOS model parameters that are almost always considered to be affected by these variations are the threshold voltage vth0 and the carrier mobility u0. These two parameters affect, among other things, the transconductance g_m of the transistor.

In digital design it is common to examine the four combinations of PMOS and NMOS minimal and maximal transconductance. They are referred to as MOS corner models and are usually named worst speed (WS), worst power (WP), worst one (WO), and worst zero (WZ). The typical model is usually denoted by typical mean (TM). See Table 7.1 for an example. The corners are obtained by perturbing the TM values of vth0 and u0 by a certain percentage.

Often these corners are employed in analog design to capture the global process variations. Corner models are usually supplied in the form of a library file, where every library section represents one corner model of NMOS and PMOS. The CMOS models from Sect. 7.3 can be transformed into a model library containing models for the variations in Table 7.1.

Suppose the library file is named cmos180n.lib. First we put the models in a separate section named model in file cmos180n.lib. The definition of the section would look like this.

```
* The model section of file cmos180n.lib
.lib model
.model nmosmod NMOS
+ Level = 53
+ Lint = 4.e-08 Tox = 4.e-09
+ Vth0 = {vth0n} Rdsw = 250
... the rest of the nmosmod model

.model pmosmod PMOS
+ Level = 53
+ Lint = 3.e-08 Tox = 4.2e-09
+ Vth0 = {vth0p} Rdsw = 450
... the rest of the pmosmod model

.endl
```

Note how the vth0 and u0 model parameters are parametrized to use the vth0n and u0n global circuit parameters in the nmosmod model. The pmosmod model uses the vth0p and u0p global circuit parameters for this purpose.

Table 7.1 Parameter variations for CMOS corner models

Corner name	NMOS			PMOS		
	g_m	vth0	u0	g_m	vth0	u0
TM	Average	0%	0%	Average	0%	0%
WP	Maximal	−10%	+6%	Maximal	−12%	+10%
WS	Minimal	+10%	−6%	Minimal	+12%	−10%
WO	Maximal	−10%	+6%	Minimal	+12%	−10%
WZ	Minimal	+10%	−6%	Maximal	−12%	+10%

To make things more neat, we put the TM values and the relative variations of these parameters in a separate section named `dfl`.

```
* The dfl section of file cmos180n.lib
.lib dfl
* Mean parameter values
.param vth0nt=0.3999
.param u0nt=3.5000000E-02
.param vth0pt=-0.42
.param u0pt=8.0000000E-03

* Relative parameter variations for corner models
.param vth0nd=0.1
.param u0nd=0.06
.param vth0pd=0.12
.param u0pd=0.1
.endl
```

We will define the TM model by adding the `tm` library section to `cmos180n.lib`.

```
* Section tm of file cmos180n.lib defines the typical mean model.
.lib tm
* Include the dfl section with vth0 and u0 values and variations.
.lib 'cmos180n.lib' dfl
* Calculate vth0 and u0 for nmos and pmos model
.param vth0n=vth0nt
.param u0n=u0nt
.param vth0p=vth0pt
.param u0p=u0pt
* Include the model section
.lib 'cmos180n.lib' model
.endl
```

In a similar way other corner models can be defined. Take, for instance, the WP model section in `cmos180n.lib`.

```
* Section wp of file cmos180n.lib defines the worst power corner model.
.lib wp
.lib 'cmos180n.lib' dfl
* This time we perturb the typical mean values.
.param vth0n=vth0nt*(1-vth0nd)
.param u0n=u0nt*(1+u0nd)
.param vth0p=vth0pt*(1-vth0pd)
.param u0p=u0pt*(1+u0pd)
.lib 'cmos180n.lib' model
.endl
```

To include the WP corner model, one simply adds the following line to the circuit netlist:

```
.lib 'cmos180n.lib' wp
```

Sometimes you want to be able to switch between different corner models without reloading the circuit. This can be achieved by using netclasses and the `netclass` command. Add the following lines to your netlist:

```
* This includes WP, WS, WO, and WZ models as 4 netclasses to the circuit.

.netclass procmod wp
.lib 'cmos180n.lib' wp
.endn

.netclass procmod ws
.lib 'cmos180n.lib' ws
.endn

.netclass procmod wo
.lib 'cmos180n.lib' wo
.endn

.netclass procmod wz
.lib 'cmos180n.lib' wz
.endn
```

To select the WO model, issue the following two NUTMEG commands:

```
netclass select procmod::wo
netclass rebuild
```

Besides global process variations, a circuit's operation is also affected by varying environmental conditions like temperature, load, supply voltage, etc. These can also be taken into account by combining their extreme values with the corner models.

Take, for instance, the amplifier from Sect. 7.3. Suppose we are interested in all combinations of extreme corner models (WP, WS, WO, WZ), temperature (0 and 75°C), supply voltage (1.6 and 2.0 V), bias current (4 and 6 μA), load capacitance (0.5 and 2 pF), and load resistance (80 and 120 kΩ). This results in a total of $4 \cdot 2 \cdot 2 \cdot 2 \cdot 2 \cdot 2 = 128$ repetitions of the analyses from Sect. 7.3.

The analysis across all 128 combinations can be automated using the NUTMEG scripting language. It is recommended that such long scripts be put in a `.control` block in the netlist. This way the script is automatically executed when the netlist is loaded.

We start by creating a plot where the extreme values and the measurement results are stored.

```
* Clean up
destroy all

* Set angle units
set units=degrees

* Create a new plot and name it ctl
setplot new
nameplot ctl

* Extremes
set proc = ( wp ws wo wz )
let temp =(0;75)
let vdd  =(1.6;2.0)
let ibias=(4u;6u)
let cload=(0.5p;2p)
let rload=(80k;120k)
```

```
* Calculate the number of combinations
let n=$(#proc)*length(temp)*length(vdd)*length(ibias)*length(cload)*length(rload)

* Prepare vectors for the measurement results
let isup=vector(n)*0
let inoffs=vector(n)*0
let acgain=vector(n)*0
... the remaining measurement vectors
```

We also want to plot the response curves for all 128 combinations. For this purpose we create empty plot windows and name them dc, ac, psrr, noiseout, noisein, and tran.

```
plot create dc
plot create ac
plot create psrr
plot create noiseout
plot create noisein
plot create tran
```

Next we start six loops that go across all of the 128 combinations.

```
* Combination counter
let ctl.cnt=0

* Corner model loop
let ctl.cnt0=0
while ctl.cnt0 lt $(#proc)
  * Select the model
  netclass select procmod::$(proc[{ctl.cnt0}])
  netclass rebuild

  * Temperature loop
  let ctl.cnt1=0
  while ctl.cnt1 lt length(ctl.temp)
    * Set temperature
    set temp = {ctl.temp[ctl.cnt1]}

    * Supply voltage loop
    let ctl.cnt2=0
    while ctl.cnt2 lt length(ctl.vdd)
      * Set supply voltage and analog ground
      let @vdd[dc] = ctl.vdd[ctl.cnt2]
      let @vagnd[dc] = ctl.vdd[ctl.cnt2]/2

      * Bias current loop
      let ctl.cnt3=0
      while ctl.cnt3 lt length(ctl.ibias)
        * Set bias current
        let @ibias[dc] = ctl.ibias[ctl.cnt3]

        * Load capacitance loop
        let ctl.cnt4=0
        while ctl.cnt4 lt length(ctl.cload)
          * Set load capacitance
          let @cload[c] = ctl.cload[ctl.cnt4]

          * Load resistance loop
          let ctl.cnt5=0
          while ctl.cnt5 lt length(ctl.rload)
            * Set load resistance
            let @rload[r] = ctl.rload[ctl.cnt5]
```

We can print some diagnostic output so that the user knows what is currently being analyzed.

```
echo Corner {ctl.cnt}: model=$(proc[{ctl.cnt0}]) T={ctl.temp[ctl.cnt1]}
+ vdd={ctl.vdd[ctl.cnt2]} ibias={ctl.ibias[ctl.cnt3]}
+ cload={ctl.cload[ctl.cnt4]} rload={ctl.rload[ctl.cnt5]}
```

The analysis script from Sect. 7.3 must be modified. The following is the new version of the part that performs the operating point and the DC analysis:

```
* Operating point analysis
op
* Measure supply current
let isup=-i(vdd)
* Store the measurement result
let ctl.isup[ctl.cnt]=isup
* Destroy operating point analysis results
destroy

* DC analysis
dc vin -2 2 lin 400
* Append the response to the dc plot window
plot quickappend dc v(out,agnd) vs v(inp,inn)
* Measure input offset
let c=0
cursor c right v(out,agnd) 0
let inoffs=v(inp,inn)[%c]
* Store measurement result
let ctl.inoffs[ctl.cnt]=inoffs
* Destroy DC analysis results
destroy
```

Instead of printing the measurement results, we add them to the result vectors in the ctl plot. The combination counter ctl.cnt is the index that specifies where the measurement result is stored in the corresponding result vector.

After every analysis the circuit's response is appended to the corresponding plot window using the plot quickappend command. The quickappend mode is used to speed up the plotting. There is no need to refresh the plot until all curves are plotted.

After all measurements on the results of a particular analysis are finished and the results are stored, we destroy the results with the destroy command. This simple step significantly reduces memory consumption. An exception is the ac analysis. Its results are named acres and must be kept until the noise analysis is finished. After the noise measurements are taken, three plots must be destroyed (the ac analysis results and the two plots that were created by the noise analysis). So after the noise measurements are stored, the following commands are issued:

```
* Move back to the plot containing integrated noise
setplot next
* Destroy noise analysis results (two plots)
destroy
destroy
* Destroy ac analysis results
destroy acres
```

The loops end with counter increments. We increase the combination counter (used as the index into the measurement result vectors) in the innermost loop.

```
              * Increase combination counter
              let ctl.cnt=ctl.cnt+1

              * Close all 6 loops
              let ctl.cnt5=ctl.cnt5+1
            end
          let ctl.cnt4=ctl.cnt4+1
        end
      let ctl.cnt3=ctl.cnt3+1
    end
    let ctl.cnt2=ctl.cnt2+1
  end
  let ctl.cnt1=ctl.cnt1+1
 end
 let ctl.cnt0=ctl.cnt0+1
end
```

We are done analyzing the 128 combinations. Let us now put some labels on plot windows and adjust the view options. By using the append mode, we force the plot windows to refresh their contents.

```
plot append dc xl -10m 10m xlabel "V(+,-) [V]" ylabel "Vout [V]"
plot append ac xlog xlabel "f [Hz]" ylabel "mag [dB], phase [deg]"
plot append psrr xlog xlabel "f [Hz]" ylabel "PSRR (Vdd/Vout) [dB]"
plot append noiseout xlog ylog xlabel "f [Hz]" ylabel "Output noise at Vout [V^2/Hz]"
plot append noisein  xlog ylog xlabel "f [Hz]" ylabel "Equiv. input noise at V(+,-)
  [V^2/Hz]"
plot append tran xlabel "t [s]" ylabel "Vin+3, Vout [V]"
```

In the end we print the worst measurement values across all 128 combinations.

```
* Go to the ctl plot
setplot ctl
echo
echo Worst values:
* Maximal supply current
echo "isup=       " {max(isup)}
* Min-max difference of the input offset
echo "inoffs=     " {max(inoffs)-min(inoffs)} (range)
* Minimal ac gain
echo "acgain=     " {min(acgain)}
...
```

Finally we store the measurement results in a file named ampcor.raw.

```
write ampcor.raw
```

The resulting circuit responses are plotted in Fig. 7.18. The script prints the following worst measurement values:

```
Worst values:
isup=        0.000107675339568416
inoffs=      0.00168792397763888 (range)
acgain=      57.2722437766492
pm=          60.4944721547554
ugbw=        4316238.66496488
psrr=        26.487859419063
inoise100=   1.97880659821485e-021
inoise100k=  1.98087072738304e-021
trise=       3.77278e-007
tsetrise=    9.83959163466204e-007
slewrise=    5680816.36882201
overshoot=   0.298057008651568
tfall=       2.27739999999998e-007
tsetfall=    4.60502313461338e-007
slewfall=    6494290.58524564
undershoot=  0.083188038945233
```

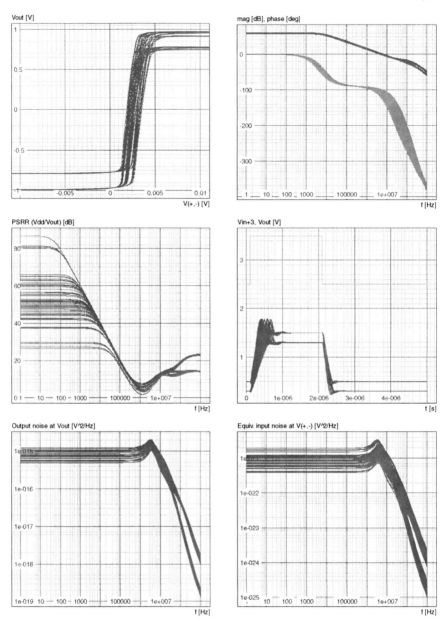

Fig. 7.18 CMOS differential amplifier simulation results across the 128 extreme combinations. DC characteristic (*top left*), AC response (*top right, top curve* is the magnitude, *bottom curve* is the phase), PSRR (*middle left*), transient response (*middle right, top curve* is input shifted up by 3 V, *bottom curve* is output), output noise power spectrum density (*bottom left*), and equivalent input noise power spectrum density (*bottom right*)

7.5 Local Variations (Mismatch) and Monte-Carlo Analysis

In Sect. 7.4 the effect of global process variation was captured by the corner MOS models (WP, WS, WO, WZ). To get a more detailed overview of the effects of these variations, we must consider that the threshold voltage and the carrier mobility are randomly distributed according to some probability distribution. Normal (Gaussian) distribution is the one that is usually assumed. The probability density function of a normally distributed random variable $X = N(a_x, \sigma_x^2)$ with mean value a_x and variance σ_x^2 is given by

$$p(x) = \frac{1}{\sigma_x \sqrt{2\pi}} e^{-(\frac{x-a_x}{2\sigma_x})^2} \tag{7.9}$$

We assume that the extreme global variations of threshold voltage and carrier mobility represent the 3σ extreme values of a normal distribution where the mean values are taken from the TM model.

Unfortunately, global variations that affect all circuit elements in the same way are not the only ones present in the manufacturing process. Due to random variations of the process, two identically designed devices exhibit different electrical behavior. Such variations are referred to as mismatch [28].

The difference is often normally distributed. The mean value of the difference is zero. The variance depends on the device size. Smaller devices exhibit greater variance. The variance of the difference in threshold voltage (V_T) and carrier mobility (μ_0) between two identically designed transistors can be modeled using Pelgrom's model [40],

$$\sigma_{V_T}^2 = \frac{A_{V_T}^2}{WL} \tag{7.10}$$

$$\frac{1}{\mu_0^2} \sigma_{\mu_0}^2 = \frac{A_{\mu_0}^2}{WL}, \tag{7.11}$$

where W and L denote the transistor channel width and length. The actual threshold voltage of a MOS transistor is modeled as a sum of two normally distributed random variables,

$$V_T = N(a_{V_T}, \sigma_{V_T,g}^2) + N(0, \frac{A_{V_T}^2}{2WL}), \tag{7.12}$$

where a_{V_T} is the mean value of the threshold voltage and $\sigma_{V_T,g}^2$ is the variance of the global variation (the one that affects all transistors in the same way). Similarly, the carrier mobility can be modeled as

$$\mu_0 = N(a_{\mu_0}, \sigma_{\mu_0,g}^2) \left(1 + N(0, \frac{A_{\mu_0}^2}{2WL})\right). \tag{7.13}$$

Table 7.2 Global and local parameter variations of MOS transistors

	$\sigma_{V_{\mathrm{T}}}/a_{V_{\mathrm{T}},g}$ (V)	$A_{V_{\mathrm{T}}}$ (V μm)	$\sigma_{\mu_0}/a_{\mu_0,g}$ (·)	A_{μ_0} (μm)
NMOS	$3.33 \cdot 10^{-2}$	$5.00 \cdot 10^{-3}$	$2.00 \cdot 10^{-2}$	$1.04 \cdot 10^{-2}$
PMOS	$4.00 \cdot 10^{-2}$	$5.49 \cdot 10^{-3}$	$3.33 \cdot 10^{-2}$	$0.99 \cdot 10^{-2}$

The division by 2 in the denominator of the mismatch variance in (7.12) and (7.13) comes from the fact that (7.10) and (7.11) represent the variance of the difference between two transistors.

Due to the random nature of the threshold voltage and carrier mobility, the circuit's performance is also distributed randomly. This distribution can be obtained by simulating a large number of random samples. For every one of the N global variation samples, M local variations are simulated. This results in NM simulations and consequently in NM values for every measurement. Such an analysis is referred to as Monte-Carlo analysis.

Sample values for a 0.18 μm IC manufacturing technology are listed in Table 7.2. In Sect. 7.4 we were able to simulate global variations as variations of the MOS device model. All transistors of the same type (NMOS or PMOS) use the same model. We can do the same now, except that the variation is random now.

Because individual MOS transistors have no device parameters for adjusting the threshold voltage and carrier mobility, we are forced to use one model per transistor if we want to model local variations. Fortunately, we can adjust the threshold voltage and carrier mobility indirectly by making some clever changes to every MOS transistor.

Variations of the threshold voltage can easily be incorporated as an independent voltage source in series with the gate electrode. The value of the source represents the deviation from the TM transistor model.

Because the carrier mobility occurs in the device equations as a factor in the product with the transistor width, we can model the relative variation of the mobility by applying the same relative variation to the transistor width. In the end we have the following subcircuit model for the NMOS and the PMOS transistors:

```
* NMOS model subcircuit
.subckt submodn drain gate source bulk param: w l m=1 vtmm=0 uOmm=0
* Calculate channel area
.param sqrtarea=sqrt(w*l)
* Calculate threshold voltage variation
.param vvt=vtmm*(5e-3*1e-6/sqrtarea/sqrt(2))
* Calculate relative carrier mobility variation
.param vuOr=uOmm*(1.04e-2*1e-6/sqrtarea/sqrt(2))
* Voltage source in series with gate
vgmm (gate gate_int) dc={vvt}
* Transistor
m0 (drain gate_int source bulk) nmosmod w={w*(1+vuOr)} l={1} m={m}
+ ad={w*0.18u} as={w*0.18u}
+ pd={2*(w+0.18u)} ps={2*(w+0.18u)}
+ nrd={0.18u/w} nrs={0.18u/w}
.ends

* PMOS model subcircuit
.subckt submodp drain gate source bulk param: w l m=1 vtmm=0 uOmm=0
.param sqrtarea=sqrt(w*l)
```

```
.param vvt=vtmm*(5.49e-3*1e-6/sqrtarea/sqrt(2))
.param vu0r=u0mm*(0.99e-2*1e-6/sqrtarea/sqrt(2))
vgmm (gate gate_int) dc={vvt}
m0 (drain gate_int source bulk) pmosmod w={w*(1+vu0r)} l={l} m={m}
+ ad={w*0.18u} as={w*0.18u}
+ pd={2*(w+0.18u)} ps={2*(w+0.18u)}
+ nrd={0.18u/w} nrs={0.18u/w}
.ends
```

Mismatch is modeled with the normalized mismatch parameters vtmm and u0mm of MOS subcircuits. If a normalized mismatch parameter is set to a, it changes the threshold voltage (or carrier mobility) by $a\sigma$ where σ is obtained from (7.10) and (7.11).

Now we are ready to perform a Monte-Carlo analysis that will estimate the effects of the manufacturing process variations on the performance of the circuit from Sect. 7.3. The script is long, so it is a good idea to put it in a .control block. We start by cleaning up, setting the angle units, and setting up the random generator.

```
destroy all

* Set angle units
set units=degrees
* Truncate normal distribution at 8 sigma
set gausstruncate=8
* Reset random generator
set rndinit=-1
```

Next we prepare a plot named ctl to hold the control vectors and the results.

```
* Create new plot and name it ctl
setplot new
nameplot ctl

* Relative global variation (sigma) of vth0 and u0 for nmos and pmos
let vth0nprocsig=0.1/3
let u0nprocsig=0.06/3
let vth0pprocsig=0.12/3
let u0pprocsig=0.1/3

* Store the nominal value of vth0 and u0 for nmos and pmos
let vth0n=@@nmosmod[vth0]
let u0n=@@nmosmod[u0]
let vth0p=@@pmosmod[vth0]
let u0p=@@pmosmod[u0]

* Number of global variations (lot) and local variations (case)
let nlot=50
let ncase=100

* Total number of variations
let n=nlot*ncase

* Prepare vectors for the results
let isup=vector(n)*0
let inoffs=vector(n)*0
let acgain=vector(n)*0
... the remaining measurement vectors
```

Next we list all normalized mismatch parameters in a variable named devpar and set the circuit's operating conditions to their nominal values.

```
* Form a list of normalized mismatch parameters
set devpar = (
```

```
+ @xm2:xamp[vtmm]   @xm2:xamp[uOmm]
+ @xm5:xamp[vtmm]   @xm5:xamp[uOmm]
+ @xm6:xamp[vtmm]   @xm6:xamp[uOmm]
+ @xm7:xamp[vtmm]   @xm7:xamp[uOmm]
+ @xm8:xamp[vtmm]   @xm8:xamp[uOmm]
+ @xm9:xamp[vtmm]   @xm9:xamp[uOmm]
+ @xm10:xamp[vtmm]  @xm10:xamp[uOmm]
+ @xm11:xamp[vtmm]  @xm11:xamp[uOmm] )

* Set circuit's operating conditions to nominal values
set temp       = 27
let @vdd[dc]   = 1.8
let @vagnd[dc] = 0.9
let @ibias[dc] = 5u
let @cload[c]  = 1p
let @rload[r]  = 100k
```

Now we are prepared to start the outer (global variation) and the inner (local variation) loop. We use the rndgauss() function to obtain the $N(0, 1)$ distribution.

```
* Global variation loop
let ctl.cntlot=0
while ctl.cntlot lt ctl.nlot
  setplot ctl
  * Create a global variation of vth0 and u0.
  * Change the vth0 and u0 parameter of the nmos and pmos model.
  * Multiply relative global variation with N(0,1)
  * to obtain the relative variation.
  let @@nmosmod[vth0]=vth0n*(1+vth0nprocsig*rndgauss(1))
  let @@nmosmod[u0]=u0n*(1+u0nprocsig*rndgauss(1))
  let @@pmosmod[vth0]=vth0p*(1+vth0pprocsig*rndgauss(1))
  let @@pmosmod[u0]=u0p*(1+u0pprocsig*rndgauss(1))

  * Now start the local variation loop
  let ctl.cntcase=0
  while ctl.cntcase lt ctl.ncase
    setplot ctl
    * Calculate the serial number of the variation.
    let cnt=cntlot*ncase+cntcase
    * Output some diagnostic messages
    echo Lot={cntlot} case={cntcase} : Var {cnt}

    * Create a local variation for all normalized mismatch parameters.
    * Use the N(0,1) distribution.
    let cnt1=0
    while cnt1 lt $(#devpar)
      let $(devpar[{cnt1}])=rndgauss(1)
      let cnt1=cnt1+1
    end
```

The rest of the loop is the same as in Sect. 7.4. The only difference is that now we do not plot the circuit's response (displaying 5,000 curves in every plot window would take up a lot of memory and slow down the computation). We list the operating point and the dc analysis as examples.

```
op
let isup=-i(vdd)
let ctl.isup[ctl.cnt]=isup
destroy

dc vin -2 2 lin 400
let c=0
cursor c right v(out,agnd) 0
let inoffs=v(inp,inn)[%c]
let ctl.inoffs[ctl.cnt]=inoffs
destroy
```

We conclude the inner and the outer Monte-Carlo loop with the following four lines of code:

```
   let ctl.cntcase=ctl.cntcase+1
  end
  let ctl.cntlot=ctl.cntlot+1
end
```

We end up with a bunch of vectors. Every one of them contains 5,000 values of some measurement corresponding to the 5,000 simulated random variations. We are interested in the standard deviation of every measurement. To make the expressions simple, we first define a function that calculates the standard deviation of a vector.

```
define sigma(vec) sqrt(mean((vec-mean(vec))^2))
```

And then we print the standard deviations.

```
setplot ctl
echo
echo Standard deviations:
echo "isup=      " {sigma(isup)}
echo "inoffs=    " {sigma(inoffs)}
echo "acgain=    " {sigma(acgain)}
...
```

We conclude by storing the Monte-Carlo results in a file for later processing.

```
write ampmm.raw
```

The script produces the following output:

```
Standard deviations:
isup=        3.63133772896295e-006
inoffs=      0.00453318797895707
acgain=      0.215458301021385
pm=          1.32175042323953
ugbw=        216870.859232791
psrr=        7.08047505024434
inoise100=   3.42975567617476e-023
inoise100k=  3.43399874329935e-023
trise=       1.30816963837019e-008
tsetrise=    2.60522290433449e-008
slewrise=    221354.204936615
overshoot=   0.0193408810033554
tfall=       8.33837325267468e-009
tsetfall=    1.24645621163439e-008
slewfall=    266532.921089193
undershoot=  0.00532661794168685
```

If we take into account only local variations (relative global variations set to 0), we get

```
Standard deviations:
isup=        3.63194415470566e-006
inoffs=      0.00450047331564304
acgain=      0.179483247246928
pm=          1.3165592500079
ugbw=        205868.520568674
psrr=        6.78368758909365
inoise100=   3.30046872977103e-023
inoise100k=  3.304612585859e-023
trise=       1.27937696471997e-008
tsetrise=    2.44662435461574e-008
slewrise=    217714.951024718
overshoot=   0.0147195772071247
```

```
tfall=        8.32785509158349e-009
tsetfall=     1.2470753762169e-008
slewfall=     265944.568525271
undershoot=   0.00529660583478486
```

From the comparison of both results, we can see that a large part of the standard deviations can be explained by the mismatch between transistors. Global variations play only a minor role. In fact, mismatch is a limiting factor of integrated circuit accuracy.

7.6 Drawing Histograms, Yield Analysis

The usual way to present Monte-Carlo analysis results is the histogram. A histogram shows the percentage of the circuits for which some property lies within some interval. The following script creates a histogram for the `inoffs` measurement from the example in Sect. 7.5:

```
* Clean up
destroy all

* Load the Monte-Carlo results (global+local variation)
load ampmm.raw

* Create a histogram of inoffs

* First bin start
let bin0=-10m
* Bin width
let binw=1m
* Last bin end
let binend=15m

* Get the vector for which we are plotting the histogram
let binvar=inoffs

* Bin count
let binn=floor((binend-bin0)/binw)+1
* Bin center position vector
let binx=vector(binn)*0
* Bin hit count
let biny=vector(binn)*0

* Go through all bins
let cnt=0
while cnt lt binn
  * Calculate new bin center and store it
  let bincenter=bin0+cnt*binw
  let binx[cnt]=bincenter
  * Select all circuits for which outoffs is within +-binw/2 of bincenter
  let mask=(binvar ge (bincenter-binw/2)) and (binvar lt (bincenter+binw/2))
  * Count them and store the count
  let biny[cnt]=sum(mask)
  * Go to next bin
  let cnt=cnt+1
end

* Set comb plot style, linewidth 5
set plottype=comb
set linewidth=5
```

```
* Plot the histogram, set y axis limits and label the plot
plot 100*biny/length(binvar) vs binx
+ yl {-100*max(biny)/length(binvar)*0.03} {100*max(biny)/length(binvar)*1.05}
+ title "Input offset histogram" xlabel "Voffs [V]" ylabel "Relative frequency [%]"
```

In a similar manner we can create a histogram of `psrr0`. We simply replace `inoffs` with `psrr0`. We set `bin0`, `binend`, and `binw` to 40, 90, and 2, respectively. Both histograms are depicted in Fig. 7.19.

By putting a requirement on a measurement, we can easily obtain the number of the circuits for which this requirement is satisfied. Take, for instance, input offset. If we require the absolute value to be less than 10 mV, we can obtain the percentage of the circuits that satisfy the requirement using the following script:

```
let yinoffs = sum(abs(inoffs) lt 10m)/length(inoffs)
echo "Input offset within 10mV: " {yinoffs*100}
```

We get the result that 95.08% of all circuits have input offset between -10 and 10 mV. The yield of a circuit [20] is defined as the percentage of circuits that satisfy a whole set of requirements on their performance measures. Take the following example where we require the supply current to be below 90 μA, the absolute value of input offset below 10 mV, AC gain above 60 dB, phase margin above 70°, unity-gain bandwidth above 5 MHz, PSRR at 0 Hz above 0 dB equivalent input noise spectrum density at 100 Hz and 100 kHz below $0.8 \cdot 10^{-21}$ V^2/Hz, rise and fall time below 250 ns, slew rate above 3 MV/s, overshoot below 22.5%, undershoot below 7.5%, and settling time below 600 ns.

```
* Load Monte-Carlo results
load ampmm.raw

* Check if a measurement satisfies the corresponding requirement
let yisup      = isup lt 90u
let yinoffs    = (inoffs gt -10m) and (inoffs lt 10m)
let yacgain    = acgain gt 60
let ypm        = pm gt 70
```

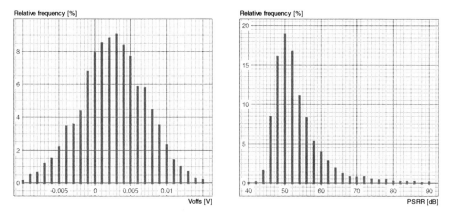

Fig. 7.19 Input offset voltage (*left*) and PSRR at 0 Hz (*right*) histograms for the CMOS differential amplifier. Global and local (mismatch) variations were taken into account in the Monte-Carlo analysis

```
let yugbw        = ugbw gt 5meg
let ypsrr0       = psrr0 gt 0
let yinoise100   = inoise100 lt 0.8e-21u
let yinoise100k  = inoise100k lt 0.8e-21u
let ytrise       = trise lt 0.25u
let ytfall       = tfall lt 0.25u
let yslewrise    = slewrise gt 3meg
let yslewfall    = slewfall gt 3meg
let yover        = over lt 0.225
let yunder       = under lt 0.075
let ytsetrise    = tsetrise lt 0.6u
let ytsetfall    = tsetfall lt 0.6u


* Calculate total yield (logical and of partial requirements)
let ytotal       = yisup and yinoffs and yacgain and ypm and yugbw
+ and ypsrr0 and yinoise100 and yinoise100k and ytrise and ytfall
+ and yslewrise and yslewfall and yover and yunder and ytsetrise
+ and ytsetfall

let n=length(ytotal)

* Print all partial yields and the total yield
echo "Yield [%]"
echo "isup        : " {100*sum(yisup)/n}
echo "inoffs      : " {100*sum(yinoffs)/n}
echo "acgain      : " {100*sum(yacgain)/n}
...
echo "--------------------"
echo "Total       : " {100*sum(ytotal)/n}
```

We obtain the following output with the partial yields and the total yield:

```
Yield [%]
isup         :  91.34
inoffs       :  95.08
acgain       :  78.84
pm           :  95.9
ugbw         :  100
psrr0        :  100
inoise100    :  99.7
inoise100k   :  99.7
trise        :  98.88
tfall        :  100
slewrise     :  99.9
slewfall     :  100
over         :  84.2
under        :  100
tsetrise     :  96.28
tsetfall     :  100
--------------------
Total        :  60.3
```

Although every partial yield is above 75%, the total yield is only a little more than 60%. Also we took into account only manufacturing process variations; we did not even account for varying environmental conditions like the supply voltage and temperature. If we did, the yield would be even lower.

7.7 Logic Gates

Besides inverters, NAND and NOR gates are the simplest building blocks of CMOS digital circuits [26]. In this section we are going to simulate the response of such gates to input signals.

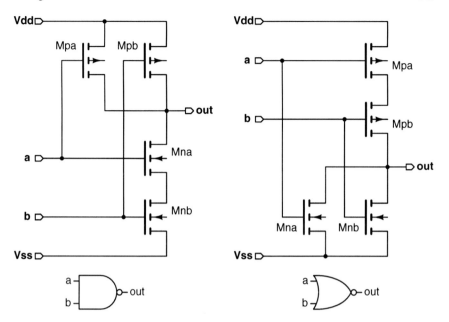

Fig. 7.20 Schematic of a CMOS NAND gate (*left*) and a CMOS NOR gate (*right*) with the corresponding symbols. Note that the Vdd and Vss inputs are missing from the symbol. Both of them are implicitly assumed to be connected to the power supply and the ground

The schematic of a 2-input NAND gate and NOR gate is depicted in Fig. 7.20. First we define subcircuits that describe these gates. We are going to use the submodp and submodn subcircuit models from Sect. 7.3.

```
* NAND gate
.subckt nand a b out vdd vss
* PMOS
xmpa (out a vdd vdd) submodp w=1.0u l=0.18u
xmpb (out b vdd vdd) submodp w=1.0u l=0.18u
* NMOS
xmna (out a t   vss) submodn w=0.5u l=0.18u
xmnb (t   b vss vss) submodn w=0.5u l=0.18u
.ends

* NOR gate
.subckt nor a b out vdd vss
* PMOS
xmpa (t   a vdd vdd) submodp w=1u l=0.18u
xmpb (out b t   vdd) submodp w=1u l=0.18u
* NMOS
xmna (out a vss vss) submodn w=0.5u l=0.18u
xmnb (out b vss vss) submodn w=0.5u l=0.18u
.ends
```

The test circuit for the NAND gate is

```
* NAND
xnand (a b out vdd 0) nand
```

```
* Inputs
va (a 0) dc=0 pulse=(0 1.8 0.5n 20p 20p 1n 2n)
vb (b 0) dc=0 pulse=(0 1.8 1n 20p 20p 2n 3n)

* Power supply
vdd (vdd 0) dc=1.8
```

The test circuit for the NOR gate is identical, except that xnand is replaced by xnor and the excitation is inverted.

```
* NOR
xnor (a b out vdd 0) nor

* Inputs
va (a 0) dc=0 pulse=(1.8 0 0.5n 20p 20p 1n 2n)
vb (b 0) dc=0 pulse=(1.8 0 1n 20p 20p 2n 3n)

* Power supply
vdd (vdd 0) dc=1.8
```

To avoid numeric integration artefacts (e.g., ringing), we use the Gear algorithm. Figure 7.21 compares the supply current obtained by simulating the circuit with the trapezoidal algorithm and the Gear algorithm.

The following set of commands can be put into the .control block to analyze the NAND gate. The same script can also be used for the NOR gate.

```
set method=gear
tran 0.01n 7n
plot v(a)+6 v(b)+4 v(out)+2 -i(vdd)/200u
+ xlabel "Time [s]" ylabel "Va+6, Vb+4, Vout+2 [V], Isup/200uA"
+ title "2-input NAND gate"
```

The results are shown in Fig. 7.22.

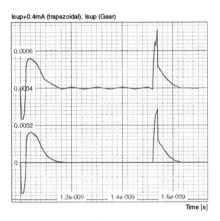

Fig. 7.21 The power supply current of the NAND gate obtained with trapezoidal integration and Gear integration

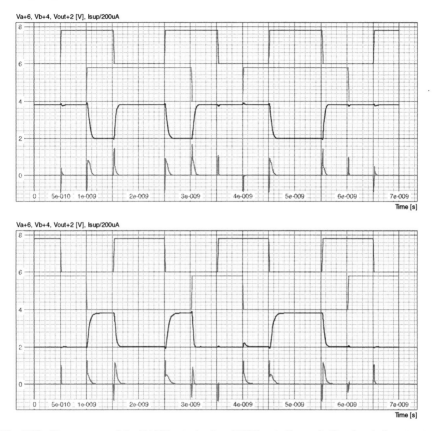

Fig. 7.22 The response of the NAND gate (*top*) and NOR gate (*bottom*). The signals from top to bottom are the a and the b inputs, the gate output, and the supply current

7.8 Flip-Flops

A flip-flop is a circuit with multiple stable DC solutions that can be used as a memory element. The synchronous D flip-flop stores the value of the input signal d at time points defined by the clock signal clk. Depending on this time point, the flip-flop can be either level triggered or edge triggered.

The state of a flip-flop is reflected by the q output. The qn output is the inverse of the q output. Many types of flip-flops have asynchronous set/reset inputs that change the state of the flip-flop at any given time regardless of the clock signal.

A rising edge-triggered flip-flop (Fig. 7.23, left) can be built using 2-input NAND gates (Sect. 7.7) and 3-input NAND gates (Fig. 7.23, right). Multiple stable DC solutions are obtained with the feedbacks in the digital circuit. A feedback is illustrated by a diagonal connection. There are eight feedback connections in the flip-flop in Fig. 7.23. Pulling down the asynchronous reset input rstn sets the state of the flip-flop (the q output) to a logic low value.

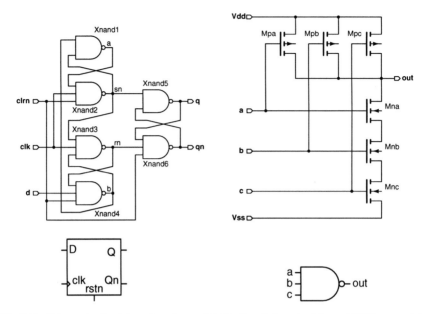

Fig. 7.23 Schematic of a rising edge-triggered D flip-flop (*left*) and a 3-input NAND gate (*right*) with the corresponding symbols. Note that the Vdd and Vss inputs are missing from the symbol. Both of them are implicitly assumed to be connected to the power supply and the ground

The subcircuit of a 3-input NAND gate and a D flip-flop are defined as follows:

```
* 3-input NAND gate
.subckt nand3 a b c out vdd vss
* PMOS
xmpa (out a vdd vdd) submodp w=1u l=0.18u
xmpb (out b vdd vdd) submodp w=1u l=0.18u
xmpc (out c vdd vdd) submodp w=1u l=0.18u
* NMOS
xmna (out a t1  vss) submodn w=0.5u l=0.18u
xmnb (t1  b t2  vss) submodn w=0.5u l=0.18u
xmnc (t2  c vss vss) submodn w=0.5u l=0.18u
.ends


* Rising clock edge triggered d flip flop with async. reset (active low)
.subckt dff clk d clrn q qn vdd vss
xnand1 (b  sn       a  vdd vss) nand
xnand2 (a  clk clrn sn vdd vss) nand3
xnand3 (sn clk b    rn vdd vss) nand3
xnand4 (rn d   clrn b  vdd vss) nand3
xnand5 (sn qn       q  vdd vss) nand
xnang6 (q  rn  clrn qn vdd vss) nand3
.ends
```

The test circuit is defined with the following netlist:

```
* dff
xdff (clk d rstn q qn vdd 0) dff

* Inputs
vclk (clk 0) dc=0 pulse=(0 1.8 5n 20p 20p 2n 4n)
vd (d 0) dc=0 pulse=(0 1.8 6n 20p 20p 12n 18n)
vcln (rstn 0) dc=0 pulse=(1.8 0 30n 20p 20p 0.5n)

* Power supply
vdd (vdd 0) dc=1.8
```

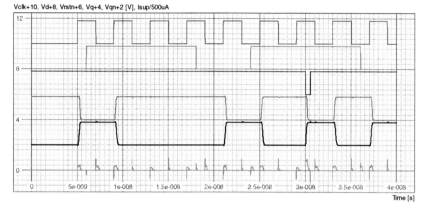

Vclk+10, Vd+8, Vrstn+6, Vq+4, Vqn+2 [V], Isup/500uA

Fig. 7.24 The response of a D flip-flop. The signals from top to bottom are the clock, the d, and the rstn inputs, the q and qn outputs, and the supply current

The D flip-flop can be analyzed with the following short NUTMEG program:

```
set method=gear
tran 0.01n 40n
plot v(clk)+10 v(d)+8 v(rstn)+6 v(q)+4 v(qn)+2 -i(vdd)/500u
+ydel 4 xlabel "Time [s]" ylabel "Vclk+10, Vd+8, Vrstn+6, Vq+4, Vqn+2 [V], Isup/500uA"
+title "D flip flop"
```

The ydel parameter of the plot command sets the tick spacing for the y-axis. The results of the simulation are depicted in Fig. 7.24. We can see that the d input is copied to the q output at every rising edge of the clk signal. The qn output is the negation of the q output. The state of the flip-flop is set to logic low whenever the rstn signal is pulled low.

Another type of flip-flop (T flip-flop) that is often used in counters can be obtained if the D flip-flop is combined with an XOR gate. The schematic of an XOR gate is depicted in Fig. 7.25.

The Miap and Mian transistors generate the inverse of the a input. The same operation is performed on the b input by Mibp and Mibn. The remaining transistors calculate the XOR of the a and the b input $(a \oplus b = \overline{a} \cdot b + a \cdot \overline{b} = \overline{(\overline{a} + b) \cdot (a + \overline{b})})$ where $\cdot$ denotes logical AND, $+$ denotes logical OR, and $\overline{a}$ denotes negation.

By feeding the d input of a D flip-flop with the XOR of the t input and the q output, a T flip-flop is obtained (Fig. 7.26). A T flip-flop inverts its state at a rising clock edge if its t input is high. If the t input is low, the state remains unchanged. The test circuit for a T flip-flop is similar to the test circuit of a D flip-flop.

```
* tff
xtff (clk t rstn q qn vdd 0) tff

* Inputs
vclk (clk 0) dc=0 pulse=(0 1.8 5n 20p 20p 2n 4n)
vd (t 0) dc=0 pulse=(0 1.8 6n 20p 20p 12n 18n)
vcln (rstn 0) dc=0 pulse=(1.8 0 27n 20p 20p 0.5n)

* Power supply
vdd (vdd 0) dc=1.8
```

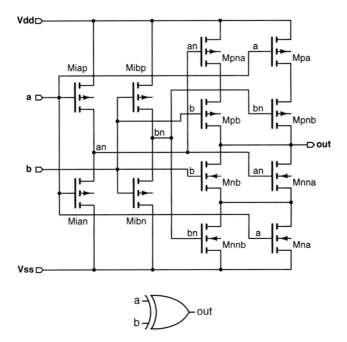

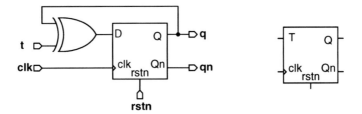

Fig. 7.25 Schematic of 2-input XOR gate and the corresponding symbol

Fig. 7.26 Schematic of a T flip-flop (*left*) and the corresponding symbol (*right*)

The script that analyzes the T flip-flop can be entered in the .control block.

```
set method=gear
tran 0.01n 40n
plot v(clk)+10 v(t)+8 v(rstn)+6 v(q)+4 v(qn)+2 -i(vdd)/500u
+ydel 4 xlabel "Time [s]" ylabel "Vclk+10, Vt+8, Vrstn+6, Vq+4, Vqn+2 [V], Isup/500uA"
+ title "T flip flop"
```

The simulation results are depicted in Fig. 7.27.

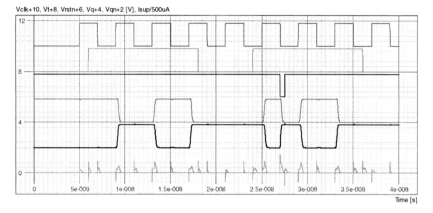

Fig. 7.27 The response of a T flip-flop. The signals from top to bottom are the clock, the t, and the rstn inputs, the q and qn outputs, and the supply current

7.9 Counters and Frequency Dividers

By pulling the t input of a T flip-flop high and using a periodic clock signal with frequency f, we obtain a new signal at the q output. The frequency of this signal is exactly $f/2$ and the duty cycle is 50%.

If we connect multiple T flip-flops to the same clock signal and feed them with appropriate t signals, we obtain a binary counter. The schematic of a 3-bit binary counter is depicted in Fig. 7.28. We can see that the first t input is pulled high, the second one is connected to the q output of the first flip-flop, and the third one is obtained as the logical AND of the q outputs of the first and the second flip-flops.

The counter can be expanded by adding another T flip-flop. The t input of the n-th flip-flop can be obtained as the logical AND of q1, q2, ..., q(n-1). The frequency of the signal at output qm is $f/2^m$, where f is the clock frequency.

The inverter and the 3-bit synchronous binary counter (Fig. 7.28) subcircuit are defined as

```
* Inverter
.subckt inv in out vdd vss
xmp (out in vdd vdd) submodp w=1.0u l=0.18u
xmn (out in vss vss) submodn w=0.5u l=0.18u
.ends

* 3-bit synchronous counter
.subckt cnt3bit clk q1 q2 q3 vdd vss
* t flip flops
xtff1 (clk vdd vdd q1 qn1 vdd vss) tff
xtff2 (clk q1  vdd q2 qn2 vdd vss) tff
xtff3 (clk t3  vdd q3 qn3 vdd vss) tff
* t signal for the 3rd flip flop
xnand (q1 q2 t3x vdd vss) nand
xinv  (t3x t3 vdd vss) inv
.ends
```

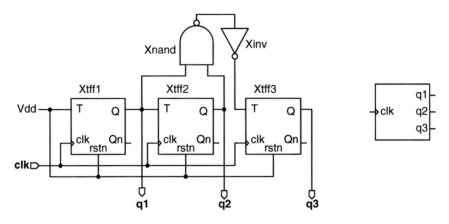

Fig. 7.28 Schematic of a 3-bit binary counter (*left*) and the corresponding symbol (*right*)

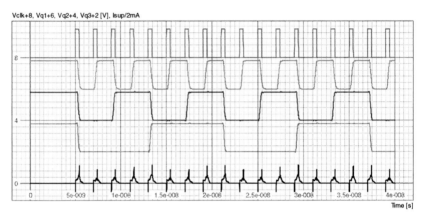

Fig. 7.29 The response of a 3-bit counter. The signals from top to bottom are the clock, the q1, q2, and q3 outputs, and the supply current

The logical AND function is implemented using a NAND gate and an inverter connected in series. The test circuit generates a clock signal with a 20% duty cycle.

```
* cnt3bit
xcnt (clk q1 q2 q3 vdd 0) cnt3bit

* Input
vclk (clk 0) dc=0 pulse=(0 1.8 5n 20p 20p 0.4n 2n)

* Power supply
vdd (vdd 0) dc=1.8
```

The analysis is conducted using the following NUTMEG script:

```
set method=gear
tran 0.01n 40n
plot v(clk)+8 v(q1)+6 v(q2)+4 v(q3)+2 -i(vdd)/2m
+ ydel 4 xlabel "Time [s]" ylabel "Vclk+8, Vq1+6, Vq2+4, Vq3+2 [V], Isup/2mA"
+ title "Synchronous 3-bit counter"
```

The results are depicted in Fig. 7.29. We can see that all of the counter's outputs produce signals with a 50% duty cycle and frequencies $f/2$, $f/4$, and $f/8$.

7.10 Phase Frequency Detector

A phase frequency detector (PFD) [46] is used when we are interested in the phase
difference between two signals. The circuit is depicted in Fig. 7.30.

The following netlist describes the PFD:

```
* Phase frequency detector
.subckt pfd a b up dn vdd vss
xdff1 (a vdd rstn qup qupn vdd vss) dff
xdff2 (b vdd rstn qdn qdnn vdd vss) dff
xnand (qup qdn rstn vdd vss) nand
xnorup (qupn qdn up vdd vss) nor
xnordn (qdnn qup dn vdd vss) nor
.ends
```

The output of the Xdff1 flip-flop goes high when a rising edge occurs in signal a.
The same operation is performed on signal b by Xdff2. When both flip-flop outputs
are high, the Xnand gate resets them. The output of the Xnorup gate is high from
the time when a rising edge occurs in signal a to the time when the flip-flops are
reset (provided that the rising edge of a precedes the rising edge of b). If the rising
edge of b precedes the rising edge of a, up remains low. The dn output operates in
the same way except that a and b switch roles.

By subtracting the average of dn from the average of up, we get a signal that is
proportional to the phase difference between a and b. If a is ahead of b the difference
of averages is positive and proportional to the phase difference. If b is ahead of a
the difference is negative.

We can test the pfd by using the following test circuit:

```
* PFD
xpfd (a b up dn vdd 0) pfd

* Inputs
va (a 0) dc=0 pulse=(0 1.8 0n 20p 20p 2n 4n)
vb (b 0) dc=0 pulse=(0 1.8 1n 20p 20p 2n 4n)

* Power supply
vdd (vdd 0) dc=1.8
```

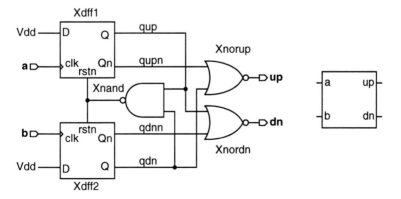

Fig. 7.30 Schematic of the PFD (*left*) and the corresponding symbol (*right*)

The tests are performed by running the following script:

```
* A is ahead of B
let @va[pulse][2]=0n
let @vb[pulse][2]=1n
tran 0.1n 8n
plot v(a)+8 v(b)+6 v(up)+4 v(dn)+2 -i(vdd)/1m
+ ydel 4 xlabel "Time [s]" ylabel "Va+8, Vb+6, Vup+4, Vdn+2 [V], Isup/1mA"
+ title "Phase frequency detector, Va early"

* B is ahead of A
let @va[pulse][2]=1n
let @vb[pulse][2]=0n
tran 0.1n 8n
plot v(a)+8 v(b)+6 v(up)+4 v(dn)+2 -i(vdd)/1m
+ ydel 4 xlabel "Time [s]" ylabel "Va+8, Vb+6, Vup+4, Vdn+2 [V], Isup/1mA"
+ title "Phase frequency detector, Vb early"

* A and B have the same phase
let @va[pulse][2]=0n
let @vb[pulse][2]=0n
tran 0.1n 8n
plot v(a)+8 v(b)+6 v(up)+4 v(dn)+2 -i(vdd)/1m
+ ydel 4 xlabel "Time [s]" ylabel "Va+8, Vb+6, Vup+4, Vdn+2 [V], Isup/1mA"
+ title "Phase frequency detector, equal phase"

* A is almost 180 degrees ahead of B.
let @va[pulse][2]=0n
let @vb[pulse][2]=1.8n
tran 0.1n 8n
plot v(a)+8 v(b)+6 v(up)+4 v(dn)+2 -i(vdd)/1m
+ ydel 4 xlabel "Time [s]" ylabel "Va+8, Vb+6, Vup+4, Vdn+2 [V], Isup/1mA"
+ title "Phase frequency detector, Va even earlier"

* Now what happens if A and B have different frequencies.
let @va[pulse][2]=0n
let @vb[pulse][2]=0n
let @va[pulse][6]=1/270meg
let @vb[pulse][6]=1/250meg
tran 0.1n 50n
plot v(a)+8 v(b)+6 v(up)+4 v(dn)+2 -i(vdd)/1m
+ ydel 4 xlabel "Time [s]" ylabel "Va+8, Vb+6, Vup+4, Vdn+2 [V], Isup/1mA"
+ title "Phase frequency detector, fa=270MHz fb=250MHz"
```

The script produces the results shown in Figs. 7.31 and 7.32.

From Fig. 7.32 it can be seen that if the frequency of a is higher than the frequency of b, the dn output is low all the time.

By adding a charge pump circuit to the PFD and a lowpass filter we get a phase detector that measures the phase difference between two signals. A charge pump pulls a current into its output if the dn input is high and delivers current at its output if the up input is high. By filtering the resulting current using a lowpass filter, we obtain a signal that is proportional to the phase difference. Figure 7.33 depicts a simple charge pump and a lowpass filter along with their corresponding symbols. Figure 7.34 depicts a phase detector.

The charge pump and the lowpass filter are defined in the following two subcircuits. The filter response is shown in Fig. 7.35 (left). The cutoff frequency is about 9 MHz.

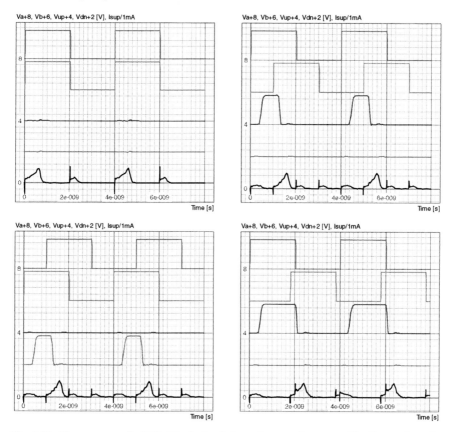

Fig. 7.31 The response of a PFD if both signals have the same frequency. The figures show the signals when both signals have the same phase (*top left*), a is ahead of b (*top right*), b is ahead of a (*bottom left*), and a is almost 180° ahead of b (*bottom right*). The signals from top to bottom are the a and b inputs, the up and dn outputs, and the supply current

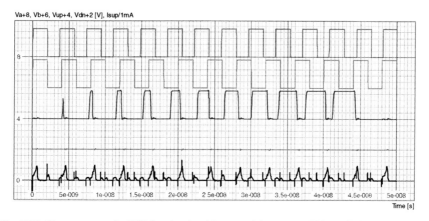

Fig. 7.32 The response of a PFD for signals with unequal frequency (270 and 250 MHz)

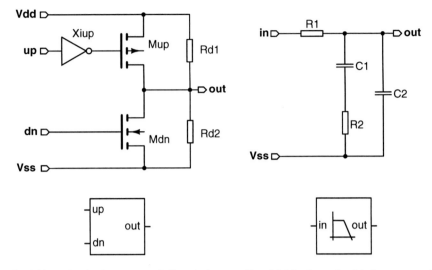

Fig. 7.33 A simple charge pump (*left*) and a lowpass filter (*right*) schematic with the corresponding symbols

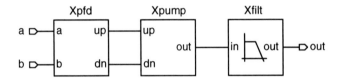

Fig. 7.34 Phase detector

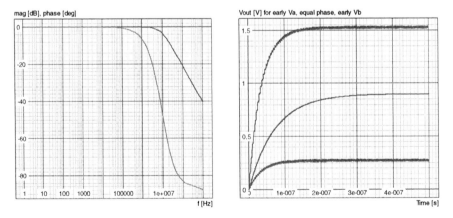

Fig. 7.35 The response of the lowpass filter (*left*) and the phase detector (*right*). The *top curve* represents the response for the case when a is almost 180° ahead of b. The *bottom curve* is for the opposite case when b is ahead of a. The *curve* in the *middle* represents the response when a and b have equal phase

```
* Simple charge pump
.subckt chgpump up dn out vdd vss
xmup (out upn vdd vdd) submodp w=0.90u l=0.18u
xmun (out dn  vss vss) submodn w=0.36u l=0.18u
xiup (up upn vdd vss) inv
rd1 (vdd  out) r=100k
rd2 (out vss)  r=100k
.ends

* Filter (high cutoff)
.subckt filterfast in out vss
r1 (in out)  r=16k
c1 (out a)   c=0.1p
r2 (a vss)   r=5k
c2 (out vss) c=1p
.ends
```

The phase detector is tested using the following test circuit:

```
* PFD
xpfd (a b up dn vdd 0) pfd

* Charge pump
xpump (up dn chg vdd 0) chgpump

* Filter
xfilt (chg out 0) filterfast


* Inputs
va (a 0) dc=0 pulse=(0 1.8 0n 100p 100p 2n 4n)
vb (b 0) dc=0 pulse=(0 1.8 0n 100p 100p 2n 4n)

* Power supply
vdd (vdd 0) dc=1.8
```

The frequency of the two input signals is 250 MHz. Three analyses are performed. Every analysis simulates a different phase delay between the two inputs.

```
plot create phout

let @va[pulse][2]=0n
let @vb[pulse][2]=1.8n
tran 2n 0.5u 0n 100n uic
plot append phout v(out)

let @va[pulse][2]=0n
let @vb[pulse][2]=0n
tran 2n 0.5u 0n 100n uic
plot append phout v(out)

let @va[pulse][2]=1.8n
let @vb[pulse][2]=0n
tran 2n 0.5u 0n 100n uic
plot append phout v(out)

plot append phout xlabel "Time [s]"
+ ylabel "Vout [V] for early Va, equal phase, early Vb"
+ title "Phase detector response"
```

The obtained response for three different phase delays (early a, equal phase, and early b) is depicted in Fig. 7.35 (right).

7.11　Voltage-Controlled Oscillator

The oscillation frequency of a voltage-controlled oscillator (VCO) can be set by setting its input voltage. Several different approaches exist for implementing a VCO. One of them is the use of a ring oscillator (see Sect. 1.3). By starving the oscillator of its current (adjusting the supply current for the inverters), we can change the oscillation frequency. The schematic of a VCO based on a ring oscillator is depicted in Fig. 7.36.

Transistors Mp1-Mp3 and Mn1-Mn3 form the ring of three inverters. The supply current of the NMOS transistors in the inverters is adjusted with a current mirror (Mnref and Mnm1-Mnm3). Similarly, the supply current of the PMOS transistors in the inverters is adjusted with a current mirror (Mpref and Mpm1-Mpm3). Transistors Mpb1, Mpb2, Mnb1, and Mnb2 represent two serially connected inverters that form a buffer for the output signal.

The VCO is described using the following subcircuit:

```
* Current-starved 3-stage ring-oscillator (VCO)
.subckt ringvco ctl out vdd vss
* Ring
xmp1 (b a vdd1 vdd) submodp w=6u l=0.3u
xmn1 (b a vss1 vss) submodn w=2u l=0.3u
xmp2 (c b vdd2 vdd) submodp w=6u l=0.3u
xmn2 (c b vss2 vss) submodn w=2u l=0.3u
xmp3 (a c vdd3 vdd) submodp w=6u l=0.3u
xmn3 (a c vss3 vss) submodn w=2u l=0.3u
* Pmos current mirrors
```

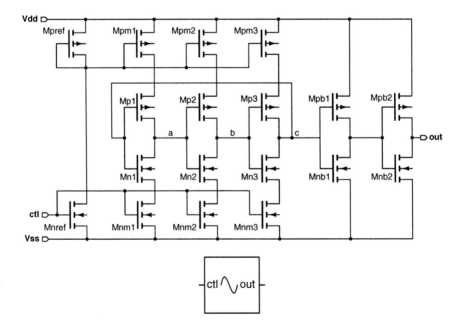

Fig. 7.36 Schematic of a VCO (*top*) and the corresponding symbol (*bottom*)

```
xmpref (ctlp ctlp vdd vdd) submodp w=2.4u l=0.2u
xmpm1  (vdd1 ctlp vdd vdd) submodp w=2.4u l=0.2u
xmpm2  (vdd2 ctlp vdd vdd) submodp w=2.4u l=0.2u
xmpm3  (vdd3 ctlp vdd vdd) submodp w=2.4u l=0.2u
* Nmos current mirrors
xmnref (ctlp ctl  vss vss) submodn w=0.8u l=0.2u
xmnm1  (vss1 ctl  vss vss) submodn w=0.8u l=0.2u
xmnm2  (vss2 ctl  vss vss) submodn w=0.8u l=0.2u
xmnm3  (vss3 ctl  vss vss) submodn w=0.8u l=0.2u
* Output buffer (2 inverters)
xmpb1  (c1  c   vdd vdd) submodp w=1.08u l=0.18u
xmnb1  (c1  c   vss vss) submodn w=0.36u l=0.18u
xmpb2  (out c1 vdd vdd) submodp w=1.08u l=0.18u
xmnb2  (out c1 vss vss) submodn w=0.36u l=0.18u
.ends
```

Because the ring with three inverters is unstable, the circuit starts to oscillate without any needed startup circuit. The numerical noise of the simulator is enough to induce oscillations. The following lines of a SPICE netlist describe the test circuit for the VCO:

```
* VCO
xvco (ctl out vdd 0) ringvco

* Control voltage
vctl (ctl 0) dc=0.5

* Power supply
vdd (vdd 0) dc=1.8
```

We are interested in the startup of the oscillator. The following script simulates it for 0.6 V input voltage:

```
let @vctl[dc]=0.6
tran 0.1n 50n 0n 10n uic
plot v(out) a:xvco+2 b:xvco+2 c:xvco+2
+ xlabel "Time [s]" ylabel "Va+2 Vb+2 Vc+2 Vout [V]"
+ title "VCO signals at Vctl=0.6V"
```

The response is depicted in Fig. 7.37. An important item is the graph of the steady-state frequency with respect to the input voltage. When the control voltage

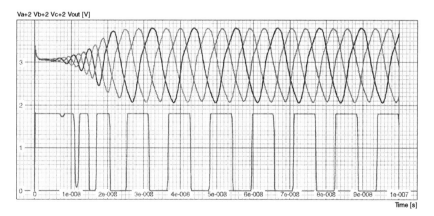

Fig. 7.37 The startup of the VCO. The three signals on the *top* are the outputs of the inverters forming the ring. The *bottom* signal is the output of the buffer (VCO output)

approaches 0, the inverter currents approach 0 and the same holds for the oscillation magnitude. Therefore, we are also interested in the steady-state envelope of all the outputs of the three inverters from the ring. The following script analyzes the VCO for different values of the input voltage:

```
* Prepare a plot holding the control vectors
setplot new
nameplot ctl
* Input voltage sweep range and the number of steps
let vstart=0.4
let vstop=1.8
let n=100

* Initial simulation time
let tsim=800n

* Prepare the vector of input voltages
let vctl=vector(n+1)/n*(vstop-vstart)+vstart
* Prepare the vectors for the frequency and the envelope min-max
let f=vctl*0
let vmaxa=vctl*0
let vmina=vctl*0
let vmaxb=vctl*0
let vminb=vctl*0
let vmaxc=vctl*0
let vminc=vctl*0

* Start the sweep loop
let cnt=0
while ctl.cnt lt length(ctl.vctl)
  setplot ctl
  * Set the input voltage
  let @vctl[dc]=vctl[cnt]

  * Simulate from tsim/2 to tsim
  tran {tsim/500} {tsim} {tsim/2} {tsim/10} uic

  * Measure the envelope min-max
  let va=v(a:xvco)
  let vmaxa=max(va)
  let vmina=min(va)
  let vb=v(b:xvco)
  let vmaxb=max(vb)
  let vminb=min(vb)
  let vc=v(c:xvco)
  let vmaxc=max(vc)
  let vminc=min(vc)

  * Measure the frequency
  let c1=0
  let c2=0
  * Move to tsim
  cursor c1 right time
  cursor c2 right time
  * Move left to the endpoints of the last two periods
  * Use envelope average (min+max )/2 as the level for the cursor
  cursor c1 left vc (vmaxc+vminc)/2 2 rising
  cursor c2 left vc (vmaxc+vminc)/2 1 rising
  * Calculate period and frequency
  let per=time[%c2]-time[%c1]
  let freq=1/per

  * Store results
  let ctl.f[ctl.cnt]=freq
  let ctl.vmaxa[ctl.cnt]=vmaxa
  let ctl.vmina[ctl.cnt]=vmina
```

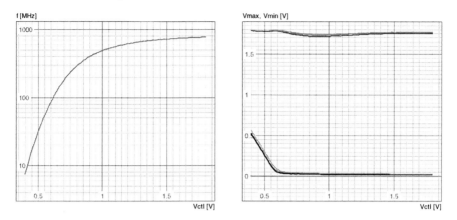

Fig. 7.38 The steady-state frequency (*left*) and the envelope of the output signals from the inverters (*right*) for a VCO

```
let ctl.vmaxb[ctl.cnt]=vmaxb
let ctl.vminb[ctl.cnt]=vminb
let ctl.vmaxc[ctl.cnt]=vmaxc
let ctl.vminc[ctl.cnt]=vminc

* Print a diagnostic message
echo {ctl.cnt}: "Vctl=" {ctl.vctl[ctl.cnt]} " T=" {per*1e9} "ns f=" {freq/1e6} "MHz"

* The simulation time for the next sweep point is
*   5 periods of the signal at this sweep point
let ctl.tsim=per*5

* Delete old simulation results
destroy

let ctl.cnt=ctl.cnt+1
end

setplot ctl
* Plot frequency vs input voltage
plot f/1e6 vs vctl ylog yl 4 1000 xlabel "Vctl [V]" ylabel "f [MHz]"
+ title "VCO frequency"

* Plot the envelope of a, b, and c with respect to input voltage
plot vmaxa vmaxb vmaxc vs vctl
+ vmina vminb vminc vs vctl
+ yl -0.2 1.8 xlabel "Vctl [V]" ylabel "Vmax, Vmin [V]"
+ title "VCO swing voltage"
```

The results are plotted in Fig. 7.38.

7.12 Phase-Locked Loop

A phase-locked loop (PLL) [46] is a frequency generator where the ratio of the output frequency and the reference frequency is kept constant by means of feedback based on the phase difference. A simple PLL is depicted in Fig. 7.39.

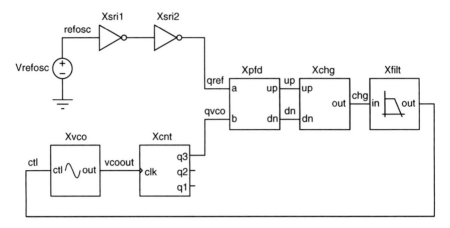

Fig. 7.39 Schematic of a PLL

The reference 20 MHz sinusoidal signal is generated by the Vrefosc source. The signal is shaped into a rectangular signal using two inverters (Xsri1 and Xsri2), resulting in the qref signal. qref is usually generated using a crystal oscillator. The frequency of such an oscillator is very stable.

The output frequency is generated by a VCO (vcoout). The frequency of the VCO signal is divided by 8 using a binary counter (Xcnt), resulting in the qvco signal. The phase of the qref signal is compared to the phase of the qvco signal using a phase detector (Xpfd, Xchg, and Xfilt). The output of the phase detector (ctl) is connected to the control input of the VCO. The feedback forces the qvco signal to have the same frequency as the qref signal. This means that the VCO runs at eight times the reference frequency (160 MHz).

All of the components of the PLL in Fig. 7.39 were described in Sects. 7.9–7.11. The cutoff frequency of the filter (11.5 kHz) is lower than the cutoff frequency in Sect. 7.10. The filter used in the PLL is described by the following netlist:

```
.subckt filter in out vss
r1 (in out)   r=655.2k
c1 (out a)    c=1p
r2 (a vss)    r=197.2k
c2 (out vss) c=20p
.ends
```

The following is the PLL netlist:

```
* Reference signal, 20MHz
vrefosc (refosc 0) dc=0 sin=(0.9 0.8 20meg)
* Shape
xsri1 (refosc a1 vdd 0) inv
xsri2 (a1 qref vdd 0) inv

* VCO
xvco (ctl vcoout vdd 0) ringvco
* Counter, modulo 8
xcnt (vcoout q1 q2 qvco vdd 0) cnt3bit

* PFD
xpfd (qref qvco up dn vdd 0) pfd
* Charge pump
```

```
xchg (up dn chg vdd 0) chgpump
* Filter
xfilt (chg ctl 0) filter

* Power supply
vdd (vdd 0) dc=1.8
```

The PLL takes a fairly long time to enter its locked state where `qref` and `qvco` have the same frequency. The simulation is performed using the following script. After the simulation is finished, the results are stored in a `.raw` file for later processing. The file takes around 300 MB of disk space.

```
tran 2n 50u 0n 100n uic
nameplot pll
* This took really long. Store it.
* The .raw file is 300MB is size.
set filetype=binary
write pll.raw
```

After the simulation is finished (or the results are loaded using `load pll.raw`), we can plot the results.

```
* Plot pll signals
plot refosc+8 qref+6 vcoout+4 qvco+2 ctl xl 2u 5u yl 0 10
+ xlabel "Time [s]" ylabel "Vrefosc+8, Vqref+6, Vvcoout+4, Vqvc+2, Vctl [V]"
+ title "VCO response (startup)"

plot refosc+8 qref+6 vcoout+4 qvco+2 ctl xl 49.89u 50u yl 0 10
+ xlabel "Time [s]" ylabel "Vrefosc+8, Vqref+6, Vvcoout+4, Vqvc+2, Vctl [V]"
+ title "VCO response (steady state)"

* PFD signals
plot up+6 dn+4 ctl*5 xl 0 20u
+ xlabel "Time [s]" ylabel "Vup+4, Vdn+2, Vctl*5 [V]"
+ title "PFD output and filter output (startup)"

plot up+6 dn+4 ctl*5 xl 49.2u 50u yl 0 8
+ xlabel "Time [s]" ylabel "Vup+4, Vdn+2, Vctl*5 [V]"
+ title "PFD output and filter output (steady state)"

* Vctl ripple (source of phase jitter)
plot ctl xl 7.2u 50u
+ xlabel "Time [s]" ylabel "Vctl [V]"
+ title "Vctl ripple (startup)"

plot ctl xl 49.2u 50u
+ xlabel "Time [s]" ylabel "Vctl [V]"
+ title "Vctl ripple (steady state)"
```

The response of the PLL is depicted in Fig. 7.40. We can see that when the locked state (steady state) is reached, the VCO frequency (160 MHz) is exactly eight times the reference frequency (20 MHz).

The response of the PFD is shown in Fig. 7.41. We can see that the initial ripple in the VCO control input eventually dies out as the locked state is reached. From the up and dn outputs of the PFD we can see that the phase difference between `qref` and `qvco` is constant in the steady state.

Actually, the phase difference is not completely constant. PLLs are plagued by a phenomenon called phase jitter. Phase jitter occurs because the VCO's control signal is not constant. We can see that during the startup the control signal has a frequency component at around 455 kHz (Fig. 7.42, left). This component dies out as the PLL approaches steady state. By looking at the steady state more closely, we can see that

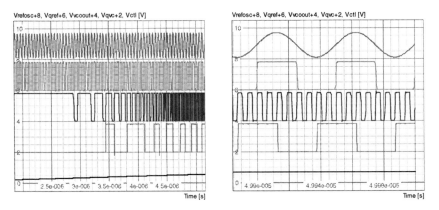

Fig. 7.40 The response of the PLL soon after startup (*left*) and in steady state (*right*). The signals from top to bottom are the output of the reference oscillator, the shaped reference signal (`qref`), the output of the VCO, the output of the counter, and the VCO control signal (`ctl`)

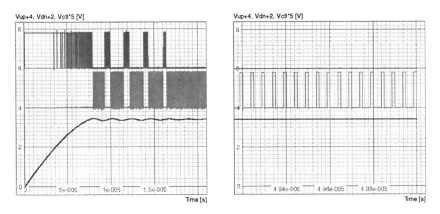

Fig. 7.41 The response of the phase detector soon after startup (*left*) and in steady state (*right*). The signals from top to bottom are the `up` and the `dn` outputs of the PFD and the VCO control signal (`vctl`)

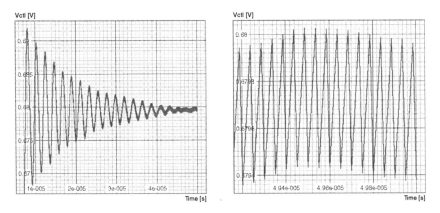

Fig. 7.42 The control signal of the VCO (`ctl`) soon after startup (*left*) and in steady state (*right*)

it contains a signal oscillating at 20 MHz with 0.3 mV magnitude (Fig. 7.42, right). This is the residual of the reference signal that is not completely removed by the filter and is the origin of the phase jitter.

A rational frequency ratio between the reference signal and the VCO output can be achieved if another counter is inserted between the reference signal and qref. If programmable digital counters are used, the output frequency of the PLL can be adjusted digitally.

References

1. K. E. Atkinson: *An Introduction to Numerical Analysis*, 2nd ed., Wiley, New York, 1989.
2. R. J. Baker: *CMOS Circuit Design, Layout, and Simulation*, IEEE Press, Hoboken, NJ, 2005.
3. *BSIM3 Homepage*, http://www-device.eecs.berkeley.edu/~bsim3/latenews.html, December 2007.
4. *BSIM4 Homepage*, http://www-device.eecs.berkeley.edu/~bsim3/bsim4.html, December 2007.
5. *BSIMSOI Homepage*, http://www-device.eecs.berkeley.edu/~bsimsoi, December 2007.
6. L. S. Bobrow: *Elementary Linear Circuit Analysis (Oxford Series in Electrical and Computer Engineering)*, Oxford University Press, Oxford, UK, 1996.
7. R. N. Bracewell: *The Fourier Transform and Its Applications*, 3rd ed., McGraw-Hill, Boston, MA, 2000.
8. M. Bucher, C. Lallement, C. Enz, F. Théodoloz, F. Krummenacher: *The EPFL-EKV MOS-FET Model Equations for Simulation*, Electronics Laboratories, Swiss Federal Institute of Technology (EPFL), Lausanne, Switzerland, 1997.
9. Yuhua Cheng, Chenming Hu: *MOSFET Modeling and BSIM3 User's Guide*, Springer, New York, 1999.
10. L. O. Chua, P. M. Lin: *Computer-Aided Analysis of Electronic Circuits*, Prentice-Hall, Englewood Cliffs, NJ, 1975.
11. F. L. Cox, III, W. B. Kuhn, J. P. Murray, S. D. Tynor: XSPICE: code-level modeling in XSPICE, *Proceedings IEEE International Symposium on Circuits and Systems*, 1992 (ISCAS 92), vol. 2, pp. 871–874, 1992.
12. D. A. Divekar: *FET Modeling for Circuit Simulation*, Kluwer, Dordrecht, 1988.
13. J. S. Duster, M. C. Jeng, P. K. Ko, C. Hu: *User's Guide for the BSIM2 Parameter Extraction Program and the SPICE3 with BSIM Implementation*, Industrial Liaison Program, Software Distribution Office, University of California, Berkeley, May 1990.
14. C. C. Enz, E. A. Vittoz: *Charge-Based MOS Transistor Modeling: The EKV Model for Low-Power and RF IC Design*, Wiley, New York, 2006.
15. *EPFL-EKV MOSFET Model Web Site*, http://legwww.epfl.ch/ekv, December 2007.
16. J. G. Fossum et al.: *UFSOI MOSFET Models User's Guide*, University of Florida, Gainesville, FL, Dec. 1997; Rev. Nov. 1998; Rev. Nov. 1999; Rev. Jun. 2001; Rev. Mar. 2002.
17. J. G. Fossum et al.: *Process-Based UFET MOSFET Model User's Guide*, University of Florida, Gainesville, FL, Jan. 1997; Rev. Apr. 1999.
18. D. Frohman-Bentchkowsky, A. S. Grove: The output conductance of MOST's in saturation, *IEEE Transactions on Electron Devices*, vol. ED-16, pp. 108–115, 1969.
19. S. Fung et. al.: *BSIM3SOI v1.0 Manual*, Department of EECS, University of California, Berkeley, 1997.
20. H. E. Graeb: *Analog Design Centering and Sizing*, Springer, New York, 2007.
21. H. K. Gummel, H. C. Poon: An integral charge control model of bipolar transistors, *Bell System Technical Journal*, vol. 49, pp. 827–852, 1970.

22. *HSPICE Simulation and Analysis User's Guide*, Synopsis, Mountain View, CA, 2003.

23. A. Ioinovici: *Computer-Aided Analysis of Active Circuits*, CRC Press, Boca Raton, FL, 1990.

24. A. Iserles: *A First Course in the Numerical Analysis of Differential Equations*, Cambridge University Press, Cambridge, UK, 1996.

25. *IsSpice4 User's Guide*, Intusoft, San Pedro, CA, 1994.

26. S. Kang, Y. Leblebici: *CMOS Digital Integrated Circuits Analysis & Design*, McGraw-Hill, New York, 2002.

27. W. Kaplan: *Advanced Calculus*, 3rd ed., Addison Wesley, Reading, MA, 1984.

28. P. R. Kinget: Device mismatch and tradeoffs in the design of analog circuits, *IEEE Journal of Solid-State Circuits*, vol. 40, pp. 1212–1224, 2005.

29. K. S. Kundert: *The Designer's Guide to SPICE and SPECTRE*, Kluwer, Dordrecht, 1995.

30. M. S. L. Lee, B. M. Tenbroek, W. Redman-White, J. Benson, M. J. Uren: A physically based compact model of partially depleted SOI MOSFETs for analog circuit simulation, *IEEE Journal of Solid-State Circuits*, vol. 36, pp. 110–121, 2001.

31. G. Massobrio, P. Antognetti: *Semiconductor Device Modeling with SPICE*, McGraw-Hill, New York, 1998.

32. D. E. Muller: A method for solving algebraic equations using an automatic computer, *Mathematical Tables and Other Aids to Computation*, vol. 10, pp. 208–215, 1956.

33. L. W. Nagel: *A Computer Program to Simulate Semiconductor Circuits*, Memorandum no. ERL-M5520, University of California, Berkeley, 1975.

34. L. W. Nagel: *The Life of SPICE*, A talk given at BCTM'96. http://www.designers-guide.org/Perspective/life-of-spice.pdf, January 2008.

35. A. R. Newton: *Presentation of the 1995 Phil Kaufman Award to Professor Donald O. Pederson*, November 1995, http://www.eecs.berkeley.edu/~newton/Presentations/Kaufman/DOPPresent.html, January 2008.

36. A. R. Newton, D. O. Pederson, A. Sangiovanni-Vincentelli: *SPICE3 Version 3f3 User's Manual*, Department of Electrical Engineering and Computer Sciences, University of California, Berkeley, 1993.

37. A. Papoulis, S. U. Pillai: *Probability, Random Variables and Stochastic Processes*, McGraw-Hill, New York, 2001.

38. A. E. Parker, D. J. Skellern: A realistic large-signal MESFET model for SPICE, *IEEE Transactions on Microwave Theory and Techniques*, vol. 45, pp. 1563–1571, 1997.

39. D. O. Pederson: A historical review of circuit simulation, *IEEE Transactions on Circuits and Systems*, vol. 31, pp. 103–111, 1984.

40. M J. M. Pelgrom, A. C. J. Duinmaijer, A. P. G. Welbers: Matching properties of MOS transistors, *IEEE Journal of Solid-State Circuits*, vol. 24, pp. 1433–1440, 1989.

41. *Predictive Technology Model*, http://www.eas.asu.edu/~ptm/, December 2007.

42. *PSpice User's Guide*, Cadence Design Systems, San Jose, CA, 2003.

43. J. S. Roychowdhury, A. R. Newton, D. O. Pederson: Algorithms for the transient simulation of lossy interconnect, *IEEE Transactions on Computer Aided Design of Integrated Circuits and Systems*, vol. 13, pp. 96–104, 1994.

44. B. Sheu, D. Scharfetter, C. Hu, D. Pederson: A compact IGFET charge model, *IEEE Transactions on Circuits and Systems*, vol. 31, pp. 745–748, 1984.

45. H. Shichman, D. Hodges: Modeling and simulation of insulated-gate field-effect transistor switching circuits, *IEEE Journal of Solid-State Circuits*, vol. SC-3, pp. 285–289, 1968.

46. K. Shu, E. Sánchez-Sinencio: *MOS PLL Synthesizers: Analysis and Design*, Springer, New York, 2005.

47. W. McC. Siebert: *Circuits, Signals, and Systems*, MIT Press, Cambridge, MA, 1985.

48. C. G. Someda: *Electromagnetic Waves*, CRC Press, Boca Raton, FL, 2006.

49. *SPECTRE Circuit Simulator User's Guide*, Cadence Design Systems, San Jose, CA, 2002.

50. *SPICE*, http://en.wikipedia.org/wiki/SPICE, January 2008.

51. *Spice OPUS Homepage*, http://www.fe.uni-lj.si/spice, December 2007.

52. H. Statz et. al.: GaAs FET device and circuit simulation in SPICE, *IEEE Transactions on Electron Devices*, vol. 34, pp. 160–169, 1987.

53. G. B. Thomas Jr., R. L. Finney: *Calculus and Analytic Geometry*, 9th ed., Addison-Wesley, Reading, MA, 1996.
54. G. Vasilescu: *Electronic Noise and Interfering Signals: Principles and Applications*, Springer, New York, 2005.
55. A. Vladimirescu, S. Liu: *The Simulation of MOS Integrated Circuits Using Spice2*, Memorandum no. UCB/ERL M80/7, Electronics Research Laboratory, College of Engineering, University of California, Berkeley, 1980.

Index

Printed in the United States
150944LV00002B/5/P